Air Pollution Control Technology Handbook

Second Edition

Air Pollution Control Technology Handbook

Second Edition

Karl B. Schnelle, Jr.
Russell F. Dunn
Mary Ellen Ternes

CRC Press is an imprint of the
Taylor & Francis Group, an **informa** business

CRC Press
Taylor & Francis Group
6000 Broken Sound Parkway NW, Suite 300
Boca Raton, FL 33487-2742

First issued in paperback 2017

ISBN-13: 978-1-4822-4560-8 (hbk)
ISBN-13: 978-1-138-74766-1 (pbk)

Library of Congress Cataloging-in-Publication Data

Schnelle, Karl B., Jr., author.
Air pollution control technology handbook / Karl B. Schnelle, Jr., Mary Ellen Ternes, and Russell Dunn. -- Second edition.
pages cm
Includes bibliographical references and index.
ISBN 978-1-4822-4560-8 (hardcover : alk. paper) 1. Air quality management--Handbooks, manuals, etc. 2. Air--Pollution--Handbooks, manuals, etc. 3. Air--Purification--Equipment and supplies--Handbooks, manuals, etc. I. Ternes, Mary Ellen, author. II. Dunn, Russell, author. III. Title.

TD883.S283 2016
628.5'3--dc23 2015026122

Visit the Taylor & Francis Web site at
http://www.taylorandfrancis.com

and the CRC Press Web site at
http://www.crcpress.com

Dedication

The authors would like to dedicate this book to all the students who have been in their air pollution control courses at Vanderbilt University, Western Washington University, Gonzaga University, Auburn University, and Education Services of the American Institute of Chemical Engineers. We appreciate your attendance, your attentiveness, and your desire to improve our environment.

Contents

Preface.....xxi
Acknowledgments.....xxiii
Authors.....xxv

Chapter 1 Historical Overview of the Development of Clean Air Regulations.....1

1.1 A Brief History of the Air Pollution Problem.....1
1.2 Federal Involvement in Air Pollution Control.....3
1.3 Characterizing the Atmosphere.....4
1.4 Recipe for an Air Pollution Problem.....6
1.4.1 Sources of Air Pollution.....8
1.4.2 Meteorological Parameters Affecting Transport of Pollutants.....8
1.4.3 Effects of Air Pollution—A Comparison of London Smog and Los Angeles Smog.....10

Chapter 2 Clean Air Act.....13

2.1 History of the Clean Air Act.....13
2.1.1 1970 Clean Air Act Amendments.....13
2.1.1.1 National Ambient Air Quality Standards.....14
2.1.1.2 New Source Performance Standards.....14
2.1.1.3 Hazardous Air Pollutants.....14
2.1.1.4 Citizen Suits.....15
2.1.2 1977 Clean Air Act Amendments.....15
2.1.2.1 Prevention of Significant Deterioration.....16
2.1.2.2 Offsets in Nonattainment Areas.....16
2.2 1990 Clean Air Act Amendments.....17
2.2.1 Title I: Provisions for Attainment and Maintenance of NAAQS.....18
2.2.1.1 NAAQS Revisions.....19
2.2.2 Title II: Mobile Sources.....21
2.2.3 Title III: Hazardous Air Pollutant Program.....22
2.2.3.1 Source Categories.....22
2.2.3.2 Establishing MACT Standards.....24
2.2.3.3 Risk Management Plans.....34
2.2.4 Title IV: Acid Deposition Control.....34
2.2.5 Title V: Operating Permits.....36
2.2.6 Title VI: Stratospheric Ozone and Global Climate Protection.....36
2.2.7 Title VII: Enforcement.....36
2.2.8 Title VIII: Miscellaneous Provisions.....37

2.2.9 Title IX: Research ... 37
2.2.10 Title X: Disadvantaged Business ... 37
2.2.11 Title XI: Employment Transition Assistance ... 37

Chapter 3 Air Permits for New Source ... 39

3.1 Elements of a Permit Application ... 39
3.1.1 Applicability ... 40
3.1.1.1 Potential to Emit ... 41
3.1.1.2 Fugitive Emissions ... 41
3.1.1.3 Secondary Emissions ... 41
3.1.2 Significant Emission Rates ... 41
3.1.3 Modification ... 42
3.1.4 Emissions Netting ... 43
3.1.4.1 Netting Example ... 44
3.2 Best Available Control Technology ... 44
3.2.1 Step 1: Identify Control Technologies ... 45
3.2.2 Step 2: Eliminate Technically Infeasible Options ... 46
3.2.3 Step 3: Rank Remaining Options by Control Effectiveness ... 46
3.2.4 Step 4: Evaluate Control Technologies in Order of Control Effectiveness ... 47
3.2.4.1 Energy Impacts ... 47
3.2.4.2 Environmental Impacts ... 47
3.2.4.3 Economic Impacts and Cost-Effectiveness ... 48
3.2.5 Step 5: Select BACT ... 48
3.3 Air Quality Analysis ... 48
3.3.1 Preliminary Analysis ... 49
3.3.2 Full Analysis ... 50
3.4 NSR Reform ... 50

Chapter 4 Atmospheric Diffusion Modeling for Prevention of Significant Deterioration Permit Regulations and Regional Haze ... 51

4.1 Introduction—Meteorological Background ... 51
4.1.1 Inversions ... 51
4.1.1.1 Surface or Radiation Inversions ... 51
4.1.1.2 Evaporation Inversion ... 52
4.1.1.3 Advection Inversion ... 52
4.1.1.4 Subsidence Inversion ... 52
4.1.2 Diurnal Cycle ... 52
4.1.3 Principal Smoke-Plume Models ... 53
4.2 Tall Stack ... 54
4.3 Classifying Sources by Method of Emission ... 55
4.3.1 Definition of Tall Stacks ... 56
4.3.2 Process Stacks ... 57

4.4 Atmospheric-Diffusion Models ... 57
4.4.1 Other Uses of Atmospheric-Diffusion Models ... 58
4.5 Environmental Protection Agency's Computer Programs for Regulation of Industry ... 59
4.5.1 Industrial Source Complex Model ... 60
4.5.2 Screening Models ... 60
4.5.3 New Models ... 61
4.6 Source-Transport-Receptor Problem ... 61
4.6.1 Source ... 61
4.6.2 Transport ... 62
4.6.2.1 Effective Emission Height ... 62
4.6.2.2 Bulk Transport of Pollutants ... 63
4.6.2.3 Dispersion of Pollutants ... 63
4.6.3 Receptor ... 63

Chapter 5 Source Testing ... 65

5.1 Introduction ... 65
5.2 *Code of Federal Regulations* ... 65
5.3 Representative Sampling Techniques ... 65
5.3.1 Gaseous Pollutants ... 65
5.3.2 Velocity and Particulate Traverses ... 68
5.3.3 Isokinetic Sampling ... 70

Chapter 6 Ambient Air Quality and Continuous Emissions Monitoring ... 73

6.1 Ambient Air-Quality Sampling Program ... 73
6.2 Objectives of a Sampling Program ... 73
6.3 Monitoring Systems ... 73
6.3.1 Fixed versus Mobile Sampling ... 74
6.3.2 Continuous versus Integrated Sampling ... 74
6.3.3 Selection of Instrumentation and Methods ... 75
6.4 Federal Reference Methods and Continuous Monitoring ... 75
6.5 *Complete* Environmental Surveillance and Control System ... 78
6.6 Typical Air Sampling Train ... 78
6.7 Integrated Sampling Devices for Suspended Particulate Matter ... 80
6.8 Continuous Air-Quality Monitors ... 81
6.8.1 Electroconductivity Analyzer for SO_2 ... 82
6.8.2 Coulometric Analyzer for SO_2 ... 83
6.8.3 Nondispersive Infrared Method for CO ... 84
6.8.4 Flame Photometric Detection of Total Sulfur and SO_2 ... 85
6.8.5 Hydrocarbons by Flame Ionization ... 86
6.8.6 Fluorescent SO_2 Monitor ... 86

6.8.7 Chemilumenescence for Detection of Ozone and Nitrogen Oxides....86
6.8.8 Calibration of Continuous Monitors....88
6.8.8.1 Specifications for Continuous Air-Quality Monitors....88
6.8.8.2 Steady-State Calibrations....88

Chapter 7 Cost Estimating....91

7.1 Time Value of Money....91
7.1.1 Annualized Capital Cost....93
7.1.2 Escalation Factors....93
7.2 Types of Cost Estimates....94
7.3 Air Pollution Control Equipment Cost....95
7.3.1 *OAQPS Control Cost Manual*....95
7.3.2 Other Cost-Estimating Resources....95

Chapter 8 Process Design and the Strategy of Process Design....97

8.1 Introduction to Process Design....97
8.2 Strategy of Process Design....97
8.2.1 Process Flowsheets....99
8.3 Mass and Energy Balances....99
8.3.1 Mass-Balance Example....101
8.3.2 Energy-Balance Example....102
8.4 Systems-Based Approaches to Design....104

Chapter 9 Profitability and Engineering Economics....109

9.1 Introduction—Profit Goal....109
9.2 Profitability Analysis....109
9.2.1 Mathematical Methods for Profitability Evaluation....109
9.2.2 Incremental Rate of Return on Investments as a Measure of Profitability....110
9.2.2.1 Example of IROI Comparing Two Cases....111
9.2.2.2 Example of IROI with Four Cases....112
9.3 Effect of Depreciation....113
9.3.1 Example....114
9.4 Capital Investment and Total Product Cost....115
9.4.1 Design Development....116

Chapter 10 Introduction to Control of Gaseous Pollutants 119

10.1 Absorption and Adsorption 121
10.1.1 Fluid Mechanics Terminology 123
10.1.2 Removal of Hazardous Air Pollutants and Volatile Organic Compounds by Absorption and Adsorption 124
10.2 Process Synthesis Technology for the Design of Volatile Organic Compounds Recovery Systems 125

Chapter 11 Absorption for Hazardous Air Pollutants and Volatile Organic Compounds Control 127

11.1 Introduction 127
11.1.1 Description 127
11.1.2 Advantages 128
11.1.3 Disadvantages 128
11.2 Aqueous Systems 128
11.3 Nonaqueous Systems 129
11.4 Types and Arrangements of Absorption Equipment 129
11.5 Design Techniques for Countercurrent Absorption Columns 132
11.5.1 Equilibrium Relationships 135
11.5.2 Ideal Solutions—Henry's Law 136
11.5.3 Countercurrent Absorption Tower Design Equations 138
11.5.4 Origin of Volume-Based Mass-Transfer Coefficients 140
11.5.4.1 Steady-State Molecular Diffusion ... 140
11.5.5 Whitman Two-Film Theory 142
11.5.6 Overall Mass-Transfer Coefficients 143
11.5.7 Volume-Based Mass-Transfer Coefficients 144
11.5.8 Determining Height of Packing in the Tower: Height of a Transfer Unit Method 145
11.5.9 Dilute Solution Case 146
11.5.10 Using Mass Exchange Network Concepts to Simultaneously Evaluate Multiple Mass Separating Agent (Absorbent) Options 147
11.6 Countercurrent Flow Packed Absorption Tower Design 151
11.6.1 General Considerations 151
11.6.2 Operations of Packed Towers 151
11.6.3 Tower Packings 153
11.6.3.1 Random or Dumped Packing 153
11.6.3.2 Types of Random Packing 154
11.6.3.3 Structured Packing 156

11.6.3.4 Types of Structured Packing 158
11.6.3.5 Grid-Type Packing 161
11.6.4 Packed Tower Internals 162
11.6.4.1 Packing Support Plate 162
11.6.4.2 Liquid Distributors 162
11.6.4.3 Liquid Redistributors 163
11.6.4.4 Bed Limiter 163
11.6.5 Choosing a Liquid–Gas Flow Ratio 164
11.6.6 Determining Tower Diameter—Random Dumped Packing 165
11.6.7 Determining Tower Diameter—Structured Packing 168
11.6.8 Controlling Film Concept 169
11.6.9 Correlation for the Effect of *L/G* Ratio on the Packing Height 169
11.6.10 Henry's Law Constants and Mass-Transfer Information 171
11.6.11 Using Henry's Law for Multicomponent Solutions 174
11.7 Sample Design Calculation 174
11.7.1 Dumped Packing 174
11.7.2 Flooding 179
11.7.3 Structured Packing 180
11.7.3.1 Flooding 182

Chapter 12 Adsorption for Hazardous Air Pollutants and Volatile Organic Compounds Control 185

12.1 Introduction to Adsorption Operations 185
12.1.1 Description 185
12.1.2 Advantages 187
12.1.3 Disadvantages 187
12.2 Adsorption Phenomenon 187
12.3 Adsorption Processes 187
12.3.1 Stagewise Process 188
12.3.2 Continuous Contact, Steady-State, Moving-Bed Adsorbers 188
12.3.3 Unsteady-State, Fixed-Bed Adsorbers 188
12.3.4 Newer Technologies 189
12.3.4.1 Rotary Wheel Adsorber 189
12.3.4.2 Chromatographic Adsorption 189
12.3.4.3 Pressure Swing Adsorption 190
12.4 Nature of Adsorbents 190
12.4.1 Adsorption Design with Activated Carbon 192
12.4.1.1 Pore Structure 192
12.4.1.2 Effect of Relative Humidity 192

12.5 Theories of Adsorption ... 192
12.6 Data of Adsorption ... 194
12.7 Adsorption Isotherms ... 195
12.7.1 Freundlich's Equation ... 195
12.7.2 Langmuir's Equation ... 196
12.7.3 Brunauer, Emmett, Teller, or BET Isotherm ... 196
12.7.3.1 Adsorption without Capillary Condensation ... 196
12.7.3.2 Adsorption with Capillary Condensation ... 197
12.8 Polanyi Potential Theory ... 198
12.8.1 Hexane Example of the Polanyi Potential Theory ... 198
12.9 Unsteady-State, Fixed-Bed Adsorbers ... 200
12.10 Fixed-Bed Adsorber Design Considerations ... 201
12.10.1 Safety Considerations ... 202
12.11 Pressure Drop through Adsorbers ... 203
12.11.1 Pressure Drop Example ... 203
12.12 Adsorber Effectiveness, Regeneration, and Reactivation ... 204
12.12.1 Steam Regeneration ... 205
12.12.2 Hot Air or Gas Regeneration ... 205
12.12.3 Reactivation ... 208
12.13 Breakthrough Model ... 208
12.13.1 Mass Transfer ... 209
12.13.2 Breakthrough Curve Example ... 211
12.13.3 Second Breakthrough Curve Example: Hexane Problem ... 213
12.14 Regeneration Modeling ... 218
12.14.1 Steam Regeneration Example ... 218
12.15 Using Mass Exchange Network Concepts to Simultaneously Evaluate Multiple Mass-Separating Agent (Absorbent and Adsorbent) Options ... 219

Chapter 13 Thermal Oxidation for Volatile Organic Compounds Control ... 221

13.1 Combustion Basics ... 221
13.2 Flares ... 223
13.2.1 Elevated, Open Flare ... 224
13.2.2 Smokeless Flare Assist ... 225
13.2.3 Flare Height ... 226
13.2.4 Ground Flare ... 227
13.2.5 Safety Features ... 228
13.3 Incineration ... 229
13.3.1 Direct Flame Incineration ... 230
13.3.2 Thermal Incineration ... 231

13.3.3 Catalytic Incineration 235
13.3.4 Energy Recuperation in Incineration 237

Chapter 14 Control of Volatile Organic Compounds and Hazardous Air Pollutants by Condensation 239

14.1 Introduction 239
14.1.1 Description 240
14.1.2 Advantages 240
14.1.3 Disadvantages 240
14.2 Volatile Organic Compounds Condensers 240
14.2.1 Contact Condensers 241
14.2.2 Surface Condensers 241
14.2.2.1 Example—Design Condensation Temperature to Achieve Desired Volatile Organic Compounds Recovery 243
14.3 Coolant and Heat Exchanger Type 244
14.3.1 Example—Heat Exchanger Area and Coolant Flow Rate 244
14.4 Mixtures of Organic Vapors 247
14.4.1 Example—Condensation of a Binary Mixture 248
14.5 Air as a Noncondensable 251
14.6 Systems-Based Approach for Designing Condensation Systems for Volatile Organic Compounds Recovery from Gaseous Emission Streams 252
Appendix 14A: Derivation of the Area Model for a Mixture Condensing from a Gas 256
Appendix 14B: Algorithm for the Area Model for a Mixture Condensing from a Gas 257

Chapter 15 Control of Volatile Organic Compounds and Hazardous Air Pollutants by Biofiltration 259

15.1 Introduction 259
15.2 Theory of Biofilter Operation 260
15.3 Design Parameters and Conditions 261
15.3.1 Depth and Media of Biofilter Bed 262
15.3.2 Microorganisms 262
15.3.3 Oxygen Supply 263
15.3.4 Inorganic Nutrient Supply 263
15.3.5 Moisture Content 263
15.3.6 Temperature 264
15.3.7 pH of the Biofilter 264

15.3.8 Loading and Removal Rates 264
15.3.9 Pressure Drop 265
15.3.10 Pretreatment of Gas Streams 265
15.4 Biofilter Compared to Other Available Control Technology 265
15.5 Successful Case Studies 266
15.6 Further Considerations 266

Chapter 16 Membrane Separation 267

16.1 Overview 267
16.1.1 Description 267
16.1.2 Advantages 268
16.1.3 Disadvantages 268
16.2 Polymeric Membranes 268
16.3 Performance 268
16.4 Applications 269
16.5 Membrane Systems Design 270

Chapter 17 NO_x Control 271

17.1 NO_x from Combustion 271
17.1.1 Thermal NO_x 271
17.1.2 Prompt NO_x 274
17.1.3 Fuel NO_x 274
17.2 Control Techniques 274
17.2.1 Combustion Control Techniques 274
17.2.1.1 Low Excess Air Firing 275
17.2.1.2 Overfire Air 275
17.2.1.3 Flue Gas Recirculation 275
17.2.1.4 Reduce Air Preheat 275
17.2.1.5 Reduce Firing Rate 275
17.2.1.6 Water/Steam Injection 276
17.2.1.7 Burners out of Service 276
17.2.1.8 Reburn 276
17.2.1.9 Low NO_x Burners 277
17.2.1.10 Ultra Low NO_x Burners 278
17.2.2 Flue Gas Treatment Techniques 279
17.2.2.1 Selective Noncatalytic Reduction 279
17.2.2.2 Selective Catalytic Reduction 280
17.2.2.3 Low Temperature Oxidation with Absorption 281
17.2.2.4 Catalytic Absorption 282
17.2.2.5 Corona-Induced Plasma 283

Chapter 18 Control of SO_x ... 285

18.1 H_2S Control ... 285
18.2 SO_2 (and HCL) Removal ... 287
18.2.1 Reagents ... 287
18.2.1.1 Calcium-Based Reactions ... 287
18.2.1.2 Calcium-Based Reaction Products ... 288
18.2.1.3 Sodium-Based Reactions ... 289
18.2.1.4 Sodium-Based Reaction Products ... 290
18.2.2 Capital versus Operating Costs ... 290
18.2.2.1 Operating Costs ... 290
18.2.3 SO_2 Removal Processes ... 291
18.2.3.1 Wet Limestone ... 291
18.2.3.2 Wet Soda Ash or Caustic Soda ... 293
18.2.3.3 Lime Spray Drying ... 294
18.2.3.4 Circulating Lime Reactor ... 296
18.2.3.5 Sodium Bicarbonate/Sodium Sesquicarbonate Injection ... 298
18.2.3.6 Other SO_2 Removal Processes ... 300
18.2.4 Example Evaluation ... 300
18.3 SO_3 and Sulfuric Acid ... 300
18.3.1 SO_3 and H_2SO_4 Formation ... 300
18.3.2 Toxic Release Inventory ... 304

Chapter 19 Fundamentals of Particulate Control ... 305

19.1 Particle Size Distribution ... 305
19.2 Aerodynamic Diameter ... 310
19.3 Cunningham Slip Correction ... 310
19.4 Collection Mechanisms ... 311
19.4.1 Basic Mechanisms: Impaction, Interception, and Diffusion ... 311
19.4.1.1 Impaction ... 312
19.4.1.2 Interception ... 313
19.4.1.3 Diffusion ... 313
19.4.2 Other Mechanisms ... 313
19.4.2.1 Electrostatic Attraction ... 313
19.4.2.2 Gravity ... 313
19.4.2.3 Centrifugal Force ... 313
19.4.2.4 Thermophoresis ... 314
19.4.2.5 Diffusiophoresis ... 314

Chapter 20 Hood and Ductwork Design 315

20.1 Introduction 315
20.2 Hood Design 316
20.2.1 Flow Relationship for Various Types of Hoods ... 317
20.2.1.1 Enclosing Hoods 317
20.2.1.2 Rectangular or Round Hoods 317
20.2.1.3 Slot Hoods 317
20.2.1.4 Canopy Hoods 318
20.3 Duct Design 318
20.3.1 Selection of Minimum Duct Velocity 319
20.3.2 Mechanical Energy Balance 320
20.3.2.1 Velocity Head 321
20.3.2.2 Friction Head 321
20.4 Effect of Entrance into a Hood 325
20.5 Total Energy Loss 325
20.6 Fan Power 325
20.7 Hood–Duct Example 326

Chapter 21 Cyclone Design 329

21.1 Collection Efficiency 329
21.1.1 Factors Affecting Collection Efficiency 331
21.1.2 Theoretical Collection Efficiency 333
21.1.3 Lapple's Efficiency Correlation 334
21.1.4 Leith and Licht Efficiency Model 335
21.1.5 Comparison of Efficiency Model Results 336
21.2 Pressure Drop 337
21.3 Saltation 338

Chapter 22 Design and Application of Wet Scrubbers 341

22.1 Introduction 341
22.2 Collection Mechanisms and Efficiency 342
22.3 Collection Mechanisms and Particle Size 342
22.4 Selection and Design of Scrubbers 344
22.5 Devices for Wet Scrubbing 344
22.6 Semrau Principle and Collection Efficiency 344
22.7 Model for Countercurrent Spray Chambers 347
22.7.1 Application to a Spray Tower 351

22.8 A Model for Venturi Scrubbers 357
22.9 Calvert Cut Diameter Design Technique 357
22.9.1 Example Calculation 360
22.9.2 Second Example Problem 361
22.10 Cut–Power Relationship 362
Appendix 22A: Calvert Performance Cut Diameter Data 363

Chapter 23 Filtration and Baghouses 367

23.1 Introduction 367
23.2 Design Issues 367
23.3 Cleaning Mechanisms 368
23.3.1 Shake/Deflate 368
23.3.2 Reverse Air 369
23.3.3 Pulse Jet (High Pressure) 370
23.3.4 Pulse Jet (Low Pressure) 372
23.3.5 Sonic Horns 372
23.4 Fabric Properties 373
23.4.1 Woven Bags 374
23.4.2 Felted Fabric 374
23.4.3 Surface Treatment 374
23.4.4 Weight 374
23.4.5 Membrane Fabrics 375
23.4.6 Catalytic Membranes 375
23.4.7 Pleated Cartridges 375
23.4.8 Ceramic Candles 376
23.5 Baghouse Size 376
23.5.1 Air-to-Cloth Ratio 376
23.5.2 Can Velocity 377
23.6 Pressure Drop 377
23.7 Bag Life 379
23.7.1 Failure Modes 379
23.7.2 Inlet Design 380
23.7.3 Startup Seasoning 380
23.8 Baghouse Design Theory 380
23.8.1 Design Considerations 382
23.8.2 Number of Compartments 383
23.8.3 Example Problem for a Baghouse Design 386

Chapter 24 Electrostatic Precipitators 391

24.1 Early Development 391
24.2 Basic Theory 391
24.2.1 Corona Formation 392
24.2.2 Particle Charging 392

24.2.3 Particle Migration ... 394
24.2.4 Deutsch Equation ... 395
24.2.4.1 Sneakage ... 396
24.2.4.2 Rapping Re-Entrainment ... 397
24.2.4.3 Particulate Resistivity ... 397
24.2.4.4 Gas-Flow Distribution ... 399
24.3 Practical Application of Theory ... 400
24.3.1 Effective Migration Velocity ... 400
24.3.2 Automatic Voltage Controller ... 400
24.4 Flue Gas Conditioning ... 401
24.4.1 Humidification ... 401
24.4.2 SO_3 ... 401
24.4.3 Ammonia ... 402
24.4.4 SO_3 and Ammonia ... 403
24.4.5 Ammonium Sulfate ... 403
24.4.6 Proprietary Additives ... 403
24.5 Using V-I Curves for Troubleshooting ... 403

References ... 407

Index ... 419

Preface

This book has been written to serve as a reference handbook for the practicing engineer or scientist who needs to prepare the basic process engineering and cost estimation required for the design of an air pollution control system. The user of this book should have a fundamental understanding of the factors resulting in air pollution and a general knowledge of the techniques used for air pollution control. The topics presented in this handbook are covered in sufficient depth so the user can proceed with the basic equipment design using the methods and design equations presented. Although moving sources, especially those powered by internal combustion engines, are serious contributors to the air pollution problem, this book will focus on stationary sources. Furthermore, this handbook will not consider nuclear power plants or other radioactive emissions. Therefore, the major audience for this book will be engineers and scientists in the chemical- and petroleum-processing industry and steam power plant and gas turbine industry.

Using this book, the air pollution control systems designer may

- Begin to select techniques for control
- Review alternative design methods
- Review equipment proposals from vendors
- Initiate cost studies of control equipment

This book is certainly suitable for anyone with an engineering or science background who needs a basic introduction to air pollution control equipment design. It can also be used as a textbook or reference in a continuing education program or a university classroom.

Karl B. Schnelle, Jr.
and
Russell F. Dunn
Vanderbilt University

Mary Ellen Ternes
Crowe and Dunlevy

Acknowledgments

Professor Schnelle acknowledges the assistance of his former student Dr. Partha Dey and his wife, Anita Dey, as well as Atip Laungphairojana, another former student, for the preparation of 70 drawings, which are part of the chapters he wrote. There would have been no book without their help. Furthermore, Professor Schnelle is forever grateful to his wife, Mary Dabney Schnelle, who read every word he wrote, including the equations. She helped remove many errors of grammar and spelling and corrected errors of algebra and definition that the author did not find.

Dr. Dunn acknowledges the support and encouragement of his wife, Donna Dunn, their children, Elizabeth and Catherine, and his parents, Gene and Carole Dunn. Dr. Dunn also acknowledges the continual inspiration that has been provided by Professor Mahmoud El-Halwagi, Texas A&M University.

Mary Ellen Ternes acknowledges the support of her children, Joshua and Jake, and her parents, William and Agnes Crowley.

All of the authors acknowledge Charles Brown for his significant contribution as an author on the first edition of this book.

Authors

Karl B. Schnelle, Jr., PhD, PE, is professor emeritus of chemical and environmental engineering; he has been a member of the Vanderbilt University faculty for more than 55 years. He has served as chair of the Environmental and Water Resources Engineering Program and the Chemical Engineering Department for a total of 14 years. He has extensive publications in the chemical engineering and environmental area. Dr. Schnelle is an emeritus member of both the American Institute of Chemical Engineers and the Air and Waste Management Association. He is a fellow of the American Institute of Chemical Engineers, a board certified environmental engineer of the American Academy of Environmental Engineers, and a life member of the American Society for Engineering Education. He was a lecturer in the American Institute of Chemical Engineers' continuing education program for more than 30 years, where he taught designing air pollution control systems and atmospheric dispersion modeling courses.

Dr. Schnelle is a licensed professional engineer in the state of Tennessee. He has been an environmental consultant to the World Health Organization, the Environmental Protection Agency, the U.S. State Department, and the Tennessee and Nashville air pollution control agencies as well as to numerous private corporations. He has served two terms as a member of the Air Pollution Control Board of the state of Tennessee and continues teaching part time at Vanderbilt University.

Russell F. Dunn, PhD, PE, is a professor of the practice of chemical and environmental engineering at Vanderbilt University. He also has prior academic experience teaching chemical engineering courses at Auburn University and the Technical University of Denmark. He has authored numerous publications and presentations on chemical and environmental engineering design, in addition to having over 30 years of professional experience in industry, consulting, and academia. His industrial experience includes being appointed as fellow when he worked at Solutia. Dr. Dunn is the founder and president of Polymer and Chemical Technologies, LLC, a company that provides chemical process and product failure analysis, in addition to developing environmental and energy-based process designs for large chemical plants. Through his company, Dr. Dunn has been a consultant to numerous private corporations and legal firms and has served as an expert witness on well over 200 chemical product and process engineering investigations.

Dr. Dunn is a licensed professional engineer in the state of Florida. Before starting his engineering consulting company, Dr. Dunn had industrial experience in various research, technology, and management positions at General Electric Company, Monsanto Chemical Company, Solutia, and Ampex Corporation. He has served a three-year term as a director of the Environmental Division of the American Institute of Chemical Engineers and currently teaches chemical engineering design and laboratory courses at Vanderbilt University.

Mary Ellen Ternes, BE ChE, JD, is a director of Crowe & Dunlevy, in the law firm's Environmental, Energy and Natural Resources practice group. For more than 20 years, Ternes has advised clients regarding Clean Air Act permitting, compliance strategies, enforcement defense, as well as federal and state litigation. Ternes has published and lectured extensively on environmental law, particularly the Clean Air Act. She is former chair of the American Bar Association's Section of Environment, Energy and Resources Air Quality Committee, the Climate Change, the Sustainable Development and Ecosystems Committee, and the Annual Conference on Environmental Law. Ternes began her career as a chemical engineer for the U.S. Environmental Protection Agency and then managed permitting and compliance for the hazardous waste incineration industry. She is a senior member of the American Institute of Chemical Engineers, and a founder and initial chair of that institute's Chemical Engineering and the Law Forum and Public Affairs and Information Committee.

Ternes earned a JD (high honors) at the University of Arkansas at Little Rock and a BE in chemical engineering at Vanderbilt University. Ternes has been admitted to practice law in Oklahoma, Arkansas, South Carolina, and the District of Columbia. Before law school, Ternes served as a Summer Honors Associate for the U.S. Environmental Protection Agency's Office of General Counsel Air and Radiation Division. She is listed in the *Chambers USA Guide* to America's Leading Lawyers for Business; The Best Lawyers in America, Super Lawyers, and International Who's Who of Environment Lawyers; and she is a fellow of the American College of Environmental Lawyers, serving on the Board of Regents.

1 Historical Overview of the Development of Clean Air Regulations

1.1 A BRIEF HISTORY OF THE AIR POLLUTION PROBLEM

Media reports about air pollution might lead us to think of air pollution as being something that developed in the second half of the twentieth century. But this is not so. The kind of air pollution to which human beings have been exposed has changed with time, as well as what we recognize as *air pollution*, but air pollution has been known in larger cities at least from the twentieth century, when people first started using coal for heating their homes.[1]

In England, during the reign of Edward I, there was a recorded protest by the nobility about the use of *sea* coal, which burned in an unusually smoky manner. Under his successor, Edward II (1307–1327), a man was put to torture for filling the air with a "pestilential odor" through the use of coal. Under the reign of Richard III and Henry V, England undertook to restrict the use of coal through taxation. Nevertheless, the situation continued to grow worse in the larger cities, so much so that during the reign of Elizabeth I (1533–1603; Queen, 1558–1603) Parliament passed a law forbidding the use of coal in the city of London while Parliament was in session. While this may have eased the pollution for the parliamentarians, it did very little to actually solve the problem.

As cities grew and the Industrial Revolution developed, the spread of coal smoke grew. In 1686, a paper was presented to the Royal Philosophical Society: "An Engine That Consumes Smoke." To this day, we have been working on this same problem, as yet to no avail. Legislation that was introduced often ignored the technical aspects of the problem, and hence was unenforceable. For example, a law passed by Parliament in 1845 stated that locomotives must consume their own smoke, which would be grand but, of course, it is not realizable.

The air pollution problem in the United States was first recognized as being due to coal smoke. In 1881, Chicago adopted a smoke control ordinance. St. Louis, Cincinnati, and other cities also adopted smoke ordinances in the years that followed. In these early years, it was established that the responsibility rested with the state and local governments.

Nashville, Tennessee, had a population of 80,865 in 1900, and it was a typical community of that period that depended on bituminous coal for heating. A short story written by O'Henry describes his visit to Nashville in 1900 as follows:

A Municipal Report

Nashville—a city, port of delivery, and the capital of the state of Tennessee, is on the Cumberland River and on the N.C. & St. L. and the L&N Railroads. This city is regarded as the most important educational centre in the South. I stepped off the train at 8 P.M. Having searched thesaurus in vain for adjectives, I must, as a substitution, hie me to comparison in the form of a recipe:

Take [a] London fog, thirty parts; malaria, ten parts; gas leaks, twenty parts; dewdrops gathered in a brick-yard at sunrise, twenty-five parts; odor of honeysuckle, fifteen parts. Mix. The mixture will give you [an] approximate conception of a Nashville drizzle. It is not so fragrant as a moth ball, nor as thick as pea soup; but 'tis enough—'twill serve.

From 1930 to 1941, the focus of air pollution was on smoke control laws. Public protest groups from Chicago, St. Louis, Cincinnati, and Pittsburgh had some success. However, air pollution was not recognized as the health hazard we know it to be today.

In 1941, war broke out with the Axis powers of Germany and Japan and their allies, and from late 1941 until 1945, there was an all-out effort to defeat these countries. This effort allowed no time or materials for air pollution control. Smoke levels reached new highs as the national effort rallied to the war. Finally, with a return to peace, action on pollution control was initiated.

In the prewar era, Pittsburgh had enacted a stringent new control regulation. In October 1946, a regulation that centered on the type of coal used was put into effect. Then, in October of 1948, tragedy struck at Donora, Pennsylvania.[2] Weather conditions were perfect for a stagnating inversion. As the inversion deepened, people in Donora became ill, and 20 died from the effects of the excessive air pollution that was prevalent. The result was an awakening to the health hazards of air pollution. Other such incidents were recognized throughout the world and are indicated in Table 1.1 compiled from several sources.[2–4]

October 1948 marks the start of a more vigorous program of air pollution control in the United States. For example, on May 1, 1949, the Pittsburgh smoke ordinance was extended to the whole of Allegheny County. Air pollution abatement was soon to attract the public's eye and money, but it was not until the advent of the Clean Air Act of 1963 that there was a national awakening to the value of our air environment.

During World War II, a new type of air pollution had been discovered in the Los Angeles atmosphere. New effects were manifest in the form of eye and skin irritation and plant damage not evident from simple smoke pollution. It was the result of a photochemical smog that was at first attributed to the oil refineries and storage facilities. When controls of these facilities did not result in a significant reduction of the problem, it was then discovered that the internal combustion engine was a major cause of this new type of pollution. The result of photochemical oxidation is seen in the brown haze apparent in the upper layer of the atmosphere. The brown haze is a mixture of particulates, oxides of nitrogen, sulfuric acid mist condensed from the oxidation of sulfur dioxide, and particles produced from photochemical reaction in the atmosphere. The haze limits visibility, decreases the amount of sunlight reaching the earth, results in an increase in the amount of cloudy weather present, and, when it accumulates, results in all the unpleasant effects associated with air pollution.

Perhaps it was in response to these visual signs of air pollution that people could see occurring that the nation decided to act. Most certainly, it was at this point that

TABLE 1.1
Horrible Total—Man Breathes, Coughs, and Dies

Location	Date	Deaths[a]	Reported Illness	Common Conditions
Meuse Valley, Belgium	12/1/30	63	6000	Low atmospheric dilution
Donora, Pennsylvania	10/26/48	18	5900 (43%)	Fog and gaseous materials
London	11/26/48	700–800		
Poza Rica, Mexico	11/21/50	22	>320	
London	12/5/52	3500–4000	Unknown	
New York	11/22/53	175–260	Unknown	
London	11/56	1000	Unknown	
London	12/2/57	700–800		
London	1/26/59	200–250		
London	12/5/62	700	Unknown	
London	1/7/63	700		
New York	1/9/63	200–400		
New York	11/23/66	170		

[a] Number of deaths above expected average death rate.

the federal government entered the picture with statutes and regulations governing conventional pollutants such as particulate and acid gases. More recently, the federal government began to regulate air pollution that people cannot see and with broader effects, specifically, ozone-depleting chemicals and greenhouse gases,[5] to protect stratospheric ozone and mitigate global climate change.

1.2 FEDERAL INVOLVEMENT IN AIR POLLUTION CONTROL

After World War II and the advent of the air pollution episodes in Donora and London, it became apparent that more concerted federal action was required. Congress first passed an air pollution law in 1955. At this time, Congress was particularly reluctant to interfere in states' rights, and early laws were not strong. These laws more or less defined the role of the federal government in research and training in air pollution effects and control. The following brief summary leads to the 1970 amendments to the Clean Air Act, beginning with the following:

The Air Pollution Control Act of 1955, Public Law 84–159, July 14, 1955

- Left states principally in charge of prevention and control of air pollution at the source
- Recognized the danger to the population of the growing problem
- Provided for research and training in air pollution control

Air Pollution Control Act Amendments of 1960, Public Law 86–493, June 6, 1960 and Amendments of 1962, Public Law 87–761, October 9, 1962

- Directed the Surgeon General to conduct a thorough study of the effects of motor vehicle exhausts on human health

The Clean Air Act of 1963, Public Law 88–206, December 1963

- Encouraged state and local programs for the control and abatement of air pollution while reserving federal authority to intervene in interstate conflicts
- Required development of air quality criteria which would be used as guides in setting ambient and emission standards
- Provided research authority to develop methods for removal of sulfur from fuels

Motor Vehicle Air Pollution Control Act of 1965, Public Law 89–272, October 20, 1965

- Recognized the technical feasibility of setting automobile emission standards
- Determined that such standards must be national standards and relegated automotive emission control to the federal government
- Gave the state of California waivers to develop standards more appropriate to the local situation

The Air Quality Act of 1967, Public Law 90–148, November 21, 1967

- Designated air quality control regions (AQCRs) within the United States, either inter-or intrastate
- Required issuance of air quality criteria
- Required states to established air quality standards consistent with air quality criteria in a fixed time schedule
- Gave states primary responsibility for action, but a very strong federal authority was provided
- Required development and issuance of information on recommended air pollution control technique

The Clean Air Amendments Act of 1970, Public Law 91–604, December 31, 1970

- Created the Environmental Protection Agency (EPA)
- Required states to prepare implementation plans on a given time schedule
- Set automotive emission standards
- Set the following basic control strategy to be employed, establishing
  - National Ambient Air Quality Standards (NAAQS)
  - Standards of performance for new stationary sources
  - National emission standards for hazardous pollutants

The Clean Air Act of 1970, subsequent amendments in 1990 and 1997, and judicial decisions followed by greenhouse gas regulations are discussed in more detail in Chapter 2.

1.3 CHARACTERIZING THE ATMOSPHERE

The atmosphere seems boundless, but of course it isn't! If we consider the relative amount of all living matter of mass equivalent to 1.0 unit, then the atmosphere, that is, all gases as we know them, would be about 300 units, and the hydrosphere, all waters, oceans, lakes, rivers, streams, ponds, and so on, would be about 70,000 units. Figure 1.1 illustrates the layers of the atmosphere.

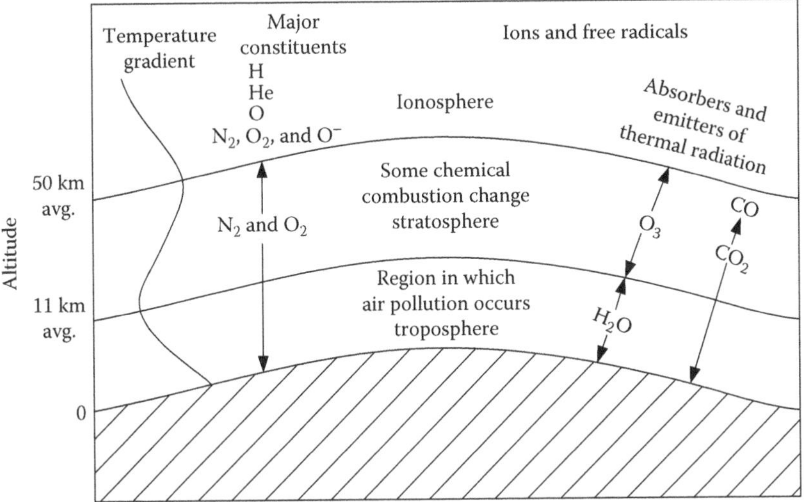

FIGURE 1.1 Layers of the atmosphere.

- Ionosphere (above 50 km)—ions and activated molecules produced by ultraviolet radiation such as in the following reaction:

$$O_2 = h\nu \rightarrow O^- + O^-$$

- Stratosphere (11–50 km)—the layer in which chemical composition changes takes place as illustrated by the following chain reaction:

$$NO_2 = h\nu \rightarrow NO + O^-$$
$$O + O_2 + M \rightarrow O_3 + M$$
$$O_3 + NO \rightarrow NO_2 + O_2$$

  Sinks above polar caps may be responsible for these actions. Temperature variation with altitude is small.
- Troposphere (0–11 km)—area of major concern in air pollution:
  Temperature decreases with altitude.
  Micrometeorological processes control the amount of pollution as it spreads and reaches ground level.

Table 1.2 records the chemical composition of air. Air normally contains water vapor, which would be somewhere around 1% by volume of the total mixture. The concentrations in Table 1.2 remain nearly constant or vary slowly. The following are variable in their concentration:

1. Water > variable 1.0% by volume
2. Meteoric dust

TABLE 1.2
Chemical Composition of Normal Air

Substance	Percent by Volume in Dry Air
N_2	78.09
O_2	20.94
Ar	0.93
CO_2	0.03
Ne	0.0018
He	0.00052
CH_4	0.00022
Kr	0.00010
N_2O	0.00010
H_2	0.00005
Xe	0.00008

Note: 1 ppm by volume = 0.0001% by volume.

3. Sodium chloride
4. Soil
5. NO_2 formed by electric discharge
6. O_3 formed by electric discharge
7. Pollen
8. Bacteria
9. Spores
10. Condensation nuclei
11. SO_2 volcanic oxygen
12. HCl volcanic origin
13. HF of volcanic origin

When doing combustion calculations, it is usual to assume that dry air contains 21% by volume of O_2 and 79% by volume of N_2. Table 1.3 compares concentrations of what could be considered pure air to concentrations in polluted air, including concentrations of CO_2 in 1750, before the industrial revolution, compared to 2010.

Table 1.4 is a historical record of concentrations of pollutants in cities in the United States in 1956, compiled by H. C. Wohlers and G. B. Bell at the Stanford Research Institute.

1.4 RECIPE FOR AN AIR POLLUTION PROBLEM

To have an air pollution incident, such as the one that occurred in Donora, or to have a problem, such as the one in Nashville, there are three factors that must occur simultaneously. There must be sources, a means of transport, and receptors.

TABLE 1.3
Comparison of Pure Air and a Polluted Atmosphere

Component	Considered to Be Pure Air	Typical Polluted Atmosphere
Particulate matter	10–20 μg/m^3	260–3200 μg/m^3
Sulfur dioxide	0.001–0.01 ppm	0.02–3.2 ppm
Carbon dioxide	300–330 ppm	350–700 ppm
Carbon monoxide	1 ppm	2–300 ppm
Oxides of nitrogen	0.001–0.01 ppm	0.30–3.5 ppm
Total hydrocarbons	1 ppm	1–20 ppm
Total oxidant	0.01 ppm	0.01–1.0 ppm

TABLE 1.4
Ranges of Concentrations of Gaseous Pollutants—A Historical Record from 1956

Pollutant	Range of Average Concentrations (ppm)	Range of Maximum Concentrations (ppm)	Number of Cities from Which Data Was Compiled
Aldehyde (as formaldehyde)	0.02–0.2	0.03–2.0	8
Ammonia	0.02–0.2	0.05–3.0	7
Carbon monoxide	2.0–10.0	3.0–300	8
Hydrogen fluoride	0.001–0.02	0.005–0.08	7
Hydrogen sulfide	0.002–0.1	Up to 1.0	4
Nitrogen oxides	0.02–0.9	0.03–3.5	8
Ozone	0.009–0.3	0.03–1.0	8
Sulfur dioxide	0.001–0.7	0.02–3.2	50

Source: Wohlers, H.C. and Bell, G.B., Stanford Research Institute Project No. SU-1816, 1956.

Figure 1.2 illustrates the process. Air pollution sources are relatively common knowledge. Their strength, type, and location are important factors. By transport, reference is made to the meteorological conditions, and the topography and climatology of a region, which are the important factors in dispersion—that is, in getting the material from the sources to the receptors. The receptors include human beings, other animals, materials, and plants. We also know that air pollution can affect visibility and can endanger our lives simply by making it difficult to travel on the highways and difficult for planes to land. The dollar cost of air pollution is the subject of much debate. However, it must be an astronomical figure, especially when you add such things as the extra dry cleaning and washing, houses that need more paintings than they should, and so on. The dollars lost to poor crops is a costly item in our economy, notwithstanding the impairment to shrubs, flowers, and trees.

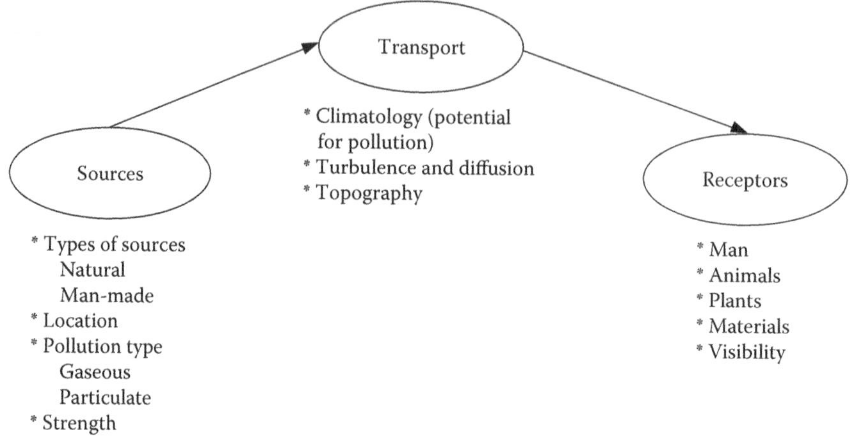

FIGURE 1.2 The trilogy: sources—transport—receptors.

1.4.1 Sources of Air Pollution

Sources of air pollution are either man-made or natural. Man-made sources are what we focus on because we may be able to effect some control on these sources. Both gaseous and particulate sources are troublesome. We have set standards for concentrations for both these materials in the atmosphere and in emission from chimneys for those sources we have recognized to be harmful. The concentration and the flow rate of the emissions are information required to determine the downwind transport of the pollutants. Knowing the location of the source relative to the receptor would allow us to calculate the concentration at a particular downwind receptor using a dispersion model. Chapter 2 details more about the pollutants that we are trying to control.

The EPA estimates the quantities of pollutants emitted each year. This information and many other facts are available at EPA's Air Trends website and other air pollution tracking websites available at: http//www.epa.gov/ttnchief1/trends/. Figure 1.3 records the annual production of air pollution by categories from Air Trends and the Inventory of U.S. Greenhouse Gas Emissions and Sinks.

1.4.2 Meteorological Parameters Affecting Transport of Pollutants

While ozone-depleting chemicals rise to the stratosphere, and greenhouse gases are recognized as being well-mixed in earth's atmosphere, the same is not true for conventional pollutants. The meteorological characteristics of the Los Angeles and Donora areas combine with the topographical features to form a container that traps conventional air pollution contaminants. The mountains in Los Angeles and the river valley in Donora form the walls of the containers that hamper horizontal air flow through these areas. A high-pressure area over the region forms the lid of such a container. A temperature inversion occurs, and the air becomes thermally stable, which has the effect of stopping vertical air flow, reinforcing the *lid* effect. In this way, the polluted air is not allowed to flow up and over the mountains or hills.

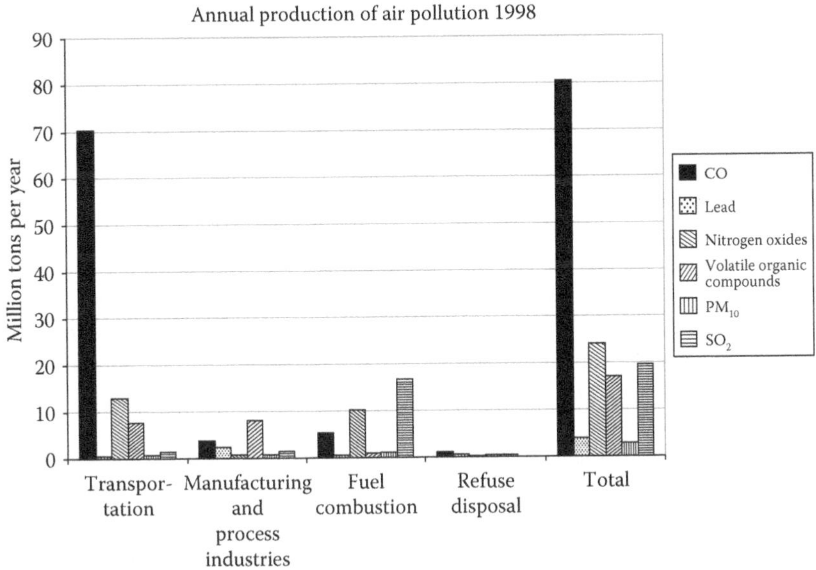

FIGURE 1.3 Annual production of pollutants by categories.

Meteorologists have long known that the amount of vertical motion of the atmosphere depends to an important extent on how the temperature varies with altitude. Near the ground, air temperature normally decreases with height. When the rate of decrease is rapid, there is a pronounced tendency toward vertical air mixing. On the other hand, when the air temperature increases with height, vertical air motions are suppressed. This temperature structure is called a *temperature inversion* because it is *inverted* from the normal condition of temperature decreasing with height.

Under adiabatic conditions, the temperature of dry air decreases at 5.45°F for each gain in altitude of 1000 ft. This temperature gradient is known as the *dry adiabatic lapse rate.* However, when the temperature of the air increases as the altitude increases; a condition known as *inversion* is present. An inversion may take place at the surface or in the upper air. A surface or a radiation inversion usually occurs on clear nights with low wind speed. In this situation, the ground cools the surface air by nature of its own cooling due to long wave radiation of heat to the outer atmosphere. The surface air becomes cooler than the upper layers, and vertical air flow is halted (hot air rises and cool air sinks). A parcel of warm air trying to rise finds the air above it hotter than itself. Thus, it will not rise because the temperature gradient is inverted. This type of inversion is common and is broken up as the sun once again heats the ground the next morning.

In Los Angeles, California, the typical inversion occurs in the upper air. There is an almost permanent high-pressure area centered over the north Pacific near the city. The axis of this high is inclined in such a way that air reaching the California coast is descending or subsiding. During the subsidence, the air is heated by compression, creating an inversion of the temperature gradient in the upper atmosphere over the city as the air moves from the sea over the land. This is termed as *subsidence inversion.*

Since the surface air in the Los Angeles area usually results from the sea breeze, the temperature difference between the upper layer and the surface is increased. The water is relatively cold and so the surface air that moves on over the land as the sea breeze is also cold. One might think that the daily cycle of sea and land breezes would break up the inversion, but this is not the case. The sea breezes only serve to raise and lower the altitude of the inversion layer. In the Los Angeles area, the effect of large air masses overrides the effect of the less powerful local heating from the surface.

In Donora, Pennsylvania, in the fall of 1948, as in London four years later, the weather was the wicked conspirator. A high-pressure area moved over western Pennsylvania on October 26, 1948, and remained fairly stationary for the next five days. Winds in the lowest 2000 ft of the atmosphere were quite weak. Mostly they were between a dead calm and 3 mph, but for brief periods they were slightly higher. The air was *thermally stable*, a formal description implying that there was very little vertical motion of the air.

Donora, lying near the bottom of a steep valley, is about 500 ft below the surrounding terrain. During the period October 26–31, 1948, an inversion capped the valley. Pilot reports and weather balloons showed that the cap, at least part of the time, was less than 1000 ft above the town. Thus, smoke fed to the atmosphere was largely confined within the valley walls and the inversion top. The air near the ground was very humid. Fog formed in the night, and in some low-lying areas of western Pennsylvania, it persisted during the day. At Donora, the visibility, cut by smoke and fog, ranged from about 0.6 to 1.5 miles. The overall weather conditions were similar to those that occurred in London, and the consequences tragically alike.

1.4.3 Effects of Air Pollution—A Comparison of London Smog and Los Angeles Smog

What makes a Los Angeles smog different from a London smog? A few conditions that apply to both cities are as follows. They are both similar in that they result in community air pollution, and the major source is the combustion of fuels. In London, it is coal and many hydrocarbons. In Los Angeles, it is primarily hydrocarbons. The peak time in London is early morning. In Los Angeles, the peak time is midday. In London, the temperature is 30°F–40°F; in Los Angeles, 75°F–90°F. The humidity is high with fog in London. Generally speaking, in Los Angeles, pollution occurs on a relatively clear day with low humidity. The inversions in London are at the surface; in Los Angeles, inversions are overhead. Visibility is severely reduced in London, but only partially reduced in Los Angeles. The effects in London are to produce bronchial irritation, whereas in Los Angeles, the effects are to produce eye and skin irritations. In Los Angeles, the smog is primarily produced through photochemical oxidation of the hydrocarbons by the ozone and nitrogen oxides that are in the atmosphere. The product of this photochemical reaction is an organic type molecule that causes plant damage and reduced visibility, and irritates skin and eyes very badly. The London type chokes us. We get a feeling of being in the midst of a big smoke, because that is primarily what it is—smoke and fumes mixed with moisture in the air. A summary of these conditions is given in Table 1.5. Both problems result

TABLE 1.5
Comparison of London Smog and Los Angeles Smog

Condition	London	Los Angeles
Fuel	Coal and hydrocarbons	Hydrocarbons
Season during year	December–January	August–September
Peak time	Early morning	Midday
Temperature	30°F–40°F	75°F–90°F
Humidity	High with fog	Low—relatively clear sky
Wind speed	Calm	<5 mph
Inversions	Surface (radiation)	Overhead (subsidence)
Visibility	Severely reduced (<1000 ft)	Reduced (½ to 1 mile)
Principal constituents	Sulfur compounds, carbon monoxide, and particulates	Ozone, nitrogen, oxides, organics, peroxyacetyl nitrate, and carbon monoxide
Effects on humans	Bronchial irritation	Eye and skin irritation
Chemical effects	Reducing	Oxidizing

from community air pollution. However, Los Angeles is different from London, Pittsburgh, and St. Louis.

Undesirable effects in Los Angeles are due to photochemical reactions irradiated by solar radiation. Due to oxidant formation, such as peroxyacetyl nitrate, photochemical smog produces undesirable effects such as eye and skin irritation, plant damage, and reduced visibility. The sources of the organic compounds and oxides of nitrogen largely are from the internal combustion engine.

It is the effects of air pollution on human beings, animals, plants, materials, and visibility that the EPA must consider when setting standards. Scientific evidence must be reviewed to support the standards selected. Therefore, the EPA has studied a great number of research documents and supported research to furnish the evidence needed for setting the standards.

Effects in all cases are dependent on a concentration–time effect, which can be called the *dosage*. All effects, especially health effects, are difficult to quantify. The problem is that people differ in their makeup and sensitivity to the various pollutants. Williamson[6] made an early attempt to quantify health effects, as can be seen in Figure 1.4. Here, the effects are shown as a function of concentration and exposure time. For example, the increased death rate in London occurred when people were exposed to a concentration of 0.30 ppm SO_2 for a period of one day. It is interesting to note that this time period was shorter than the other data showing mortality in humans. As discussed previously, we do know that the SO_2 emissions in London were accompanied by particulate matter, which formed the London smog. Experiments with exposure of individuals to controlled concentrations of SO_2 in otherwise pure air indicated that the human body can stand higher levels of SO_2 without the detriment experienced in the Donora incident. This incident also was accompanied by a high concentration of particulate matter. When the experiments were carried out with particulates mixed in the air along with the SO_2, the level of tolerance dropped well below the value, which can be tolerated for SO_2 alone.

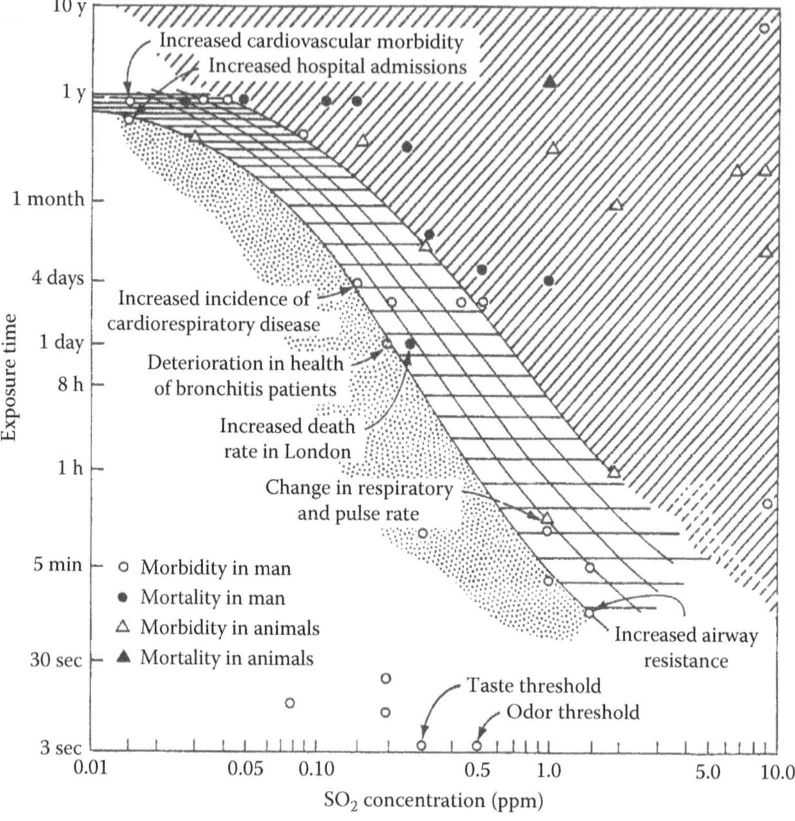

FIGURE 1.4 Health effects due to exposure to SO_2. Shaded area—range of exposures where excess deaths have been reported, grid area—range of exposures in which increased morbidity has been reported, speckled area—range of exposures where health effects are suspected. (Data from Williamson, S., *Fundamentals of Air Pollution,* Addison-Wesley, Reading, MA, 1973. With permission.)

These results can be summarized by noting that two plus two is greater than four when dealing with air pollution effects. Single contaminants in significant concentrations produce effects, but the effect is more generally associated with the mixture. Furthermore, the effect is intensified when a mixture is in the air. Thus, the total effect is greater than the sum of the effects of each individual pollutant. In other words, air pollution effects are synergistic.

2 Clean Air Act

2.1 HISTORY OF THE CLEAN AIR ACT

The development and maturation of the Clean Air Act (CAA) in the United States demonstrates an increasing federal role in the regulation of air pollution, including a broader scope to address the global issues, stratospheric ozone protection, and climate change. Until the 1950s, air pollution generally was perceived as a local and regional problem to be regulated at the local and state levels. California was the first state to act against air pollution, primarily because of deteriorating air quality in the already highly populated Los Angeles basin area with its unique geography and meteorology that exacerbated the problem.

The Air Pollution Control Act of 1955 authorized air pollution research and training programs and technical assistance to state and local governments, but the responsibility for air pollution control was left to state and local governments. The Air Pollution Control Act was amended in 1961 and again in 1962 to authorize special studies for health effects associated with motor vehicle pollutants.

The CAA was passed in 1963, establishing for the first time federal responsibility for air pollution control. In addition to authorizing grant research and training and technical assistance, federal enforcement authority was granted to abate interstate air pollution control problems. Also, air quality criteria were required to be developed for the protection of public health and welfare. The Motor Vehicle Air Pollution Control Act of 1965 authorized the Department of Health, Education, and Welfare to promulgate emission standards for motor vehicles, and the first federal emission standards for light motor vehicles were established for the 1968 model year.

2.1.1 1970 Clean Air Act Amendments

The 1970 Clean Air Act Amendments (CAAA) produced a dramatic change in broadening the federal responsibility for air pollution control. It created the U.S. Environmental Protection Agency (EPA), and it established the following programs for new air quality standards:

- National Ambient Air Quality Standards (NAAQS)
- New Source Performance Standards (NSPS)
- National Emission Standards for Hazardous Air Pollutants (NESHAP)
- Citizen suits for enforcement
- New automobile emission standards

States received EPA's delegation to implement and enforce the CAA upon obtaining EPA's approval of a state's proposed state implementation plan. The EPA-approved state implementation plans are promulgated in the federal regulations at 40 CFR

Part 52, so that they are federally enforceable. However, even in CAA-delegated states, EPA retains oversight and enforcement authority.

2.1.1.1 National Ambient Air Quality Standards

The NAAQS for criteria pollutants were to be established by the EPA at two levels: (1) primary standards to protect health and (2) secondary standards to protect welfare (e.g., crops, vegetation, buildings, visibility). The following six criteria pollutants were specified:

1. Particulate
2. Carbon monoxide (CO)
3. Ozone (O_3)
4. Sulfur dioxide (SO_2)
5. Nitrogen dioxide (NO_2)
6. Lead

Note that volatile organic compounds (VOCs) are not a criteria pollutant, but they are regulated like a criteria pollutant because VOCs and nitrogen oxides are precursors to ozone, which is produced by photochemical reactions.

Specifying lead as a criteria pollutant instead of a hazardous air pollutant is interesting, and this distinction proved to be effective. At that time, tetraethyl lead was prevalent in leaded gasoline as an inexpensive way to boost the octane rating. Controlling the spread of lead molecules throughout the environment as gasoline was burned was accomplished effectively as a criteria pollutant. Conversely, the hazardous air pollutant program, NESHAP, became bogged down in establishing health-based standards.

The EPA was required to review the NAAQS every five years to ensure that new research would be considered. The standard could remain at the same level if the review proved that the standard provides sufficient protection for health and welfare.

2.1.1.2 New Source Performance Standards

The 1970 amendments established a program of technology-based NSPS, so that any new source of air pollution would be required to apply effective air pollution controls. The NSPS are codified at 40 CFR Part 60. Existing sources were *grandfathered* and were not required to retrofit pollution controls. Existing sources that are modified are considered new sources, so old plants have to upgrade their air pollution control equipment as the plant is upgraded. The definition of *modified* is a subject of controversy to this day, centering on the distinction between *modification* and *maintenance*.[1]

2.1.1.3 Hazardous Air Pollutants

The NESHAP program for hazardous air pollutants required setting pollutant-specific, health-based standards for each hazardous air pollutant. Unfortunately, this program turned out to be too cumbersome to be implemented effectively. One significant problem was that the EPA administrator was to set the standard "at the level which in his judgment provides an ample margin of safety to protect the public from such hazardous pollutant."[2] First, it takes a great deal of research to

TABLE 2.1
Hazardous Air Pollutants for Which NESHAP Standards Were Established

Asbestos
Benzene
Beryllium
Inorganic arsenic
Mercury
Radionuclides
Radon-22
Vinyl chloride
Coke oven emissions (listed, but not promulgated)

establish a human health-based standard. Second, establishing an ample margin of safety was problematic, especially for the portion of the public that is sensitive to air pollution. It required court decisions to determine that cost of control could not be a factor in determining a health-based standard, but cost could be used to establish an ample margin of safety. Finally, establishing standards for over a 100 hazardous chemicals proved to be too burdensome. NESHAP standards were promulgated and codified at 40 CFR Part 61 for only the few hazardous pollutants listed in Table 2.1.

2.1.1.4 Citizen Suits

Determining that any person has sufficient interest to protect a universal resource such as clean air, Congress established citizen suits that authorize civil action against any person, including a government entity, who is alleged to be in violation of an emission standard or limitation. It also authorized suits against the EPA administrator when the administrator is alleged to fail to perform acts or duties required by the CAA. Costs of litigation, including reasonable attorney and expert witness fees, may be awarded. A key concept associated with citizen suits is the granting of *standing* to any individual, corporation, state, or municipality, doing away with the normal requirement of *injury in fact*.

There are procedural requirements, codified at 40 CFR Part 54, to limit excessive use of citizen litigation. One is providing 60 days' notice of a violation to the EPA, state, and alleged violator prior to starting a civil action. This is intended to allow agencies and violators the opportunity to correct a problem. Another condition barring a citizen suit is that the EPA or state has not already commenced and is not diligently pursuing a civil action. Again, citizen suits come into play when the enforcement agencies fail to meet their duties.

2.1.2 1977 Clean Air Act Amendments

Two highlights of the 1977 amendments were (1) codifying of the concept of prevention of significant deterioration (PSD) and (2) establishing an emission offset policy in areas that are not in attainment of the NAAQS.

2.1.2.1 Prevention of Significant Deterioration

The concept of preventing an area with good air quality from deteriorating to a lesser quality was introduced with the 1970 amendments. However, the wording was thin and controversy erupted. In 1973, the EPA proposed a plan for defining no significant deterioration, and reproposed and adopted the concept in 1974. The 1977 amendments removed the controversy by codifying PSD.

Three classifications of geographical areas were established, with boundaries initially set by the EPA. Changes could be made by state and local officials to allow local determination of land use. Class I areas were designated to allow very little deterioration of the existing air quality. Pristine areas were to remain pristine. These areas include national parks and wilderness areas and many Indian reservations. Class II areas allow a moderate decline from existing ambient air quality as growth occurs. Class III areas are specifically designated as heavy industrial areas, where a larger amount of deterioration of the existing ambient air quality is allowed. In no case are any areas allowed to have pollution levels that exceed the NAAQS, so if the existing air quality is already poor, limited or no additional deterioration is allowed.

Each area classification allows a specified incremental increase in the ambient pollutant concentration above the ambient background level as of 1977. Thus, a permanent cap is established, which may be lower than the NAAQS.

Areas with air pollutant concentrations below the applicable NAAQS are referred to as *attainment areas*. Any new major stationary source of air emissions, or major modification of an existing stationary source, which proposes to construct in an NAAQS attainment area, or an area that is unclassified, must first obtain a PSD construction permit issued pursuant to an approved state implementation plan, where both the permit and the plan satisfy substantive requirements provided by 40 CFR §§ 51.165 and 51.166, respectively. The PSD air permit authorizes the projected increase in ambient pollutant concentration. New source review for PSD air permit applications is discussed in Chapter 3. An approved PSD air permit allots a portion of the available increment for deterioration to that source. Later, new sources wishing to locate in the area, or major modifications of existing sources, must consider the amount of the increment that has already been consumed. Delegated air permitting agencies must consider reasonable increment consumption and projected growth in the area to avoid granting the increment spuriously on a *first come, first served* basis.

2.1.2.2 Offsets in Nonattainment Areas

A new air pollution source that wishes to locate in an area that does not meet the NAAQS is precluded from doing so, unless a net decrease in air pollution can be demonstrated. The proposed new source can agree to control or shut down emission sources of like pollutant in exchange, in order to offset the new source of pollution. To ensure that progress is made toward bringing the ambient air quality closer to the NAAQS, the offset ratio must be greater than 1 to 1.

To avoid putting industry in the position of keeping old, dirty emission sources operating to preserve offsets for future expansions, and to encourage emission reductions, the offset policy was accompanied by banking. When emission reductions were made, emission reduction credits could be applied for and granted. Banked emission reduction credits could be used as offsets in the future. Under federal guidance, credits were only good for a five-year contemporaneous period, however. Some state agencies, to which air permitting authority has been delegated by the EPA, use different contemporaneous periods. This is the type of discrepancy that has led to confusion between federal and state authority and program implementation.

2.2 1990 CLEAN AIR ACT AMENDMENTS

The 1990 CAAA made major, sweeping changes to the Act. They have been hailed[3] as "one of the most significant pieces of environmental legislation ever enacted." There was significant and lengthy Congressional debate throughout the 1980s as to how to address shortcomings of the Act, especially over provisions for acid rain control and air toxics.

The previous CAAA fell short of their goal of achieving acceptable air quality throughout the United States. Indeed, every major urban area was in violation of at least one of the NAAQS, affecting over 100 million people.[4] NESHAP had been promulgated for only eight hazardous air pollutants in the 20 years since the program has been established by the 1970 amendments. The EPA estimated that an excess of 1000–3000 cancer deaths per year were occurring as a result of hazardous air pollutants.

The political mandate for the 1990 amendments was large. The U.S. House of Representatives passed the amendments by a vote of 401 to 21, and the Senate voted in favor by 89 to 11.

The 1990 amendments were organized in 11 Titles, as listed in Table 2.2. The major changes include new provisions addressing visibility to mitigate *regional haze* (Title I), a completely overhauled hazardous air pollutant program (Title III), specific provisions to control pollutants that cause acid rain (Title IV), a new operating permit program (Title V), and a specific program to control pollutants that cause stratospheric ozone depletion (Title VI).

Concerns about the effectiveness of implementing the 1990 amendments still remain. There are numerous deadlines that the EPA must meet, but few *statutory hammers* or consequences to ensure that they are met. This opens the door to resorting

TABLE 2.2
1990 Clean Air Act Amendments

Title I	Attainment and Maintenance of NAAQS
Title II	Mobile Sources
Title III	Hazardous Air Pollutants (Air Toxics)
Title IV	Acid Rain (Acid Deposition Control)
Title V	Operating Permits
Title VI	Stratospheric Ozone and Global Climate Protection
Title VII	Enforcement
Title VIII	Miscellaneous Provisions
Title IX	Research
Title X	Disadvantaged Business
Title XI	Employment Transition Assistance

to lawsuits to force the EPA to meet the deadlines, diverting the EPA's budget resources to the defense of those lawsuits.

2.2.1 Title I: Provisions for Attainment and Maintenance of NAAQS

The 1990 CAAA recognized that many urban areas were not in attainment of the NAAQS, and that there were major problems with high levels of tropospheric ozone and CO. To address the problems practically, the amendments extended the time for states to achieve compliance, but required constant progress in reducing emissions and established provisions for sanctions on areas of the country that do not meet the conditions.

The new amendments also established degrees of severity for nonattainment for O_3, CO, and particulate matter with an aerodynamic diameter of less than 10 μ (PM_{10}).

Ozone nonattainment was broken into five degrees of severity, as listed in Table 2.3. Areas of extreme severity were allowed more time to achieve attainment, but receive increased federal scrutiny for new source review of major sources. The definition of *major* stationary source, discussed further in new source review permit applications in Chapter 3, applies to smaller sources, and the emission offset ratio requirement for new sources is increased.

Carbon monoxide and particulate matter degrees of severity for nonattainment were established at two levels as shown in Table 2.4.

Also in 1990, Congress added Section 169B, *Visibility*, to better address regional haze issues, which were not sufficiently addressed with Section 169A. Pursuant to

TABLE 2.3
Degrees of Severity for Ozone Nonattainment

	1-h Avg.	Attainment Date	Major Source Threshold (tons/year)	Offset Ratio for New Sources
Marginal	0.12–0.138	11/15/1993	100	1.1
Moderate	0.138–0.16	11/15/1996	100	1.15
Serious	0.16–0.18	11/15/1999	50	1.2
Severe	0.18–0.19	11/15/2005	25	1.3
	0.19–0.28	11/15/2007		
Extreme	>0.28	11/15/2010	10	1.5

TABLE 2.4
Degrees of Severity for CO and PM Nonattainment

	CO	PM
Moderate	9–16.5 ppm	Area that can achieve attainment by November 1996
Serious	>16.5 ppm	Area that cannot achieve compliance by November 1996

Section 169B, the EPA promulgated additional visibility protection requirements within 40 CFR 51.300–309, requiring states to submit state implementation plans addressing regional haze visibility impairment no later than December 17, 2007 and creating the Grand Canyon Visibility Transport Commission.

2.2.1.1 NAAQS Revisions

Current NAAQS as of December 2014 are listed in Table 2.5. The CAA Section 109(d)(1) requires that the EPA review all its criteria published under Section 108, and NAAQS published under Section 109 every five years to determine if new studies and scientific evidence warrant revisions to the standards. In 1997, the EPA issued a new primary and secondary ozone standard of 0.08 ppm for an 8-h average in addition to the existing standard of 0.12 ppm for a 1-h average. The EPA determined that longer term exposures to lower levels of ozone caused health effects including asthma attacks, breathing and respiratory problems, loss of lung function, and possible long-term lung damage and decreased immunity to disease. This was the first update of the ozone standard in 20 years.

Also in 1997, the EPA established a new particulate matter standard for fine particulates with an aerodynamic diameter of less than 2.5 μ ($PM_{2.5}$). The standard was established at 65 μg/m^3 for a 24-h average and 15 μg/m^3 for an annual average. The new standards followed a lawsuit by the American Lung Association that the EPA missed the deadline for the required review of the particulate standard. The previous review of the particulate standard was conducted in 1987. A new court-ordered deadline of July 1997 was established to finalize the particulate standard, and the EPA finalized the ozone standard simultaneously.

In May 1999, the new ozone and particulate standards were remanded by the Court of Appeals following a lawsuit by the American Trucking Association. The key issue was neither the quality of the health-based review, the review process, nor the degree of public health concern. Indeed, the Court of Appeals agreed that there was growing

TABLE 2.5
National Ambient Air Quality Standards

	Primary		Secondary	
	Level	**Averaging Time**	**Level**	**Averaging Time**
O_3	0.075 ppm	8 h	Same	Same
PM_{10}	150 μg/m^3	24 h	Same	Same
$PM_{2.5}$	35 μg/m^3	24 h	35 μg/m^3	24 h
	12 μg/m^3	Annual	15 μg/m^3	Annual
CO	35 ppm	1 h		
	9 ppm	8 h		
SO_2	0.075 ppm	1 h	0.5 ppm	3 h
NO_2	0.100 ppm	1 h	0.053 ppm	Annual
	0.053 ppm	Annual		
Lead	0.15 μg/m^3	Rolling 3 month	Same	Same

empirical evidence demonstrating a relationship between fine particle pollution and adverse health effects that justified a new fine particle standard. Instead, the key issue centered around EPA's authority to establish a standard for a *nonthreshold* pollutant that weighs health with the cost of implementing the standard.

A threshold pollutant is one that exhibits a minimum level, or threshold, below which no health effects are observed. A nonthreshold pollutant has diminishing health effects with decreasing concentration, but there will always be some effect even at extremely low concentrations. Threshold and nonthreshold pollutant responses are illustrated in Figure 2.1. The data are the difference in mortality between mice exposed to ozone and an unexposed control group, then both subsequently exposed to *Streptococcus* bacteria.[5] Figure 2.1 also demonstrates the difficulty in collecting data to measure and interpret a threshold level. A large number of expensive data at very low concentrations and with very small responses may be required to detect a threshold.

If a pollutant indeed has no health-based threshold, then the only healthy level is zero, which is impractical. Therefore, a judgment that compromises an acceptable level of health risk with a reasonable implementation cost must be reached, and such judgment constitutes establishing policy. In *American Trucking Associations, Inc. v. U.S. EPA*, the D.C. Circuit Court of Appeals said that the EPA lacked any determinate criterion for establishing where the standard for a nonthreshold pollutant should be set, so choosing a standard was capricious and arbitrary. In a decision of 2 to 1, the Court of Appeals determined that the CAA, as applied and absent further clarification, is unconstitutional because it gives "an unconstitutional delegation of legislative power"[6] to the EPA. The dissenting opinion of the Court was that this interpretation "ignores the last half-century of Supreme Court nondelegation jurisprudence."[6]

The EPA appealed to the U.S. Supreme Court. On February 27, 2001, the Supreme Court unanimously upheld the constitutionality of EPA's interpretation in setting the 1997 NAAQS revisions. Specifically, the Supreme Court held that the CAA does not

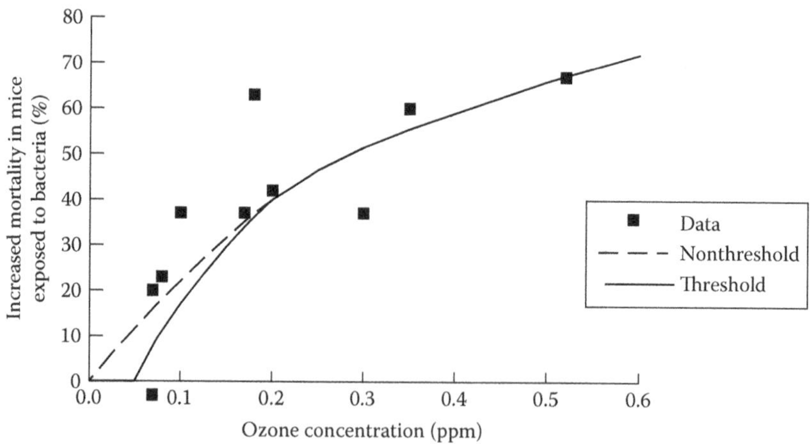

FIGURE 2.1 Threshold versus nonthreshold response. (Adapted from De Neveres, N., *Air Pollution Control Engineering*, Copyright 1995. The McGraw-Hill Companies, p. 16.)

permit the EPA to consider implementation costs in setting NAAQS, and disagreed that the EPA lacked determinate criteria for drawing the lines that guide the protection of public health.[7]

2.2.2 Title II: Mobile Sources

Cars and trucks account for almost half the emissions of the ozone precursors, VOCs and NO_x, and up to 90% of the CO emissions in urban areas. Although today's motor vehicles emit less pollution (60%–80% less, depending on the pollutant) than those built in the 1960s, there has been a rapid growth in the number of vehicles on the roadways and in the total miles driven, which has offset a large portion of the emission reductions gained from motor vehicle controls.

The 1990 amendments established standards that reduced tailpipe emissions of VOCs, NO_x, and CO on a phased-in basis beginning in model year 1994. Automobile manufacturers also are required to reduce vehicle emissions resulting from the evaporation of gasoline during refueling.

Reductions in the vapor pressure of gasoline and the sulfur content of diesel fuel also are required. Reformulated gasoline programs were initiated in 1995 for cities with the worst ozone problems. Higher levels (2.7%) of alcohol-based oxygenated fuels are produced and sold in areas that exceed the NAAQS for CO during the winter months.

The new law also established a clean fuel car pilot program in California, and a program limiting emissions from centrally fueled fleets of 10 or more vehicles in several areas of the country with severe smog problems. In 2007, ruling in favor of several states suing the EPA to regulate greenhouse gases (GHGs) emitted from mobile sources, the U.S. Supreme Court held that GHGs are *air pollutants* under the CAA.[8] In response to this decision, the EPA adopted several regulations governing GHG emissions from mobile sources and stationary sources. In 2009, the EPA published its threshold *Endangerment* and *Cause or Contribute* findings substantiating further CAA GHG rulemaking based on EPA's conclusions that current or future concentrations of six GHGs (carbon dioxide, methane, nitrous oxide, hydrofluorocarbons, perfluorocarbons, and sulfur hexafluoride) in the atmosphere endanger public health and welfare, and that GHG emissions from vehicles cause or contribute to this endangerment. In 2010, the EPA adopted GHG emission standards for mobile sources (*tailpipe rule*), triggering regulation of GHG emissions from major stationary sources for which the EPA adopted regulations addressing timing of implementation and tailoring the statutory major stationary source GHG emission thresholds for applicability purposes (*timing* and *tailoring rules*). After many years of litigation involving over 100 separate law suits challenging EPA's regulations implementing the Supreme Court's 2007 decision, the EPA retained authority to regulate GHG emissions from mobile and stationary sources. However, in 2014, the U.S. Supreme Court vacated EPA's authority to require CAA major stationary source permits based on GHG emissions alone, while affirming EPA's discretion to continue to regulate GHG emissions from stationary sources with criteria pollutants exceeding major source thresholds (*PSD anyway sources*).[9]

2.2.3 Title III: Hazardous Air Pollutant Program

The 1990 amendments took a completely new approach toward controlling hazardous air pollutants (HAPs). The HAPs are those pollutants that are hazardous to human health or the environment but are not specifically covered under another portion of the CAA. These pollutants are typically carcinogens mutagens and reproductive toxins. The original NESHAP program was based on establishing health-based standards for individual HAPs, but this approach became bogged down and unproductive with its burdensome requirements for regulatory development.

The new program requires establishment of technology-based standards for source categories that originally included 189 different HAPs, although caprolactam, used for manufacturing nylon, has been de-listed. The idea was to stop worrying about the specifics of health and risk for specific chemicals and start taking action by applying established control technologies to industrial sources. A determination of health-based *residual risk* could come later. A second major portion of Title III was requiring plans to prevent and mitigate accidental releases of HAPs.

The new technology-based NESHAP program requires the EPA to determine maximum achievable control technology (MACT) standards for major sources of HAPs. Major sources are defined as facilities with the potential to emit 10 tons/year of any single HAP or a total of 25 tons/year of all listed HAPs. The *potential to emit* is key, as opposed to actual emissions.

2.2.3.1 Source Categories

In 1992, the EPA published a list of 174 major source categories, for example, alkyd resins production; polystyrene production; petroleum refineries—catalytic cracking, catalytic reforming, and sulfur plant units—primary aluminum production; primary copper smelting; and industrial boilers. The list was updated in 1998, and again in 1999.[10] Among other changes, the latest revision added six new source categories (cellulosic sponge manufacturing, brick and structural clay products manufacturing, ceramics manufacturing, clay minerals processing, lightweight aggregate manufacturing, and wet-formed fiberglass mat production) and deleted two source categories (aerosol can-filling facilities and antimony oxides manufacturing).

Eight area source categories were listed for small stationary sources that were considered to pose a health threat on an individual or aggregate basis. Area sources are required to use generally available control technology. The area sources are listed in Table 2.6.

The original schedule for MACT and generally available control technology standards called for promulgating 40 standards by November 1992, 28 by November 1994, 28 by November 1997, and 54 by November 2000. These are designated 2-, 4-, 7-, and 10-year standards. It soon became apparent that the EPA could not meet this schedule, and a lawsuit resulted in a consent decree for court-ordered deadlines that required standards for 14 major sources and 4 area sources to be promulgated by July 1993.[11] Court-ordered deadlines tend to be missed less frequently than statutory deadlines to avoid the legal issue of being held in contempt of court.

A list of source categories for which MACT standards have been promulgated is provided in Table 2.7.

TABLE 2.6
70 Area Source Categories

Acrylic and Modacrylic Fibers Production
Ag Chemicals and Pesticides Manufacturing
Aluminum Foundries
Asphalt Processing and Asphalt Roofing Manufacturing
Autobody Refinishing
Brick and Structural Clay
Carbon Black Production
Chemical Manufacturing: Chromium Compounds
Chemical Preparations
Chromic Acid Anodizing[a]
Clay Ceramics
Commercial Sterilization Facilities[a]
Copper Foundries
Cyclic Crude and Intermediate Production
Decorative Chromium Electroplating[a]
Dry Cleaning Facilities[a]
Fabricated Metal Products, Electrical and Electronic Equipment—Finishing Operations
Fabricated Metal Products, Fabricated Metal Products, not elsewhere classified
Fabricated Metal Products, Fabricated Plate Work (Boiler Shops)
Fabricated Metal Products, Fabricated Structural Metal Manufacturing
Fabricated Metal Products, Heating Equipment, Except Electric
Fabricated Metal Products, Industrial Machinery and Equipment—Finishing Operations
Fabricated Metal Products, Iron and Steel Forging
Fabricated Metal Products, Primary Metal Products Manufacturing
Fabricated Metal Products, Valves and Pipe Fittings
Ferroalloys Production: Ferromanganese and Silicomanganese
Flexible Polyurethane Foam Production
Flexible Polyurethane Foam Fabrication
Gasoline Distribution Stage I
Halogenated Solvent Cleaners[a]
Hard Chromium Electroplating[a]
Hazardous Waste Incineration
Hospital Sterilizers
Industrial Boilers
Industrial Inorganic Chemical Manufacturing
Industrial Organic Chemical Manufacturing
Inorganic Pigments Manufacturing
Institutional/Commercial Boilers
Iron Foundries
Lead Acid Battery Manufacturing
Medical Waste Incinerators
Mercury Cell Chlor-Alkali Plants
Miscellaneous Coatings

(Continued)

TABLE 2.6 (*Continued*)
70 Area Source Categories

Miscellaneous Organic Chemical Manufacturing (MON)
Municipal Landfills
Municipal Waste Combustors
Nonferrous Foundries
Oil and Natural Gas Production
Other Solid Waste Incineration
Paint Stripping
Paints and Allie
Pharmaceutical Production
Plastic Materials and Resins Manufacturing
Plating and Polishing
Polyvinyl Chloride and Copolymers Production
Portland Cement Manufacturing
Prepared Feeds Manufacturing
Pressed and Blown Glass and Glassware Manufacturing
Primary Copper Smelting
Primary Nonferrous Metals—Zinc, Cadmium, and Beryllium
Publicly Owned Treatment Works
Secondary Copper Smelting
Secondary Lead Smelting
Secondary Nonferrous Metals
Sewage Sludge Incineration
Stainless and Non-Stainless Steel Manufacturing: Electric Arc Furnaces (EAF)
Stationary Internal Combustion Engines
Steel Foundries
Synthetic Rubber Manufacturing

[a] Notes original source category.

2.2.3.2 Establishing MACT Standards

The EPA must establish technology-based MACT standards for new and existing sources in the source categories. The minimum level of this technology is called the *MACT floor*. First, the EPA establishes the floor, then determines if the cost and benefit warrant more stringent technology, or going *beyond the floor*.

To evaluate the control technologies, the EPA collects information from companies by sending *114 Letters*, which must be answered under the authority of Section 114 of the CAA. The information includes emissions, controls, and costs. The EPA also may require source testing of designated facilities.

For existing facilities, the *floor* is the technology used by the best-performing 12% of the existing sources within a source category; unless there are fewer than

TABLE 2.7
Final MACT Standards (as of February 2015)

NESHAP (MACT) Standard Source Categories Affected	CFR Subpart(s) (40 CFR Part 63 Unless Noted)	Final Federal Register Date and Citation	Compliance Date (Check with EPA If No Date Listed)
Aerospace	GG	09/01/1995	09/01/1998
Acrylic/modacrylic fiber (area sources)	LLLLLL (6L)	07/16/2007	
Asbestos	40 CFR 61 Subpart M	40 CFR 61.140	
Asphalt processing and asphalt roofing manufacturing	LLLLL	04/29/2003	05/01/2006
Auto and light duty truck (surface coating)	IIII	04/26/2004	04/26/2007
Auto body refinishing (area sources)	HHHHHH (6H)	01/09/2008	
Benzene waste operations	40 CFR 61 Subpart FF	12/04/2003	12/04/2006
Boat manufacturing	VVVV	08/22/2001	08/22/2004
Brick and structural clay products manufacturing	JJJJJ	05/16/2003	05/16/2006
Clay ceramics manufacturing	KKKKK		
Carbon black production (area sources)	MMMMMM (6M)	07/16/2007	
Cellulose products manufacturing Miscellaneous viscose processes • Cellulose food casing • Rayon • Cellulosic sponge • Cellophane • Cellulose ethers production • Caroxymethyl cellulose • Methyl cellulose • Cellulose ethers	UUUU	06/11/2002	06/11/2005
Chemical manufacturing industry (area sources):CMAS	VVVVVV (6V)	10/29/2009	
Chemical preparations industry (area sources)	BBBBBBB (7B)	12/30/2009	
Chromium electroplating • Chromic acid anodizing • Decorative chromium electroplating • Hard chromium electroplating	N	01/25/1995	01/25/1996 deco 01/25/1997 others
Chromium compounds (area sources)	NNNNNN (6N)	07/16/2007	
Clay ceramics manufacturing (area sources)	RRRRRR (6R)	12/26/2007	12/26/2007

(Continued)

TABLE 2.7 (*Continued*)
Final MACT Standards (as of February 2015)

NESHAP (MACT) Standard Source Categories Affected	CFR Subpart(s) (40 CFR Part 63 Unless Noted)	Final Federal Register Date and Citation	Compliance Date (Check with EPA If No Date Listed)
Coke ovens: pushing, quenching, and battery stacks	CCCCC	04/14/2003	04/14/2006
Coke ovens • Charging, top side, and door leaks	L	10/27/1993	Contact project lead
Combustion sources at kraft, soda, and sulfite pulp and paper mills (pulp and paper MACT II)	MM	01/12/2001	01/12/2004
Commercial sterilizers • Commercial sterilization facilities	O	12/06/1994	12/06/1998
Degreasing organic cleaners • Halogenated solvent cleaners	T	12/02/1994	12/02/1997
Dry cleaning • Commercial dry cleaning dry-to-dry • Commercial dry cleaning transfer machines • Industrial dry cleaning dry-to-dry • Industrial dry cleaning transfer machines	M	09/22/1993	09/23/1996
Electric arc furnace steelmaking facilities (area sources)	YYYYY	12/28/2007	06/30/2008
Engine test cells/stands (combined with rocket testing facilities)	PPPPP	05/27/2003	See 68FR28774
Fabric printing, coating and dyeing	OOOO	05/29/2003	05/29/2006
Ferroalloys production (major sources)	XXX	05/20/1999	05/20/2001
Ferroalloys production (area sources)	YYYYYY (6Y)	12/23/2008	12/23/2011
Flexible polyurethane foam fabrication operation	MMMMM	04/14/2003	04/14/2004
Flexible polyurethane foam production and fabrication (area sources)	OOOOOO (6-O)	07/16/2007	
Flexible polyurethane foam production	III	10/07/1998	10/08/2001
Friction products manufacturing	QQQQQ	10/18/2002	10/18/2005
Gasoline dispensing facilities (area sources)	CCCCCC (6C)	01/10/2008	01/10/2011

(*Continued*)

TABLE 2.7 (*Continued*)
Final MACT Standards (as of February 2015)

NESHAP (MACT) Standard Source Categories Affected	CFR Subpart(s) (40 CFR Part 63 Unless Noted)	Final Federal Register Date and Citation	Compliance Date (Check with EPA If No Date Listed)
Gasoline distribution (stage 1)	R	12/14/1994	12/15/1997
Gasoline distribution bulk terminals, bulk plants, and pipeline facilities (area sources)	BBBBBB (6B)	01/10/2008	01/10/2011
General provisions	A		
Generic MACT I-acetal resins	YY UU	06/29/1999	06/29/2002
Generic MACT I-hydrogen fluoride	YY UU	06/29/1999	06/29/2002
Generic MACT I-polycarbonates production	YY UU	06/29/1999	06/29/2002
Generic MACT I-acrylic/Modacrylic fibers	YY UU	06/29/1999	06/29/2002
Generic MACT II-spandex production	YY UU	07/12/2002	07/12/2005
Generic MACT II-carbon black production	YY UU	07/12/2002	07/12/2005
Generic MACT II-ethylene processes	YY UU	07/12/2002	07/12/2005
Glass manufacturing (area sources)	SSSSSS (6S)	12/26/2007	12/26/2009 (existing sources) or upon startup (new sources)
Glass manufacturing plants—inorganic arsenic emissions	40 CFR 61, Subpart N		
Gold mine ore processing and production (area sources)	EEEEEEE (7E)	02/17/2011	02/17/2014
Hazardous waste combustion • Hazardous waste incinerators (A) • Hazardous waste incinerators (M)	Parts 63, 261 and 270	09/30/1999	09/30/2003
Hazardous organic NESHAP (synthetic organic chemical manufacturing industry)	F, G, H, I	04/22/1994	F/G-05/14/2001 H-05/12/1999 new sources 05/12/1998
Hospitals: ethylene oxide sterilizers (area sources)	WWWWW	12/28/2007	12/28/2007 (new sources) 12/28/2008 (existing sources)

(*Continued*)

TABLE 2.7 (*Continued*)
Final MACT Standards (as of February 2015)

NESHAP (MACT) Standard Source Categories Affected	CFR Subpart(s) (40 CFR Part 63 Unless Noted)	Final Federal Register Date and Citation	Compliance Date (Check with EPA If No Date Listed)
Hydrochloric acid production • Fumed silica production	NNNNN	04/17/2003	04/17/2006
Industrial, commercial, and institutional boilers and process heaters—major sources	DDDDD	09/13/2004	09/13/2007
Industrial, commercial, and institutional boilers—area sources (see also boiler compliance at area sources)	JJJJJJ (6J)	03/21/2011	03/21/2014
Industrial cooling towers	Q	09/08/1994	03/08/1995
Integrated iron and steel	FFFFF	05/20/2003	05/20/2006
Iron and steel foundries (major sources)	EEEEE	04/22/2004	04/22/2007
Iron and steel foundries (area sources)	ZZZZZ	01/02/2008	01/02/2011
Large appliances (surface coating)	NNNN	07/23/2002	07/23/2005
Lead acid battery mfg. (area sources)	PPPPPP (6P)	07/16/2007	
Leather finishing operations	TTTT	02/27/2002	02/27/2005
Lime manufacturing	AAAAA	01/05/2004	01/05/2007
Magnetic tape (surface coating)	EE	12/15/1994	without new control devices 12/15/1996 with new control device 12/15/1997
Manufacturing nutritional yeast (formerly baker's yeast)	CCCC	05/21/2001	05/21/2004
Marine vessel loading operations	Y	09/19/1995	MACT-09/19/1999 RACT-09/19/1998
Mercury cell chlor-alkali plants	IIIII	12/19/2003	12/19/2006
Metal can (surface coating)	KKKK	11/13/2003	11/13/2006
Metal coil (surface coating)	SSSS	06/10/2002	06/10/2005
Metal fabrication and finishing source nine categories (area sources)	XXXXXX (6×)	07/25/2008	07/25/2011
Metal furniture (surface coating)	RRRR	05/23/2003	05/23/2006
Mineral wool production	DDD	06/01/1999	06/01/2002
Misc. coating manufacturing	HHHHH	12/11/2003	12/11/2006
Misc. metal parts and products (surface coating) • Asphalt/coal tar application to metal pipes	MMMM	01/02/2004	01/02/2007

(*Continued*)

TABLE 2.7 (*Continued*)
Final MACT Standards (as of February 2015)

NESHAP (MACT) Standard Source Categories Affected	CFR Subpart(s) (40 CFR Part 63 Unless Noted)	Final Federal Register Date and Citation	Compliance Date (Check with EPA If No Date Listed)
Misc. organic chemical production and processes (MON) • Alkyd resins production • Ammonium sulfate production • Benzyltrimethylammonium chloride production • Carbonyl sulfide production • Chelating agents production • Chlorinated paraffins production • Ethyllidene norbomene production • Explosives production • Hydrazine production • Maleic anhydride copolymers production • Manufacture of paints, coatings, and adhesives • OBPA/1, 3-diisocyanate production • Photographic chemicals production • Phthalate plasticizers production • Polyester resins production • polymerized vinylidene chloride production • Polymethyl methacrylate resins production • Polyvinyl acetate emulsions production • Polyvinyl alcohol production • Polyvinyl butyral production • Quaternary ammonium compound production • Rubber chemicals production • Symmetrical tetrachloropyridine production	FFFF	11/10/2003	05/10/2008
Municipal solid waste landfills	AAAA	01/16/2003	
Natural gas transmission and storage	HHH	06/17/1999	06/17/2002

(*Continued*)

TABLE 2.7 (*Continued*)
Final MACT Standards (as of February 2015)

NESHAP (MACT) Standard Source Categories Affected	CFR Subpart(s) (40 CFR Part 63 Unless Noted)	Final Federal Register Date and Citation	Compliance Date (Check with EPA If No Date Listed)
Nonferrous foundries: aluminum, copper, and others (area sources)	ZZZZZZ (6Z)	06/25/2009	Existing sources—06/27/2011 new sources—*upon start-up*
Off-site waste recovery operations	DD	07/01/1996	02/01/2000
Oil and natural gas production includes area sources	HH	06/17/1999	06/17/2002
Organic liquids distribution (nongasoline)	EEEE	02/03/2004	02/03/2007
Paint stripping and miscellaneous surface coating operations—(area sources)	HHHHHH (6H)	01/09/2008	
Paper and other web (surface coating)	JJJJ	12/04/2002	12/04/2005
Pesticide active ingredient production • 4-chlror-2-methyl acid production • 2,4 salts and esters production • 4,6-dinitro-o-cresol production • Butadiene furfural cotrimer • Captafol production • Captan production • Chloroneb production • Chlorothalonil production • Dacthal™ production • Sodium pentachlorophenate production • Tordon™ acid production	MMM	06/23/1999	12/23/2003
Petroleum refineries	CC	08/18/1995	08/18/1998
Petroleum refineries • Catalytic cracking • Catalytic reforming • Sulfur plant units • Associated bypass lines	UUU	04/11/2002	04/11/2005
Pharmaceuticals production	GGG	09/21/1998	09/21/2001
Phosphoric acid	AA	06/10/1999	06/10/2002
Phosphate fertilizers	BB		
Plastic parts (surface coating)	PPPP	04/19/2004	04/19/2007
Plating and polishing operations (area sources)	WWWWWW (6W)	07/01/2008	07/01/2010

(*Continued*)

TABLE 2.7 (*Continued*)
Final MACT Standards (as of February 2015)

NESHAP (MACT) Standard Source Categories Affected	CFR Subpart(s) (40 CFR Part 63 Unless Noted)	Final Federal Register Date and Citation	Compliance Date (Check with EPA If No Date Listed)
Plywood and composite wood products (formerly plywood and particle board manufacturing)	DDDD	07/30/2004	
Polyether polyols production	PPP	06/01/1999	06/01/2002
Polymers and Resins I • Butyl rubber • Epichlorohydrin elastomers • Ethylene propylene rubber • Hypalon™ production • Neoprene production • Nitrile butadiene rubber • Polybutadiene rubber • Polysulfide rubber • Styrene-butadiene rubber and latex	U	09/05/1996	07/31/1997
Polymers and resins II • Epoxy resins production • Non-nylon polyamides production	W	03/08/1995	03/03/1998
Polymers and resins III • Amino resins • Phenolic resins	OOO	01/20/2000	01/20/2003
Polymers and resins IV • Acrylonitrile-butadiene-styrene • Methyl methacrylate-acrylonitrile+ • Methyl methacrylate-butadiene++ • Polystrene • Styrene acrylonitrile • Polyethylene terephthalate • Nitrile resins	JJJ	09/12/1996	07/31/1997
Polyvinyl chloride and copolymers production	J		
Polyvinyl chloride and copolymers production (area sources)	DDDDDD (6D)		
Portland cement manufacturing	LLL	06/14/1999	06/10/2002
Primary aluminum	LL	10/07/1997	10/07/1999
Primary copper	QQQ	06/12/2002	06/12/2005
Primary copper smelting (area sources)	EEEEEE (6E)	01/23/2007	

(*Continued*)

TABLE 2.7 (*Continued*)
Final MACT Standards (as of February 2015)

NESHAP (MACT) Standard Source Categories Affected	CFR Subpart(s) (40 CFR Part 63 Unless Noted)	Final Federal Register Date and Citation	Compliance Date (Check with EPA If No Date Listed)
Primary lead smelting	TTT	06/04/1999	05/04/2001
Primary magnesium refining	TTTTT	10/10/2003	10/10/2004
Primary nonferrous metals-zinc, cadmium, and beryllium (area sources)	GGGGGG (6G)	01/23/2007	
Printing and publishing (surface coating)	KK	05/30/1996	05/30/1999
Publicly owned treatment works (POTW)	VVV	10/26/1999	10/26/2002
Pulp and paper (noncombust) MACT	S	04/15/1998 03/08/1996	04/15/2001 04/16/2001
Reciprocating internal combustion engines includes area sources	ZZZZ	06/15/2004	06/15/2007
Refractory products manufacturing	SSSSS	04/16/2003	New or reconstructed 04/16/2003 existing 04/17/2006
Reinforced plastic composites production	WWWW	04/21/2003	04/21/2006
Rubber tire manufacturing	XXXX	07/09/2002	07/11/2005
Secondary aluminum	RRR	03/23/2000	Existing sources 03/24/2003 new sources 03/23/2000 or startup
Secondary copper smelting (area sources)	FFFFFF (6F)	01/23/2007	
Secondary lead smelters	X	06/23/1995	06/23/1997
Secondary nonferrous metals processing (brass, bronze, magnesium, and zinc) (area sources)	TTTTTT (6T)	12/26/2007	12/26/2007 (existing sources) or upon startup (new sources)
Semiconductor manufacturing	BBBBB	05/22/2003	05/22/2006
Shipbuilding and ship repair (surface coating)	II	12/15/1995	12/16/1996
Site remediation	GGGGG	10/08/2003	10/08/2006
Solvent extraction for vegetable oil production	GGGG	04/12/2001	04/12/2004
Stationary combustion turbines	YYYY	03/05/2004	03/05/2007
Steel pickling-HCL process	CCC	06/22/1999	06/22/2001

(*Continued*)

TABLE 2.7 (*Continued*)
Final MACT Standards (as of February 2015)

NESHAP (MACT) Standard Source Categories Affected	CFR Subpart(s) (40 CFR Part 63 Unless Noted)	Final Federal Register Date and Citation	Compliance Date (Check with EPA If No Date Listed)
Taconite iron ore processing	RRRRR	10/30/2003	10/30/2006
Tetrahydrobenzaldehyde manufacture (formerly butadiene dimers production)	F	05/12/1998	05/12/2001
Utility NESHAP	UUUUU		
Wet formed fiberglass mat production	HHHH	04/11/2002	04/11/2005
Wood building products (surface coating) (formerly flat wood paneling products)	QQQQ	05/28/2003	05/28/2006
Wood furniture (surface coating)	JJ	12/07/1995	11/21/1997
Wood preserving (area sources)	QQQQQQ (6Q)	07/16/2007	
Wool fiberglass manufacturing	NNN	06/14/1999	06/14/2002

Source: http://www.epa.gov/airtoxics/mactfnlalph.html.

30 sources, in which case, the standard must be at least as stringent as the average of the five best-performing facilities. This may not be as onerous as it first appears. In some industries, the same control technology is commonly used by most facilities. NSPS have been established for a long time for some source categories. It may well be that the best-performing 12% of sources use the same control technology as the best-performing 80% of the existing sources. Then, if MACT is set at the floor level, only 20% of the existing facilities would have to be upgraded to comply with MACT.

For new (and reconstructed) sources, the MACT *floor* is the single best-controlled similar facility within the source category. Again, if all of the best-performing plants use the same technology, this stringent requirement does not necessarily require that new technology will be required.

The EPA can, however, determine that available technology and the cost and health benefits warrant control technology that is more stringent than the floor technology. Typically, cost effectiveness is expressed in terms of dollars per ton of pollutant removed, and considers both capital and operating costs. Also, the EPA promotes flexibility in an attempt to enhance cost effectiveness by incorporating emissions averaging between units into the standards. This allows facilities to choose between lower cost options while still reducing hazardous air pollutant emissions.

While the technology-based MACT standards give a jump-start for action toward controlling HAPs, health-based standards have not been abandoned completely. Within eight years after promulgation of the technology-based standards, the EPA is required to review the residual risk associated with hazardous air pollutants. If the EPA determines that the remaining emissions from a facility after application of technology standards still pose a health risk, the facility may have to reduce emissions further. This may yet cause the program to fall into the same difficulties that burdened the NESHAP program prior to the 1990 amendments.

2.2.3.3 Risk Management Plans

A second feature of Title III is a program to prevent accidental releases of hazardous air pollutants and to reduce the risk of exposure to HAPs in the event of accidents. This is directly related to the accidental release of methyl isocyanate in Bhopal, India. The heart of the program is the requirement for *risk management plans.*

The program requires facilities to evaluate accidental release scenarios, mitigation action, and consequences and prepare a risk management plan. While this may seem like common sense for any facility that handles hazardous materials, it was not always done well. Hence, the regulation is intended to ensure that all facilities review, understand, mitigate, and plan for accidental releases of the HAPs.

Because many facilities used or stored hazardous chemicals, but did not emit them during normal operations, this provision brings those facilities under CAA regulations for the first time. It affects facilities that store or handle greater than threshold quantities of regulated substances from a list that is different from the list of HAPs for which MACT standards are promulgated.

2.2.4 Title IV: Acid Deposition Control

Title IV is directed at SO_2 and NO_x emissions from utility power plants to control the precursors of acid deposition, which include acid rain, acid snow, and acid fog. Coal-fired utility power plants are the largest source of man-made SO_2 emissions. The amendment repealed the requirement for a specified percentage reduction for new power plants and established a two-phase program for SO_2 and NO_x emission reductions. The amendments also established a market-based SO_2 allowance-trading program to promote cost-effective reductions.

Phase I, which became effective January 1, 1995, required 110 listed power plants of greater than 100 MW electrical capacity to reduce their emissions to a level equivalent to the product of an emissions rate of 2.5 lb of SO_2/MBTU by an average of their 1985–1987 fuel use. An additional 182 units joined Phase I of the program as substitution or compensating units, bringing the total number of affected units to 445. The success of the flexible program is reflected in that 1995 SO_2 emissions were reduced by almost 40% below the required level overall. Phase I also specified NO_x emission limits for tangentially fired boilers and dry-bottom, coal-fired boilers.

Phase II became effective January 1, 2000. It affected all electric utility units greater than 25 MW electrical capacity. The SO_2 emission rate was limited to 1.2 lb/MBTU. And a permanent cap of 8.9 million tons SO_2/year was established for the utility industry. Also, regulations for NO_x control were promulgated in order to achieve a NO_x reduction of 2 million tons/year.

For the market-based SO_2 allowance-trading program, each utility is allocated SO_2 allowances by the EPA based on an emission rate of 1.2 lb SO_2/MBTU. An allowance is permission to emit 1 ton of SO_2. Therefore, Phase II limits the number of available allowances in 2000 at 8.9 million. Each source must have sufficient allowances to cover its annual emissions. If a source's emissions exceed its allowance, it is subject to a $2000/ton excess emissions fee and a requirement to offset the excess emissions in the following year.

Allowances are fully marketable commodities that may be bought, sold, and traded by any individual, corporation, or governing body, including brokers, municipalities, environmental groups, and private citizens. There have been cases of allowances being purchased and not used in order to keep those tons of SO_2 from being emitted.

Under this program, utilities that emit less than their allocated allowances may sell the excess allowances or bank them for later use. In addition to the allocated allowances, a certain number of allowances are made available for direct sale, by EPA-sponsored auction. New power plants after 1996 are not allocated allowances, and must obtain them from the market.

To keep track of SO_2 emissions, stringent monitoring requirements have been developed. Of course, to measure the mass of SO_2 emitted, both exhaust gas flow and pollutant concentration are required. Daily calibration and record-keeping requirements must be met.

Initially, the program was met with some skepticism and cynicism. Some opponents claimed that utilities were given the ability to *buy the right to pollute*. However, the program has been a significant success. Through the end of 1999, over 80 million allowances have been traded, with approximately 62% of the trades between power plants within organizations, and 38% between organizations. The price of allowances, displayed in Figure 2.2, has fluctuated between $70 and $215, which is much lower than the anticipated $1500 value, as the program was being planned. The General Accounting Office estimated that the allowance trading system saves as much as $3 billion/year—over 50%—compared with a command and control approach typical of previous environmental protection programs.

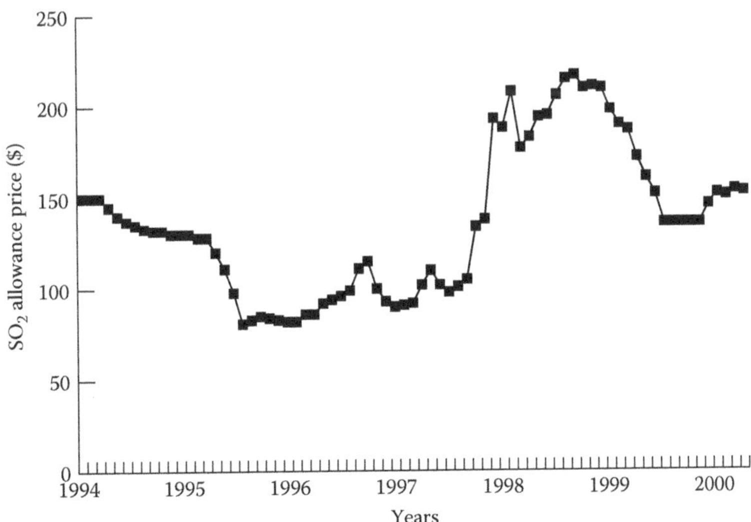

FIGURE 2.2 SO_2 allowance price history.

2.2.5 Title V: Operating Permits

The concept of an operating permit is to detail in one place a description of a major emission source, how it operates, the effect of operations on emissions, and all air emissions permit limitations and obligations. This would provide understanding and clarity for both the operator and the regulatory agency, which would aid in ensuring compliance. Currently, a source's pollution control obligations may be scattered throughout numerous hard-to-find provisions of state and federal regulations. The permit program is intended to ensure that all of a source's air pollutant obligations are being met, and that the source files periodic reports identifying the extent to which it has complied with those obligations.

The EPA struggled with a variety of monitoring programs that would ensure compliance with permit obligations. Initially, the compliance assurance monitoring rule was proposed, which emphasized continuous measurement of either pollutants or surrogate operating parameters that could be linked to pollutants. Eventually, the monitoring requirements were relaxed into periodic monitoring, which would be less stringent than continuous monitoring.

Title V also authorizes regulatory agencies to collect emission fees based on the quantity of emissions. This provision is intended to augment the resources for agencies to carry out enforcement of the CAA, as well as to provide some incentive for facilities to reduce emissions.

2.2.6 Title VI: Stratospheric Ozone and Global Climate Protection

The law requires a complete phase-out of chlorofluorocarbons and bromo-chlorofluorocarbons (i.e., halons) with interim reductions and some related changes to the existing Montreal Protocol, revised in June 1990. Under these provisions, the EPA must list all regulated substances along with their ozone depletion potential, atmospheric lifetimes, and global warming potentials within 60 days of enactment. EPA must ensure that Class I chemicals are phased out on a schedule similar to that specified in the Montreal Protocol—chlorofluorocarbons, halons, and carbon tetrachloride by 2000; methylchloroform by 2002—but with more stringent interim reductions. Class II chemicals, hydrochlorofluorocarbons (HCFCs) will be phased out by 2030. The law also requires the EPA to publish a list of safe and unsafe substitutes for Class I and Class II chemicals and to ban the use of unsafe substitutes.

2.2.7 Title VII: Enforcement

The 1990 amendments granted the EPA new authority to issue administrative penalty orders, and enhanced civil judicial penalties. Criminal penalties for knowing violations are upgraded from misdemeanors to felonies. Knowingly releasing a hazardous air pollutant that places another person in imminent danger of death or serious injury can result in a fine of up to $1,000,000 for organizations, and imprisonment of up to 15 years. Negligent release of a HAP is punishable by up to one year in prison. Making a false material statement, representation, or certification, or falsifying, tampering with, or failing to install monitoring devices can result in two years in prison.

Company officials must certify either compliance or acknowledgment of being out of compliance with regulations. If out of compliance, there must be a plan in place to achieve compliance. Certification adds personal liability and causes many managers to be much more concerned about compliance than they had been in the past.

Administrative penalties were set at a maximum potential fine up to $25,000/day, to a maximum of $200,000. These penalty maximums are increased for inflation by the Civil Monetary Penalty Inflation Rule, mandated by the Debt Collection Improvement Act of 1996. The last adjustment, in 2013, raised penalty maximums to $37,500 and $320,000, as codified by 40 CFR Part 19.When assessing penalties, EPA is required to consider the size of the business, compliance history, duration of the violation, good faith efforts to comply, payment of penalties previously assessed for the same violation, the economic advantage to the company for noncompliance, and the seriousness of the violation calculated pursuant to the Stationary Source Penalty Policy available on EPA's website at http://www2.epa.gov/enforcement.

An added legal right now allows citizens to seek penalties against violators, called *citizen suits*. Penalties are placed in a U.S. Treasury fund to be used by the EPA for compliance and enforcement activities. An award of up to $10,000 also was established for information leading to criminal conviction or administrative penalties for violations emissions standards or limitations imposed through the hazardous air pollutant, acid deposition, operating permit, or stratospheric ozone programs.

2.2.8 Title VIII: Miscellaneous Provisions

Among several other miscellaneous provisions, Title VIII requires that the EPA review the emission factors that are used to estimate quantities of CO, VOC, and NO_x emissions every three years. In addition, any person may demonstrate, after appropriate public participation, and request approval for improved emission estimating techniques.

2.2.9 Title IX: Research

Title IX provides federal funding for research programs including monitoring and modeling, health effects, ecological effects, accidental releases, pollution prevention, emissions control, acid rain, and alternative motor vehicle fuels. Ecosystem studies must be conducted for the effects of air pollutants on water quality, forests, and biological diversity.

2.2.10 Title X: Disadvantaged Business

Title X requires that at least 10% of federal research funds be made available to disadvantaged businesses.

2.2.11 Title XI: Employment Transition Assistance

Title XI provides additional unemployment benefits through the Job Training Partnership Act for workers laid off as a consequence of compliance with the CAA.

3 Air Permits for New Source

A facility that proposes to add or modify a source of air pollution may be required to obtain an approved air permit before starting construction of the new source. Local regulations define the permit requirements for small sources. Federal regulations define the air permit application requirements for major stationary sources. The process for evaluation and approval of a new major source permit application is called *new source review* (NSR). It covers permits for new major sources in both attainment, called *prevention of significant deterioration* (PSD) permits, and nonattainment area (NAA) permits. Although remaining in *draft* form for the past 25 years, when read with the current NSR regulations, Environmental Protection Agency (EPA)'s *New Source Review Workshop Manual* continues to provide excellent guidance for preparing an air permit application.[1,4]

Permit requirements are specific for each pollutant, including each criteria pollutant, precursors to tropospheric ozone (i.e., volatile organic compounds), and each hazardous air pollutant. Depending on the quantity of each pollutant emitted, each may be subject to a local regulation or to federal regulations for NSR.

3.1 ELEMENTS OF A PERMIT APPLICATION

In general, an air permit application will include the following:

- Facility information
- Description of the new source of air pollution
- Estimated quantities of each pollutant
- Applicability determination (if a major source is being considered)
- Description of air pollution control devices and justification for selection
- Effect on ambient air pollutant concentration
- Other impacts

Local regulatory agencies may provide forms that detail the required information explicitly and succinctly. A major NSR requires a large amount of discussion, so simple forms are not appropriate for the wide variety of circumstances and the long discussions of pertinent subjects. Instead, the required elements of the permit application are presented in chapters of a document with appendices for supporting information (e.g., detailed dispersion modeling output).

Although local requirements for small sources may not require as much detailed information as major sources, a similar evaluation process often is used. Many local agencies require a best available control technology (BACT) analysis for selection of controls for small sources styled in the same manner and using the same approach as required for major sources.

Interestingly, a permit application for a new major source in a NAA may require less information than a PSD permit application in an attainment area. The reason is because some of the information, such as cost-effectiveness of control equipment and PSD increment consumption, is moot. Control equipment for new sources in a NAA must meet lowest achievable emission rate, where cost-effectiveness is not a consideration.

3.1.1 Applicability

Applicability must be evaluated to determine which regulations apply to a new source. Three basic criteria are considered: (1) whether the source emissions are sufficiently large to cause the source to be considered a *major* stationary source; (2) if the source is in an attainment area; and (3) if *significant* amounts of specific pollutants will be emitted.

A stationary source is considered to be *major* if the *potential to emit* (PTE) exceeds the specified major source threshold. The threshold for industrial categories listed in Table 3.1 is 100 tons per year (tpy) of any regulated pollutant except the listed hazardous air pollutants. For all other industrial categories, the threshold is 250 tpy. However, as discussed in Chapter 2, the emission threshold for a *major* source is reduced in ozone NAAs, bringing smaller facilities into the scrutiny of federal NSR requirements in those areas.

TABLE 3.1
PSD Source Categories with Major Source Threshold of 100 tpy

Fossil-fuel-fired steam electric plants of more than 250 MBTU/h heat input
Coal cleaning plants (with thermal dryers)
Kraft pulp mills
Portland cement plants
Primary zinc smelters
Iron and steel mill plants
Primary aluminum ore reduction plants
Primary copper smelters
Municipal incinerators capable of charging more than 250 tons of refuse/day
Hydrofluoric, sulfuric, or nitric acid plants
Petroleum refineries
Lime plants
Phosphate-rock processing plants
Coke oven batteries
Sulfur recovery plants
Carbon black plants (furnace plants)
Primary lead smelters
Fuel conversion plants
Sintering plants

(Continued)

TABLE 3.1 (*Continued*)
PSD Source Categories with Major Source Threshold of 100 tpy

Secondary metal production plants
Chemical process plants (not including ethanol production by natural fermentation)
Fossil fuel boilers (or combinations thereof) totaling more than 250 MBTU/h heat input
Petroleum storage and transfer units with a total storage capacity exceeding 300,000 barrels
Taconite ore processing plants
Glass fiber processing plants
Charcoal production plants

3.1.1.1 Potential to Emit

PTE is the amount of pollution that a facility is capable of emitting at the maximum design capacity of the facility, except as constrained by federally enforceable conditions. The PTE is determined separately for each pollutant. Federally enforceable conditions are permit restrictions that can be verified, such as installed air pollution control equipment at prescribed efficiency, restrictions on hours of operation, and restrictions on the raw materials or fuel that is burned, stored, or processed. The facility must be able to show continual compliance (or noncompliance) with each limitation through testing, monitoring, and/or recordkeeping.

3.1.1.2 Fugitive Emissions

Fugitive emissions, to the extent that they are quantifiable, are included in the PTE for the sources listed in Table 3.1 and considered in subsequent impact analysis. Fugitive emissions are those that cannot be reasonably collected and passed through a stack, vent, or a similar opening. Examples include particulate matter from coal piles, roads, and quarries, and volatile organic compound emissions from leaking valves, flanges, and pump seals at refineries and organic chemical plants.

3.1.1.3 Secondary Emissions

Secondary emissions are not considered as part of the PTE. They are associated with the source, but are not emitted from the source itself. For example, a production increase from a new cement plant could require increased emissions at the adjacent quarry owned by another company. Increased secondary emissions at the quarry would not be included in the cement plant's PTE. However, if they are specific, well-defined, quantifiable, and impact the same general area as the cement plant, they would be considered in the discussion of the cement plant's ambient impacts analysis.

3.1.2 Significant Emission Rates

The significant emission rate thresholds, often called the *PSD trigger*, for federal NSR for specific pollutants are listed in Table 3.2. In attainment areas, any new or modified source emissions at a major source facility that increases the PTE of any of these pollutants by the significant emission rate triggers the requirement for a PSD permit.

TABLE 3.2
Significant Emission Rates for Federal New Source Review

Pollutant	Significant Emission Rate (tpy)
Carbon monoxide	100
Nitrogen oxides (all oxides of nitrogen)	40
Sulfur dioxide	40
Particulate matter	25
PM_{10}	15
$PM_{2.5}$	10
Ozone (as volatile organic compound or nitrogen oxides)	40
Lead	0.6
Fluorides	3
Sulfuric acid mist	7
Hydrogen sulfide	10
Total reduced sulfur compounds including H_2S	10
Municipal waste combustor organics 3.5×10^{-6}	
Municipal waste combustor metals 15	
Municipal waste combustor acid gases 40	
Municipal solid waste landfill emissions 50	

There is a second criterion for significant emission threshold. If any new major stationary source is constructed within 10 km of a Class I area and if the potential emissions would increase the 24-h average concentration of any regulated air pollutant by 1.0 μg/m^3 or more, then a PSD review is triggered.

3.1.3 Modification

A *modification* is generally a physical change or a change in the method of operation that results in an increase of air pollutant emissions, and an NSR for a modification is treated in the same way as for a new source. Existing sources that are not modified generally are *grandfathered* with no change in their existing pollution controls, although existing sources in NAAs may be required to install reasonably available control technology (RACT) for criteria pollutants, and existing sources of hazardous air pollutants may be regulated by the maximum achievable control technology standards for hazardous air pollutants as discussed in Chapter 2.

A modification is subject to PSD review only if the existing source is a *major* stationary source and if the net emissions increase of any pollutant emitted by the source is equal to or greater than the *significant emission rate*. The regulations do not define *physical change* or *change in the method of operation* precisely. However, they specifically exclude certain activities:

- Routine maintenance, repair, and replacement
- A fuel switch due to an order under the Energy Supply and Environmental Coordination Act of 1974 (or any superceding legislation) or due to a natural gas curtailment plan under the Federal Power Act

- A fuel switch due to an order or rule under Section 125 of the Clean Air Act
- A switch at a steam generating unit to a fuel derived in whole or in part from municipal solid waste
- A switch to a fuel or raw material, which (a) the source was capable of accommodating before January 6, 1975, so long as the switch would not be prohibited by any federally enforceable permit condition established after that date under a federally approved state implementation plan (SIP) (including any PSD permit condition) or a federal PSD permit or (b) the source is approved to make under a PSD permit
- Any increase in the hours or rate of operation of a source, so long as the increase would not be prohibited by any federally enforceable permit condition established after January 6, 1975, under a federally approved SIP (including any PSD permit condition) or a federal PSD permit
- A change in ownership of a stationary source

3.1.4 Emissions Netting

In determining the change in emissions that result from a modification at an existing source, other contemporaneous increases and decreases can be considered. A facility may increase NO_x emissions with one project, and decrease NO_x emissions at nearly the same time with another project. The net NO_x emissions may be less than the significant emission rate for NO_x, and the projects could *net out of PSD*. Avoiding the scrutiny of a PSD permit approval process may have advantages for the facility, including different agency jurisdiction for the review, fewer agencies involved in the process, and no PSD increment evaluation. The advantage to the environment is the incentive for reducing emissions, while allowing improvements at existing facilities to proceed.

The net emissions change equals the emissions increases associated with the proposed modification minus source-wide creditable contemporaneous emissions decreases plus source-wide creditable contemporaneous emissions increases. The increases and decreases are changes in actual emissions (not permitted or potential emissions). They must be at the same major source facility. They cannot be traded between facilities and the netting equation is pollutant specific.

The contemporaneous time period is defined by federal regulations as five years before the date that construction on the proposed modification is expected to commence. Changes older than the contemporaneous period are not considered. Many states to which jurisdiction for the PSD program has been delegated by the EPA have developed regulations with different time periods for the definition of contemporaneous. Where approved by the EPA, these time periods are used. Establishing the exact dates for the contemporaneous time period can be very important and a bit difficult to determine. Since the date that construction will commence is unknown when the permit application is prepared and reviewed, the scheduled date may be used. The date of increases and decreases is set when the changed emission unit becomes operational, which may be after a reasonable shakedown period that does not exceed 180 days.

3.1.4.1 Netting Example

A major source proposes a modification to its facility that would result increasing the PTE by 35 tpy NO_x, 80 tpy SO_2, and 25 tpy particulate matter less than 10 μm (PM_{10}). Two years ago, the facility incorporated a smaller modification that increased the PTE by 30 tpy NO_x, 15 tpy SO_2, and 5 tpy PM_{10}, all of which were less than the significant amounts that would trigger a PSD review for that modification. Three years ago, the facility made a change that decreased actual emissions by 10 tpy NO_x, 50 tpy SO_2, and 20 tpy PM_{10}. The facility applied for and received emission reduction credits for the decreased emissions.

Since the proposed increase in NO_x emissions of 35 tpy is less than the significant amount of 40 tpy, a PSD review for NO_x is not required. The net increase in potential SO_2 emissions is 45 tpy (+80 + 15 − 50), which is greater than the PSD trigger and a PSD review for SO_2 is required. The net increase in potential particulate matter emissions is +10 tpy (+25 + 5 − 20), which is less than the PSD trigger and a PSD review for PM_{10} is not required.

3.2 BEST AVAILABLE CONTROL TECHNOLOGY

The Federal NSR program requires that BACT be applied to major new sources and major modifications in attainment areas. Many local agencies also require that BACT be applied to smaller new sources of air pollution. Often, the selection of control technology does not depend on whether a PSD review is triggered because the same control technology will be required in any case.

BACT is a technology-based standard that can be met by available technology. Typically, it is specified in an air permit as a maximum emission rate or concentration that can be monitored and verified. Frequently, one will read or hear a vendor's advertisement that their technology is BACT for a type of emission source. More precisely, they mean that their technology usually is accepted by permitting agencies as the technology that will be used to control emissions to a specified limit. BACT is determined on a case-by-case basis as the technology that will provide the maximum degree of reduction taking into account energy, environmental, and economic impacts and other costs. The permit applicant proposes BACT for the proposed project and provides a detailed BACT analysis to support their proposal. The permitting agency reviews the permit application for completeness and accuracy, and may approve the proposed technology and emission limitation, reject it, or request more information.

BACT may be differentiated from other acronymed technologies. RACT applies to retrofits for existing sources in NAAs, while BACT applies to new and modified sources. Lowest achievable emission rate applies to new and modified sources in NAAs, while BACT applies to areas that are in attainment of the National Ambient Air Quality Standards (NAAQS). Maximum achievable control technology applies to hazardous air pollutants from industrial point sources, while BACT applies to criteria pollutants. Generally available control technology applies to hazardous air pollutants from area sources, which is any stationary source of hazardous air pollutants that is not a major source.

Some local agencies adopt additional distinctions for control technologies. Best available retrofit control technology is used in California for existing sources in NAAs, just like RACT. It may be difficult to distinguish between best available retrofit control technology and RACT, although both are based on judgment resulting from an evaluation of the source and situation, and best available retrofit control technology is sometimes more stringent than RACT for a given source type.[2] Toxic BACT is used by some local agencies for new sources of toxic air pollutants.

Cost-effectiveness is a very important part of BACT, whereas it is not considered when selecting the lowest achievable emission rate for new sources in NAAQS NAAs. One also frequently anticipates permitting agencies to be looking for a specific cost-effectiveness, expressed as dollars per ton of pollutant removed, when they accept BACT. Agencies, meanwhile, are reluctant to establish a firm threshold for cost-effectiveness, because a BACT analysis is more than a simple economic evaluation and needs to take other criteria into account. This makes it difficult for an applicant to judge what an agency considers to be cost-effective. Precedent, both locally and nationally, can be used as a guide.

The EPA has adopted a *top down* methodology for determining BACT. This means that the control technology with the highest degree of pollutant reduction must be evaluated first. If that technology meets all of the criteria for feasibility, cost-effectiveness, air quality impacts, and other impacts, it is determined to be BACT for the proposed new source and the analysis ends. If not, the next most effective control technology is evaluated for BACT. The *top down* approach is rigorous and is driven using the highest reduction technology that is considered cost-effective by the agency rather than more cost-effective compromises that would be desired by an applicant.

As a minimum level, BACT must at least meet the emission limitations of new source performance standards. Established new source performance standards have already been based on a thorough technology and economic evaluation. But as control technology improves, BACT may become more stringent than new source performance standards.

3.2.1 Step 1: Identify Control Technologies

First, potential control technologies are identified. The potential technologies include those that are used outside of the United States and those that are used for similar emission sources. Innovative control technologies may be included, but if a technology has not been commercially demonstrated and if it is not commercially available, then it is not considered to be a potential for *available* control technology.

This step takes some research. As a minimum, research should include EPA's Office of Air Quality Planning and Standards Technology Transfer Network Clean Air Technology Center RACT/BACT/ lowest achievable emission rate Clearinghouse (www.epa.gov/ttn/catc). Agencies use this database to post permit application emission limits, technologies, and restrictions for other agencies and applicants to use for just this purpose. Technical books and journals are another common source of technology information. Internet searches now can be quite valuable. Control equipment vendors often are invaluable sources of information, although one must bear in mind their vested interest in selling their equipment. Environmental consultants often are hired to prepare permit applications because of their prior knowledge of control technologies and permit application requirements. Less common, but acceptable, sources of information include technical conference reports, seminars, and newsletters.

Agencies typically expect a reasonably complete list, but it does not have to be exhaustive. Agency personnel use their own knowledge of the industry to check for completeness, and will request additional information for a potential technology if they believe a good possibility has been left out.

3.2.2 Step 2: Eliminate Technically Infeasible Options

Some of the potential control technologies may not work in the specific application being proposed. To be technically feasible, the control technology must be available and applicable to the source. Technical infeasibility may be based on physical, chemical, or engineering principles. Technical judgment must be used by the applicant and the reviewing agency in determining whether a control technology is applicable to the specific source.

For example, while modifying a process heater, low NO_x burners might not be able to be retrofit into the heater due to its configuration. The longer flame length produced by low NO_x burners may impinge on the back wall of the heater.

Demonstrating unresolvable technical difficulty due to the size of the unit, location of the proposed site, or operating problems related to specific circumstances could show technical infeasibility. However, when the resolution of technical difficulties is a matter of cost, the technology should be considered technically feasible, and let the economic evaluation determine if the cost is reasonable. There may be instances where huge cost can solve many problems. Judgment should be exercised to avoid wasteful effort in detailed economic evaluations that don't stand a chance of being reasonable.

3.2.3 Step 3: Rank Remaining Options by Control Effectiveness

All remaining control technologies that are not eliminated during Step 2 are ranked in order of control effectiveness for the pollutant under review. While seemingly straightforward, two key issues must be addressed in this step. The first is ensuring that a comparable basis is used for emission-reduction capability. Calculating potential emissions on a basis of mass per unit production, that is, pounds SO_2/MBTU heat input, can help.

A more difficult problem arises for control techniques that can operate over a wide range of emission performance levels, depending on factors such as size (e.g., a large electrostatic precipitator versus a small one) and operating parameters (e.g., scrubber recirculation rate and reagent concentration). Every possible level of efficiency does not have to be analyzed. Again, judgment is required. Recent regulatory decisions and performance data can help identify practical levels of control. This reduces the need to analyze extremely high levels in the range of control. By spending a huge amount of money and installing a huge electrostatic precipitator, an extremely high particulate removal efficiency is theoretically possible. However, as discussed in Chapter 24, there are practical limitations to theoretical performance. Demonstrated performance is critical to the evaluation of available technology. Also, it is generally presumed that demonstrated levels of control can be achieved unless there are source-specific factors that limit the effectiveness of the technology. This eliminates the need to analyze lower levels of a range of control.

3.2.4 Step 4: Evaluate Control Technologies in Order of Control Effectiveness

Starting with the feasible technology having the highest pollutant removal performance, the technology is evaluated for energy, environmental (other than air quality, which will be analyzed in detail in another section of a new source permit application), and economic impacts. Both adverse and beneficial effects should be discussed, and supporting information should be presented.

3.2.4.1 Energy Impacts

Many control technologies either consume or produce energy or change the amount of energy that is produced by the process. Thermal oxidation of volatile organics in a concentrated gas stream may be an example of energy production if that energy can be recovered and put to use.

Energy and energy conservation is considered by the government to be important to the economy of the nation, and the energy analysis assures that energy has been considered. However, quantitative thresholds for energy impacts have not been established, so it is difficult to reject a technology based on energy alone. There may be a circumstance where the required energy is not available at the proposed source. The cost of energy will be considered separately in the economic evaluation.

The energy evaluation should consider only direct energy consumption by the control technology. Indirect energy consumption that should not be considered includes the energy required to produce the raw materials for the control technology (e.g., the heat required to produce lime from limestone) and for transportation of raw materials and by-products.

3.2.4.2 Environmental Impacts

This section of the BACT analysis provides information about environmental impacts other than air quality. A detailed discussion of pollutant concentration in ambient air must be provided separately in the air permit application. This discussion of environmental impacts includes topics such as creation of solid or hazardous waste (e.g., as by-products of the control technology or as spent catalyst), wastewater, excess water use in an area with inadequate supply, emissions of unregulated air pollutants, and destruction of critical habitat. It also contains an evaluation of impacts on visibility. Both positive and negative impacts should be discussed.

That a control technology generates liquid or solid waste or has other environmental impacts does not necessarily eliminate the technology as BACT. Rather, if the other impacts create significant environmental problems, there may be a basis for elimination. Site-specific and local issues may define what problems are considered to be *significant*.

3.2.4.3 Economic Impacts and Cost-Effectiveness

This section tends to be more objective than Sections 3.2.4.1 and 3.2.4.2 because costs are quantified in terms of dollars spent per ton of pollutant reduced. The focus is the cost of control, not the economic situation of the facility.

The primary factors required to calculate cost-effectiveness, defined as dollars per ton of pollutant reduced, are the emission rate, the control effectiveness, the annualized capital cost of the control equipment, and the operating cost of the control equipment. With these factors, the cost-effectiveness is simply the annualized capital and operating costs divided by the amount of pollution controlled by the technology in one year.

Estimating and annualizing capital costs are discussed in Chapter 7. Total capital cost includes equipment, engineering, construction, and startup. Annual operating costs include operating; supervisory; maintenance labor, maintenance materials, and parts; reagents; utilities; overhead and administration; property tax; and insurance.

In addition to the overall cost-effectiveness for a control technology, the incremental cost-effectiveness between dominant control options can be presented. The incremental cost-effectiveness is the difference in cost between two dominant technologies divided by the difference in the pollution emission rates after applying these two technologies. An extraordinarily high incremental cost-effectiveness can be used as justification that a control technology having only slightly higher removal efficiency than the next technology, but which has significantly higher cost, is not cost-effective.

The incremental cost-effective approach sometimes can be used to establish the practical limits for technologies with variable removal efficiency. For example, an electrostatic precipitator with 98% removal efficiency may be considered by an applicant to be cost-effective. But to achieve 99.5% efficiency, a much larger and more expensive precipitator would be required. The large additional cost for a small improvement in efficiency is not cost-effective.

3.2.5 Step 5: Select BACT

Step 4 is repeated for each technology in order of the control effectiveness established in Step 3 until one of the technologies is not eliminated by the evaluation. The most effective level of control that is not eliminated by the evaluation is proposed by the applicant to be BACT. It is not necessary to continue evaluating technologies that have lower control effectiveness.

The permitting agency will review the application for completeness and accuracy, and is responsible for assuring that all pertinent issues are addressed before approving the application. The ultimate BACT decision is made by the agency after public comment is solicited and public review is held and comments are addressed.

3.3 AIR QUALITY ANALYSIS

Each PSD source or modification must perform an air quality analysis to demonstrate that its new pollutant emissions would not violate either the applicable NAAQS or the applicable PSD increment. To determine if a new industrial source of air pollution can be established in any area, the projected increase in ambient pollutant concentration must be determined using dispersion modeling. If the proposed source is large and if there is any doubt that either the allowed increment could be exceeded or that the NAAQS could be approached, air permitting agencies may require up to two years of ambient air monitoring to establish accurate background concentrations. Additional monitoring may be required after the source starts up to ensure that modeling calculations did not underpredict the effect of the new source.

3.3.1 Preliminary Analysis

A preliminary dispersion modeling analysis is first performed to: (1) determine whether the applicant can forego further air quality analyses for a particular pollutant; (2) allow the applicant to be exempted from ambient monitoring data requirements; and (3) determine the impact area within which a full impact analysis must be carried out. The levels of significance for air quality impacts in Class II areas are listed in Table 3.3. If a proposed source is located within 100 km of a Class I area, an impact of 1 μg/m^3 on a 24-h basis is significant. If the projected increase in pollutant concentration is less than the significance level for the appropriate time averaging period, then it is considered *insignificant*, and an evaluation of PSD increment consumption is not required. Also, the available increment is not reduced by projects having an insignificant impact, so future PSD increment analyses can neglect these small projects.

Sometimes a simple, conservative screening model can be used to quickly demonstrate that a new source is insignificant. The SCREEN model can be downloaded from the EPA's Office of Air Quality Planning and Standards Technology Transfer Network web page (http://www.epa.gov/ttn/). This model applies conservative default meteorological conditions so meteorological data input is not required. As a result, it calculates conservatively high ambient pollutant concentrations.

TABLE 3.3
Significant Impact Levels for Air Quality Impacts in Class II Areas

	Time Averaging Period				
Pollutant	**Annual**	**24 h**	**8 h**	**3 h**	**1 h**
SO_2	1 μg/m^3	5 μg/m^3	–	25 μg/m^3	7.8 μg/m^3
PM_{10}	1 μg/m^3	5 μg/m^3	–	–	–
$PM_{2.5}$	0.3 μg/m^3	1.2 μg/m^3	–	–	–
NO_2	1 μg/m^3	–	–	–	7.5 μg/m^3
CO	–	–	500 μg/m^3	–	2000 μg/m^3
O_3	–	–	–	–	–[a]

[a] No significant ambient impact concentration has been established. Instead, any net emissions increase of 100 tpy of volatile organic compound subject to PSD would be required to perform an ambient impact analysis.

A more complex model that uses local meteorological data can be used to calculate the impact of a source on local ambient air quality, with more realistic results than are obtained from a screening model. The ISCST and ISCLT (Industrial Source Complex Short Term and Long Term) models often are used, with AERMOD now preferred. Many commercial pollutant dispersion models use the ISC model as the base for the calculations and add input/output routines to make them user-friendly and to display results attractively.

3.3.2 Full Analysis

A full impact analysis is required when the estimated increased in ambient pollutant concentration exceeds the prescribed significant ambient impact level. The full analysis expands the scope of the dispersion modeling to include existing sources and the secondary emissions from residential, commercial, and industrial growth that accompanies the new activity at the new source or modification. The full impact analysis is used to project ambient pollutant concentrations against which the applicable NAAQS and PSD increments are compared.

3.4 NSR REFORM

In July 1996, the EPA proposed revisions to the NSR program.[4] The goals of the reform were to reduce the regulatory burden and to streamline the NSR permitting process. According to the EPA, the proposed revisions would reduce the number and types of activities at a source that would be subject to major NSR review, provide state agencies with more flexibility, encourage the use of pollution prevention and innovative technologies, and address concerns related to permitting sources near Class I areas. However, not everyone agreed with EPA's assessment of the value of the reforms. Significant debate, particularly the applicability of major NSR to modification of existing sources,[3] delayed adoption of the 1996 proposal until December 2002, followed by years of litigation, which defeated several of the reforms, but allowed several more to remain. Significant elements of NSR reform that survived judicial challenge include more flexible methods of determining baseline emission, emission increases, and permitting. Now, emission sources can determine baseline actual emissions as the emission rate during any consecutive 24-month period within 10 years prior to the proposed change. Emission increases resulting from the proposed change utilize actual emissions compared to projected actual emissions (*actual-to-projected-actual*). Also, emission sources can now create *plantwide applicability limits*, which function as facility-wide emissions caps allowing operational flexibility beneath the emissions cap.

4 Atmospheric Diffusion Modeling for Prevention of Significant Deterioration Permit Regulations and Regional Haze

4.1 INTRODUCTION—METEOROLOGICAL BACKGROUND

Reduction of ground-level concentrations from a point source can be accomplished by elevation of the point of emission above the ground level. The chimney has long been used to accomplish the task of getting the smoke from fires out of the house and above the inhabitants' heads. Unfortunately, meteorological conditions have not cooperated fully, and, thus, the smoke from chimneys does not always rise up and out of the immediate neighborhood of the emission. To overcome this difficulty for large sources where steam is produced, for example, power plants and space heating boiler facilities, taller and taller stacks have been built. These tall stacks do not remove the pollution from the atmosphere, but they do aid in reducing ground-level concentrations to a value low enough, so that harmful or damaging effects are minimized in the vicinity of the source.

4.1.1 Inversions

Inversions are the principal meteorological factor present when air pollution episodes are observed. They can be classified according to the method of formation and according to the height of the base, the thickness, and the intensity. An inversion may be based at the surface or in the upper air.

4.1.1.1 Surface or Radiation Inversions

A surface inversion usually occurs on clear nights with low wind speed. In this situation, the ground cools rapidly due to the prevalence of long-wave radiation to the outer atmosphere. Other heat transfer components are negligible, which means the surface of the earth is cooling. The surface air becomes cooler than the air above it, and vertical air flow is halted. In the morning, the sun warms the surface of the earth, and the breakup of the inversion is rapid. Smoke plumes from stacks are quite often trapped in the radiation inversion layer at night and then brought to the ground in a fumigation during morning hours. The result is high ground-level concentration.

4.1.1.2 Evaporation Inversion

After a summer shower or over an irrigated field, heat is required as the water evaporates. The result is a transfer of heat downward, cooling the upper air by convection and forming an evaporation inversion.

4.1.1.3 Advection Inversion

An advection inversion forms when warm air blows across a cooler surface. The cooling of the air may be sufficient to produce fog. When a sea breeze occurs from open water to land, an inversion may move inland, and a continuous fumigation may occur during the daytime.

4.1.1.4 Subsidence Inversion

In Los Angeles, the typical inversion is based on the upper air. This inversion results from an almost permanent high-pressure area centered over the north Pacific Ocean near the city. The axis of this high is inclined in such a way that air reaching the California coast is slowly descending or subsiding. During the subsidence, the air compresses and becomes warmer, forming an upper-air inversion. As the cooler sea breeze blows over the surface, the temperature difference increases, and the inversion is intensified. It might be expected that the sea breeze would break up the inversion but this is not the case. The sea breeze serves only to raise and lower the altitude of the upper air inversion.

4.1.2 Diurnal Cycle

On top of the general circulation a daily, or 24-h, cycle, referred to as the *diurnal cycle*, is superimposed. The diurnal cycle is highly influenced by radiation from the sun. When the sun appears in the morning, it heats the earth by radiation, and the surface of the earth becomes warmer than the air above it. This causes the air immediately next to the earth to be warmed by convection. The warmer air tends to rise and creates thermal convection currents in the atmosphere. These are the thermals, which birds and glider pilots seek out, and which allow them to soar and rise to great altitudes in the sky.

On a clear night, a process reverse of that described above occurs. The ground radiates its heat to the blackness of space, so that the ground cools off faster than the air. Convection heat transfer between the lower air layer and the ground causes the air close to the ground to become cooler than the air above, and a radiation inversion forms. Energy lost by the surface air is only slowly replaced, and a calm may develop.

These convection currents set up by the effect of radiant heat from the sun tend to add or subtract from the longer term mixing turbulence created by the weather fronts. Thus, the wind we are most familiar with, the wind close to earth's surface, tends to increase in the daytime and to die down at night.

There are significant diurnal differences in the temperature profiles encountered in a rural atmosphere and those in an urban atmosphere. On a clear sunny day in rural areas, a late afternoon normal but smooth temperature profile with temperatures decreasing with altitude usually develops. As the sun goes down, the ground begins to radiate heat to the outer atmosphere, and a radiation inversion begins to build up near the ground. Finally, by late evening, a dog-leg shaped inversion is firmly established and remains until the sun rises in the early morning. As the sun begins to warm the ground, the inversion is broken from the ground up, and the temperature profile becomes z-shaped. Smoke plumes emitted into the atmosphere under the late evening inversion tend to become trapped. Since vertical mixing is very poor, these plumes remain contained in very well-defined layers and can be readily observed as they meander downwind in what is called a *fanning fashion*. In the early morning as the inversion breaks up, the top of the thickening normally negative temperature gradient will encounter the bottom edge of the fanning plume. Since vertical mixing is steadily increasing under this temperature profile, the bottom of the fanning plume suddenly encounters a layer of air in which mixing is relatively good. The plume can then be drawn down to the ground in a fumigation, which imposes high ground-level concentrations on the affected countryside.

A similar action is encountered in the city. However, in this case, due to the nature of the surfaces and numbers of buildings, the city will hold in the daytime heat, and thus the formation of the inversion is delayed in time. Furthermore, the urban inversion will form in the upper atmosphere, which loses heat to the outer atmosphere faster than it can be supplied from the surfaces of the city. Thus, the evening urban inversion tends to form in a band above the ground, thickening both toward the outer atmosphere and toward the ground. Smoke plumes can be trapped by this upper air radiation inversion, and high ground concentrations will be found in the early morning urban fumigation.

4.1.3 Principal Smoke-Plume Models

Even though the objective of air pollution control is to reduce all smoke emissions to nearly invisible conditions, some visible plumes are likely to be with us for quite a long while. Visible plumes are excellent indicators of stability conditions. Five special models have been observed and classified by the following names:

1. Looping
2. Coning
3. Fanning
4. Fumigation
5. Lofting

All of these types of plumes can be seen with the naked eye. A recognition of these conditions is helpful to the modeler and in gaining an additional understanding of dispersion of pollutants.

In Section 4.1.2, the condition for fanning followed by fumigation has been described. Lofting occurs under similar conditions to fumigation. However, in this case, the plume is trapped above the inversion layer where upward convection is present. Therefore, the plume is lofted upward with zero ground-level concentration resulting.

When the day is very sunny with some wind blowing, radiation from the ground upward is very good. Strong convection currents moving upward are produced. Under these conditions plumes tend to loop upward and then down to the ground in what are called *looping plumes*.

When the day is dark with steady relatively strong winds, the temperature profile will be neutral, so that the convection currents will be small. Under these conditions, the plume will proceed downwind spreading in a conical shape. Hence the name *coning plume* is applied. Under these conditions, dispersion should most readily be described by Gaussian models.

4.2 TALL STACK

The Tennessee Valley Authority (TVA) has pioneered the use of tall stacks in the United States and has carried out extensive experiments, collected data, and determined the design variables and mathematical models to predict minimum ground-level concentrations. The 170 ft stacks provided at the first large steam plant constructed by TVA at Johnsonville, Tennessee, in 1952 were soon found to be inadequate. These stacks were then extended to 270 ft in 1955, and TVA stack height has crept upward ever since. As evidence, the large coal-fired power plant at Cumberland City, Tennessee, has two 1000 ft stacks, and the Kingston and Widows Creek plants, which have each a 1000 ft stack, topping the former tallest stacks at the Bull Run and Paradise plants by 200 ft.

Ever since structural steel became plentiful and strong enough to carry extreme loads, longer and taller structures have been built. Competition in this area is keen, and one wonders whether stack structures grow out of a rational need to reduce ground-level concentrations, or out of man's need to excel. Whatever the reason, it is amusing to compile and contemplate the statistics on tall structures, as listed in Table 4.1. Table 4.2 has the details of the TVA stacks at their major steam plants.

TABLE 4.1
The Size of Tall Things

Structure	Height	Structure	Height
Sears Tower (Chicago, Illinois)	1450 ft	INCO Stack (Sudbury, Ontario)	1250 ft
World Trade Center (Lower Manhattan, New York City)	1350 ft	Kennecott Company (Magna, Utah)	1215 ft
Empire State Building (Midtown Manhattan, New York City) (1475 ft to top of mast)	1250 ft	Penn Electric Company New York State Electric Co. (Homer City, Pennsylvania)	1210 ft
Chrysler Tower (Midtown Manhattan, New York City)	1040 ft	American Electric Power Mitchell Plant (Cresap, West Virginia)	1206 ft
Eiffel Tower (Paris, France)	984 ft	Keystone Group (Conemaugh, Pennsylvania)	1000 ft
Gateway Arch (St. Louis, Missouri)	630 ft	TVA-Cumberland Plant (Nashville, Tennessee)	1000 ft
Washington Monument (Washington, DC)	555 ft	TVA-Widows Creek Plant (Guntersville, Tennessee)	1000 ft
		TVA-Kingston Plant (Kingston, Tennessee)	1000 ft

TABLE 4.2
Major TVA Steam Plants

Name	First Unit in Operation	Unit No.	Rated Capacity per Unit (MW)	Total Plant (MW)	Stacks Number	Stacks Height (ft)
Cumberland	1972	1–2	1300	2600	2	1000
Bull Run	1967	1	950	950	1	800
Paradise	1963	1–2	704	2558	2	600
		3	1150		1	800
Allen	1959	1–3	330	990	3	400
Gallatin	1956	1–2	300	1255	1	500
		3–4	327.5		1	500
Colbert	1955	1–2	200	1397	2	300
		3–4	223		2	300
		5	550.5		1	500
John Sevier	1955	1–2	223	846	1	350
		3–4	200		1	350
Kingston	1954	1–4	175	1700	2	1000[a]
		5–9	200			
Shawnee	1953	1	150	1750	2	800[b]
		2–7	175			
		8	150			
		9	175			
		10	150			
Widows Creek	1952	1–2	140.6	1978	1	1000[c]
		3	150			
		4–6	140.6			
		7	575		1	500
		8	550		1	500
Johnsonville	1951	1–4	125	1485	4	270[d]
		5–6	147		2	270[d]
		7–10	173		2	400
Watts Bar	1942	1–4	60	240	4	177

[a] Original heights units 1–4, 250 ft, units 5–9, 300 ft.
[b] Replaces ten 250 ft stacks.
[c] Original six stacks 170 ft high, raised to 270 ft, then replaced.
[d] Original height 170 ft.

4.3 CLASSIFYING SOURCES BY METHOD OF EMISSION

Table 4.3 summarizes useful methods by which air pollution sources can be classified. Dispersion models exist that fit into this scheme. For stationary sources, three cases are defined: area sources, process stacks, and tall stacks.

TABLE 4.3
Classifying Air Pollution Sources by Method of Emission

Moving Sources

Transportation using fossil fuel
Internal combustion engine
Jet engine
Steam engine

Stationary Sources

Area Based
Low-level urban sources
Result of space heating and trash burning
Homes, apartments, commercial buildings
Improper firing of furnaces, poor quality coal, uncontrolled emission
Models: Require extensive source-emission information

Process Stacks
Chemical and petroleum processing
Space heating and process steam
May be result of leak or venting waste inorganic or organic gases
Heights up to about 250 ft
Low buoyancy, high velocity—could be a pure jet emission—plume rise not great
Model: Gaussian, but must evaluate the effects of stack and building downwash and surrounding topographical features

Tall Stacks
Fossil fuel burning for electrical power production
Heights up to 1250 ft
High buoyancy and velocity
Plume rise significant
Furnace heat input: 10^9 BTU/h (100 MW plant and larger)
Heat emission rate: 19,000 BTU/sec
Stack height: 2.5 times height of tallest structure near stack
Stack velocity: 1.5 times maximum average wind speed expected
Model: Gaussian, maximum concentration encountered depends upon regional meteorological conditions and topographical features

4.3.1 Definition of Tall Stacks

Adopting the TVA viewpoint to define a tall stack requires reference to the amount of furnace heat input, which should be greater than 293 MW (10^9 BTU/h). A furnace with this heat input would require 9072 kg/day (10,000 tons/day) of coal with a total heating value of 27.89×10^{10} J/kg (12,000 BTU/lb). A 100 MW plant would use about 771 kg/day (850 tons/day) and could qualify as having a tall stack. Most tall stack sources will be associated with fossil fuel-burning steam electric power-generating facilities. Another method to identify a tall stack is through the heat emission rate. This quantity should be greater than 20 MW (68.24×10^6 BTU/h) to define a tall stack.

Heat input is not the only requirement for establishing a tall stack. Such stacks produce plumes with great buoyancy, and these plumes have a high plume rise after leaving the stack. Furthermore, the exit velocity is high enough to avoid any building downwash. Rules of thumb to estimate the required exit velocity and stack height are

$$h_s > 2.5\,h_b$$

$$v_s > 1.5\bar{u}$$

where:

h_s is the stack height
v_s is the stack exit velocity
h_b is the height of tallest structure near stack
$\bar{u}$ is the maximum average wind speed that will be encountered

When a stack satisfies all these conditions, it may be considered a tall stack, and calculations are simplified.

4.3.2 Process Stacks

All other point sources differ in several ways. Most process stacks are not connected to sources with a high furnace heat input. Thus buoyancy is limited, and plume rise may be smaller. Quite often these stack plumes will have a high velocity, but little density difference, compared to ambient conditions. Thus, the plumes might be considered as jets into the atmosphere. Furthermore, since these stacks are usually shorter than 400 ft, the plumes may be severely affected by the buildings and the terrain that surround them. If stack efflux velocity is low, stack downwash may become prominent. In general, this is the kind of stack that is found in a chemical or a petroleum processing plant. Emissions from such a stack range from the usual mixture of particulates, sulfur oxides, nitrogen oxides, and excess air to pure organic and inorganic gases. To further complicate matters, these emissions usually occur within a complex of multiple point emissions; the result is that single-point source calculations are not valid. A technique for combining these process complex sources must then be devised.

4.4 ATMOSPHERIC-DIFFUSION MODELS

An atmospheric-diffusion model is a mathematical expression relating the emission of material into the atmosphere to the downwind ambient concentration of the material. The heart of the matter is to estimate the concentration of a pollutant at a particular receptor point by calculating from some basic information about the source of the pollutant and the meteorological conditions. For a detailed discussion of the models and their use, refer to the texts by Turner[1] and Schnelle and Dey.[2]

Deterministic, statistically regressive, stochastic models and physical representations in water tanks and wind tunnels have been developed. Solutions to the deterministic models have been analytical and numerical, but the complexities of analytical solution are so great that only a few relatively simple cases have been solved. Numerical solutions of the more complex situations have been carried out but require a great amount of computer time. Progress appears to be the most likely for the deterministic models. However, for the present, the stochastically based Gaussian-type model is the most useful in modeling for regulatory control of pollutants.

Algorithms based on the Gaussian model form the basis of models developed for short averaging times of 24 h or less and for long-time averages up to a year. The short-term algorithms require hourly meteorological data, while the long-term algorithms require meteorological data in a frequency distribution form. Algorithms are available for single and multiple sources, as well as single and multiple receptor situations. On a geographical scale, effective algorithms have been devised for distances up to 10–20 km for both urban and rural situations. Long-range algorithms are available but are not as effective as those for the shorter distance. Based on a combination of these conditions, the Gaussian plume model can provide at a receptor either (1) the concentration of an air pollutant averaged over time and/or space or (2) a cumulative frequency distribution of concentration exceeded during a selected time period.

4.4.1 Other Uses of Atmospheric-Diffusion Models

Atmospheric-diffusion models have been put to a variety of scientific and regulatory uses. Primarily, the models are used to estimate the atmospheric concentration field in the absence of monitored data. In this case, the model can be a part of an alert system serving to signal when air pollution potential is high, requiring interaction between control agencies and emitters. The models can serve to locate areas of expected high concentration for correlation with health effects. Real-time models can serve to guide officials in cases of nuclear or industrial accidents or chemical spills. Here, the direction of the spreading cloud and areas of critical concentration can be calculated. After an accident, models can be used in *a posteriori* analysis to initiate control improvements. The models also can be used for

- Stack-design studies
- Combustion-source permit applications
- Regulatory variance evaluation
- Monitoring-network design
- Control strategy evaluation for state implementation plans
- Fuel (e.g., coal) conversion studies
- Control-technology evaluation
- New-source review

A current frequent use for atmospheric-diffusion models is in air-quality impact analysis. The models serve as the heart of the plan for new-source reviews and the prevention of significant deterioration of air quality. Here, the models are used to calculate the amount of emission control required to meet ambient air-quality standards. The models can be employed in preconstruction evaluation of sites for the location of new industries. Models have also been used in monitoring-network design and control-technology evaluation.

4.5 ENVIRONMENTAL PROTECTION AGENCY'S COMPUTER PROGRAMS FOR REGULATION OF INDUSTRY

The Environmental Protection Agency (EPA) has developed a series of atmospheric-dispersion programs available through the Support Center for Regulatory Air Models (SCRAM), now on the Web and maintained by the EPA's Air Quality Modeling Group (AQMG). Models used for regulatory purposes were initially made available through the Users' Network of Applied Modeling of Air Pollution system in three ways: (1) executable codes on EPA's IBM mainframe at Research Triangle Park, (2) source codes for the UNISYS UNIVAC computer and test data on a magnetic tape from the National Technical Information Service, or (3) source codes and test data in packed form from EPA's Users' Network of Applied Modeling of Air Pollution Bulletin Board Service. During the summer of 1989, a new system for distribution was put in place. Source codes for models used for regulatory purposes were made available from the SCRAM Bulletin Board Service operated by EPA's Office of Air Quality Planning and Standards. Updating this service, the models and technical information concerning their use are now available on the Web, at the Technology Transfer Network Support Center for Regulatory Atmospheric Modeling established by the Office of Air Quality Planning and Standards. SCRAM is now available under this network at http://www.epa.gov/ttn/scram/, and provides links to models, data, processors, and changes to EPA's models documented through Model Change Bulletins.

Using these programs, it is possible to predict the ground-level concentrations of a pollutant resulting from a source or a series of multiple sources. These predictions are suitable evidence to submit to states when requesting a permit for new plant construction. Of course, the evidence must show that no ambient air-quality standard set by the EPA is exceeded by the predicted concentration.

The basic dispersion model employed by the EPA SCRAM programs is the Gaussian equation. Briggs plume-rise method and logarithmic wind speed-altitude equations are also used in the algorithms comprising SCRAM. SCRAM requires the source-receptor configuration to be placed in either a rectangular or a polar-type grid system. The rectangular system is keyed to the Universal Transverse Mercator grid system employed by the U.S. Geological Survey on its detailed land contours maps. This grid is indicated on the maps by blue ticks spaced 1 km apart running both North-South and East-West. Sources and receptors can be located in reference to this grid system, and the dispersion axis located from each source in reference to each of the receptor grid points. The polar grid system is used in a screening model to select worst meteorological conditions. If concentrations under the worst conditions are high enough, a more detailed study is conducted using the rectangular coordinate system. The location of the highest concentration then is determined within 100 m on the rectangular grid.

Meteorological data is obtained from on-site measurement, if possible. If not, data must be used from the nearest weather bureau station. This data can be obtained from the National Weather Records Center in Asheville, North Carolina. At the weather stations, data is recorded every hour. However, since 1964, the center in Asheville only digitizes the data every third hour. Thus, air-quality impact analysis studies can employ 1964 hourly data for short averaging time studies. However, some of the SCRAM programs have meteorological data preprocessors, which take the surface data and daily upper air data from the Asheville center and produce an hourly record of wind speed and direction, temperature, mixing depth, and stability. The meteorological data is used in the dispersion programs to calculate hourly averages, which are then further averaged to determine 3-, 8-, and so on, up to 24-h averages. Long-term modeling for monthly, seasonal, or annual averages requires use of the same data and a special program known as *stability array* (STAR). This program will compute an array of frequencies of occurrence of wind from the 16 compass directions, in one of six wind speed classes, for either five, six, or seven stability classes.

4.5.1 Industrial Source Complex Model

One of the most widely used models for estimating concentrations of nonreacting pollutants within a 10 mile radius of the source is EPA's Industrial Source Complex Short-Term, version 3 (ISCST3) program. It is a steady-state Gaussian plume model. Therefore, the parameters such as meteorological conditions and emission rate are constant throughout the calculation. There is also a long-term program, ISCLT. The time periods for the short-term program include 1, 2, 3, 4, 6, 8, 12, and 24 h. The ISCST3 program can calculate annual concentration if used with a year of sequential hourly meteorological data.

The ISCLT is a sector-averaged model, which combines basic features of several older programs prepared for the EPA. It uses statistical wind summaries and calculates seasonal or annual ground-level concentrations. ISCLT accepts stack-, area-, and volume-source types, and like the ISCST model, it uses the Gaussian-plume model.

In both of these programs, the generalized plume-rise equations of Briggs, which are common to most EPA dispersion models, are used. There are procedures to evaluate effects of aerodynamic wakes and eddies formed by buildings and other structures. A wind-profile law is used to adjust observed wind speed from measurement height to emission height. Procedures from former models are used to account for variations in terrain height over receptor grid. There are one rural and three urban options, which vary due to turbulent mixing and classification schemes. The models make the following assumptions about plume behavior in elevated terrain:

- The plume axis remains at the plume stabilization height above mean sea level as it passes over elevated or depressed terrain.
- Minimum turbulent mixing depth is set to terrain height.
- The wind speed is a function of height above the surface.
- If terrain height exceeds stack height, the ISC Programms truncate the terrain to stack height.

4.5.2 Screening Models

In scenarios where there are few sources or emissions, which are not very large, it is usually advantageous to employ a screening model. For regulatory purposes, if the concentrations predicted by the screening model exceed certain significant values, a more refined model must be employed. EPA's SCREEN3 is available for this screening operation. SCREEN3 allows a group of sources to be merged into one source, and it can account for elevated terrain, building downwash, and wind speed modifications for turbulence.

4.5.3 New Models

CALPUFF, a multilayer, multispecies, nonsteady-state dispersion model that views a plume as a series of puffs, is a new model under consideration. This model simulates space–time, varying meteorological conditions on pollutant-transport, chemical reaction, and removal. It can be applied from around 100 ft downwind up to several hundreds of miles.

The American Meteorological Society/EPA Regulatory Model (AERMOD) is a refined model currently under development by the EPA as a supplement to ISCST3 for regulatory purposes. By accounting for varying dispersion rates with height, refined turbulence based on planetary boundary layer theory, advanced treatment of turbulent mixing, plume height, and terrain effects, AERMOD improves the estimate of downwind dispersion. A version of the ISCST3 model known as *ISC-PRIME* was developed to incorporate plume-rise enhancements and the next generation of building downwash effects. *ISC-PRIME* was incorporated into AERMOD. AERMOD, along with AERMET, a meteorological data preprocessor, and AERMAP, a terrain data preprocessor, are the state-of-the-art air-quality models are now EPA's regulatory model of choice.

There are also a variety of specialized models for accidental release modeling, roadway modeling, offshore sources, and regional transport modeling.

4.6 SOURCE-TRANSPORT-RECEPTOR PROBLEM

The heart of the matter with which we are dealing is, given a source emitting a pollutant, can we estimate, by calculation, the ambient concentration of that pollutant at a given receptor point? To make the calculation, it is obvious that we must have a well-defined source and that we must know the geographic relation between the source and the receptor. But we must understand the means of transport between the source and the receptor, as well. Thus, source-transport-receptor becomes the trilogy, which we must quantitatively define in order to make the desired computation.

4.6.1 Source

Defining the source is a difficult matter in most cases. We need to consider first whether it is mobile or stationary and then whether it is emitted from a point, in a line, or more generally from an area. Then we must determine its chemical and physical properties. The properties can be determined most appropriately by sampling and analysis, when possible. It is then that we turn, for example, to estimation by a mass balance to determine the amount of material lost as pollutant. The major factors that we need to know about the source are as follows:

1. Composition, concentration, and density
2. Velocity of emission

3. Temperature of emission
4. Pressure of emission
5. Diameter of emitting stack or pipe
6. Effective height of emission

From these data, we can calculate the flow rate of the total stream and of the pollutant in question.

4.6.2 Transport

Understanding transport begins with three primary factors that affect the mixing action of the atmosphere: radiation from the sun and its effect at the surface of the earth, rotation of the earth, and the terrain or topography and the nature of the surface itself. These factors are the subject of basic meteorology.

The way in which atmospheric characteristics affect the concentration of air pollutants after they leave the source can be viewed in three stages:

1. Effective emission height
2. Bulk transport of the pollutants
3. Dispersion of the pollutants

4.6.2.1 Effective Emission Height

After a hot or buoyant effluent leaves a properly designed source, such as a chimney, it keeps on rising. The higher the plume goes, the lower will be the resultant ground-level concentration. The momentum of the gases rising up the chimney initially forces these gases into the atmosphere. This momentum is proportional to the stack gas velocity. However, stack gas velocity cannot sustain the rise of the gases after they leave the chimney and encounter the wind, which eventually will cause the plume to bend over. Thus, mean wind speed is a critical factor in determining plume rise. As the upward plume momentum is spent, further plume rise is dependent upon the plume density. Plumes that are heavier than air will tend to sink, while those with a density less than that of air will continue to rise until the buoyancy effect is spent. The buoyancy effect in hot plumes is usually the predominate mechanism. When the atmospheric temperature increases with altitude, an inversion is said to exist. Loss of plume buoyancy tends to occur more quickly in an inversion. Thus, the plume may cease to rise at a lower altitude, and be trapped by the inversion.

Many formulas have been devised to relate the chimney and the meteorological parameters to the plume rise. The most commonly used model is credited to Briggs. See Reference 2 in Chapter 8 for a detailed discussion of the models including Briggs model. The plume rise that is calculated from the model is added to the actual height of the chimney and is termed the *effective source height*. It is this height that is used in the concentration-prediction model.

4.6.2.2 Bulk Transport of Pollutants

Pollutants travel downwind at the mean wind speed. Specification of the wind speed must be based on data usually taken at weather stations separated by large distances. Since wind velocity and direction are strongly affected by the surface conditions, the nature of the surface, predominant topologic features such as hills and valleys, and the presence of lake, rivers, and buildings, the exact path of pollutant flow is difficult to determine. Furthermore, wind patterns vary in time, for example, from day to night. The Gaussian concentration model does not take into account wind speed variation with altitude, and only in a few cases are there algorithms to account for the variation in topography. For the future, progress in modeling downwind concentrations will come through increased knowledge of wind fields and numerical solutions of the deterministic models.

4.6.2.3 Dispersion of Pollutants

Dispersion of pollutants depends on the mean wind speed and atmospheric turbulence. The dispersion of a plume from a continuous elevated source increases with increasing surface roughness and with increasing upward convective air currents. Thus, a clear summer day produces the best meteorological conditions for dispersion, and a cold winter morning with a strong inversion results in the worst conditions for dispersion.

4.6.3 Receptor

In most cases, legislation will determine the ambient concentrations of pollutant to which the receptor is limited. Air-quality criteria delineate the effects of air pollution and are scientifically determined dosage-response relationships. These relationships specify the reaction of the receptor or the effects when the receptor is exposed to a particular level of concentration for varying periods of time. Air-quality standards are based on air-quality criteria and set forth the concentration for a given averaging time. Regulations have been developed from air-quality criteria and standards, which set the ambient quality limits. Thus, the objective of our calculations will be to determine if an emission will result in ambient concentrations, which meet air-quality standards that have been set by reference to air-quality criteria.

Usually, in addition to the receptor, the locus of the point of maximum concentration, or the contour enclosing an area of maximum concentration, and the value of the concentration associated with the locus or contour should be determined. The short-time averages that are considered in regulations are usually 3 min, 15 min, 1 h, 3 h, or 24 h. Long-time averages are one week, one month, a season, or a year.

5 Source Testing

5.1 INTRODUCTION

Source testing is measuring the air pollutant concentration and/or quantity at a source or stack. The terminology implies using test methods to measure concentration on a one-time or snapshot basis, as opposed to continuously monitoring the source, as discussed in Chapter 6. Source testing may be performed to provide design data or to measure performance of a process, or it may be prescribed on a periodic basis to demonstrate compliance with air permit emission limitations. Most source test procedures require labor to set up the test, collect samples, and analyze the results.

5.2 *CODE OF FEDERAL REGULATIONS*

The gospel of source testing is the *Code of Federal Regulations*, 40 CFR Part 60, Appendix A.[1] Very specific test procedures with step-by-step instructions are detailed in the *Code of Federal Regulations.* A list of source test methods that are described in the *Code of Federal Regulations* is provided in Table 5.1. These procedures have been tested, reviewed, and adopted by the Environmental Protection Agency (EPA) as the reference source test methods for a number of pollutants in a variety of applications. The universality of these test methods makes it easy for those practicing in the field to know just how a source test was conducted, and understand its limitations, just by giving a shorthand reference such as *Method 8.*

Of course, the *Code of Federal Regulations* test methods cannot be applied to all possible applications, because it would have been impossible to evaluate and fund the research required for all circumstances. It would be an inefficient use of public funds for the EPA to sponsor research for test methods that cover unusual operating conditions for a unique process, and it is impossible to predict future processes, conditions, and improvements. Sometimes experience, judgment, and skill are needed to modify the test method to overcome a limitation that arises in a specific application. In such cases, test reports can reference the test method and describe the modification.

5.3 REPRESENTATIVE SAMPLING TECHNIQUES

5.3.1 Gaseous Pollutants

One of the fundamentals in source testing for air contaminants is to obtain a representative sample of the gas stream. This is easy to do for gaseous pollutants, since molecules of gas can be assumed to be evenly distributed throughout the gas stream

TABLE 5.1
***Code of Federal Regulations* Source Test Methods**

Method	Description
1	Sample and velocity traverses for stationary sources
1A	Sample and velocity traverses for stationary sources with small stacks or ducts
2	Determination of stack gas velocity and volumetric flow rate (type-S pitot tube)
2A	Direct measurement of gas volume through pipes and small ducts
2B	Determination of exhaust gas volume flow rate from gasoline vapor incinerators
2C	Determination of stack gas velocity and volumetric flow rate in small stacks or ducts (standard pitot tube)
2D	Measurement of gas volumetric flow rates in small pipes and ducts
2E	Determination of landfill gas production flow rate
2F	Determination of stack gas velocity and volumetric flow rate with three-dimensional probes
2G	Determination of stack gas velocity and volumetric flow rate with two-dimensional probes
2H	Determination of stack gas velocity taking into account velocity decay near the stack wall
3	Gas analysis for the determination of dry molecular weight
3A	Determination of oxygen and carbon dioxide concentrations in emissions from stationary sources (instrumental analyzer procedure)
3B	Gas analysis for the determination of emission rate correction factor or excess air
3C	Determination of carbon dioxide, methane, nitrogen, and oxygen from stationary sources
4	Determination of moisture content in stack gases
5	Determination of particulate matter emissions from stationary sources
5A	Determination of particulate emissions from the asphalt processing and asphalt roofing industry
5B	Determination of nonsulfuric acid particulate matter from stationary sources
5C	Reserved
5D	Determination of particulate emissions from positive pressure fabric filters
5E	Determination of particulate emissions from the wool fiberglass insulation manufacturing industry
5F	Determination of nonsulfate particulate matter from stationary sources
5G	Determination of particulate emissions from wood heaters (dilution tunnel sampling location)
5H	Determination of particulate emissions from wood heaters from a stack location
5I	Determination of low level particulate matter emissions from stationary sources
6	Determination of sulfur dioxide emissions from stationary sources
6A	Determination of sulfur dioxide, moisture, and carbon dioxide emissions from fossil-fuel combustion sources
6B	Determination of sulfur dioxide and carbon dioxide daily average emissions from fossil-fuel combustion sources
6C	Determination of sulfur dioxide emissions from stationary sources (instrumental analyzer procedure)
7	Determination of nitrogen oxide emissions from stationary sources
7A	Determination of nitrogen oxide emissions from stationary sources—ion chromatographic method
7B	Determination of nitrogen oxide emissions from stationary sources (ultraviolet spectrophotometry)

(*Continued*)

TABLE 5.1 (*Continued*)
***Code of Federal Regulations* Source Test Methods**

Method	Description
7C	Determination of nitrogen oxide emissions from stationary sources—alkaline-permanganate/colorimetric method
7D	Determination of nitrogen oxide emissions from stationary sources—alkaline-permanganate/ion chromatographic method
7E	Determination of nitrogen oxide emissions from stationary sources (instrumental analyzer procedure)
8	Determination of sulfuric acid mist and sulfur dioxide emissions from stationary sources
9	Visual determination of the opacity of emissions from stationary sources
9 Alt. 1	Determination of the opacity of emissions from stationary sources remotely by LIDAR
10	Determination of carbon monoxide emissions from stationary sources
10A	Determination of carbon monoxide emissions in certifying continuous emission monitoring systems at petroleum refineries
10B	Determination of carbon monoxide emissions from stationary sources
11	Determination of hydrogen sulfide content of fuel gas streams in petroleum refineries
12	Determination of inorganic lead emissions from stationary sources
13A	Determination of total fluoride emissions from stationary sources—SPADNS zirconium lake method
13B	Determination of total fluoride emissions from stationary sources—specific ion electrode method
14	Determination of fluoride emissions from potroom roof monitors for primary roof monitors for primary aluminum plants
14A	Determination of total fluoride emissions from selected sources at primary aluminum production facilities
15	Determination of hydrogen sulfide, carbonyl sulfide, and carbon disulfide emissions from stationary sources
15A	Determination of total reduced sulfur emissions from sulfur recovery plants in petroleum refineries
16	Semicontinuous determination of sulfur emissions from stationary sources
16A	Determination of total reduced sulfur emissions from stationary sources (impinger technique)
16B	Determination of total reduced sulfur emissions from stationary sources
16C	Determination of total reduced sulfur emissions from stationary sources
17	Determination of particulate emissions from stationary sources (in-stack filtration method)
18	Measurement of gaseous organic compound emissions by gas chromatography
19	Determination of sulfur dioxide removal efficiency and particulate, sulfur dioxide, and nitrogen oxides emission rates
20	Determination of nitrogen oxides, sulfur dioxide, and diluent emissions from stationary gas turbines
21	Determination of volatile organic compound leaks
22	Visual determination of fugitive emissions from material sources and smoke emissions from flares
23	Determination of polychlorinated dibenzo-p-dioxins and polychlorinated dibenzofurans from stationary sources

(*Continued*)

TABLE 5.1 (*Continued*)
***Code of Federal Regulations* Source Test Methods**

Method	Description
24	Determination of volatile matter content, water content, density, volume solids, and weight solids of surface coatings
24A	Determination of volatile matter content and density of printing inks and related coatings
25	Determination of total gaseous nonmethane organic emissions as carbon
25A	Determination of total gaseous organic concentration using a nondispersive infrared analyzer
25B	Determination of total gaseous organic concentration using a nondispersive infrared analyzer
25C	Determination of the nonmethane organic compounds (NMOC) in MSW landfill gases
25D	Determination of the volatile organic concentration in waste samples
25E	Determination of vapor phase organic concentration in waste samples
26	Determination of hydrogen chloride emissions from stationary sources
26A	Determination of hydrogen halide and halogen emissions from stationary sources—isokinetic method
27	Determination of vapor tightness of gasoline delivery tank using pressure-vacuum test
28	Certification and auditing of wood heaters
28A	Measurement of air to fuel ratio and minimum achievable burn rates for wood-fired appliances
29	Determination of metals emissions from stationary sources
30A	Determination of total vapor phase mercury emissions from stationary sources (instrumental analyzer procedure)
30B	Determination of total vapor phase mercury emissions from coal-fired combustion sources using carbon sorbent traps

due to mixing and diffusion. It is highly unlikely that gaseous pollutants will segregate in a moving gas stream. A simple probe can be used to withdraw a sample. Due care must be used to avoid pulling a sample from a nonrepresentative location, such as just downstream of an injection point.

The primary consideration for gaseous pollutant sampling is that the sample is not contaminated, or decontaminated, by incompatibility with the materials of the sampling device or container. Teflon, stainless steel, and glass sample lines and containers often are used to avoid reactions with pollutants.

5.3.2 Velocity and Particulate Traverses

Volumetric flow rate in a duct or stack is measured using a pitot tube to detect the difference between the static and dynamic pressure difference created by the velocity head at several points in the duct. Details of the tip of a pitot tube are shown

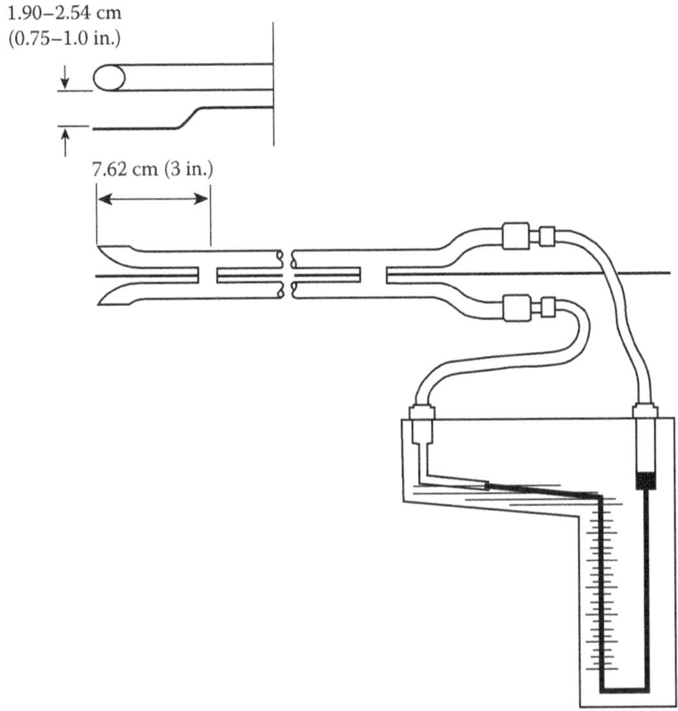

FIGURE 5.1 Pitot tube for velocity measurement.

in Figure 5.1. The measured pressure difference is used to calculate velocity. The key is to position the pitot tube at the correct points in the duct, so that the average velocity is determined. This is done by positioning the pitot tube at the centroid of equal-area segments of the duct. Method 1 (*Code of Federal Regulations*) provides tables for probe positions based on this principle. Figure 5.2 is an example of sampling points for a circular stack cross section. Figure 5.3 is an example for a rectangular duct. Note that the two lines of sampling points lie at 90°. For a circular duct, this requires at least two sampling ports at 90°. For small diameter stacks, the pitot tube can reach across the stack to pick up the points on the far side. For large diameter stacks, it is easier to reach no more than half way across the stack, so four sampling ports are provided to allow shorter sampling probes.

Similarly, because each sampling position is representative of a small area of the duct or stack, particulate samples are withdrawn at the same traverse points at which velocity measurements are made. However, because particles do not necessarily follow the streamlines of gas flow and because gravity can act on particles in a horizontal duct, Method 1 recommends more traverse points for particulate sampling than for a simple velocity traverse. The minimum number of sample points for traverses depends on the proximity of the test port to flow disturbances in the

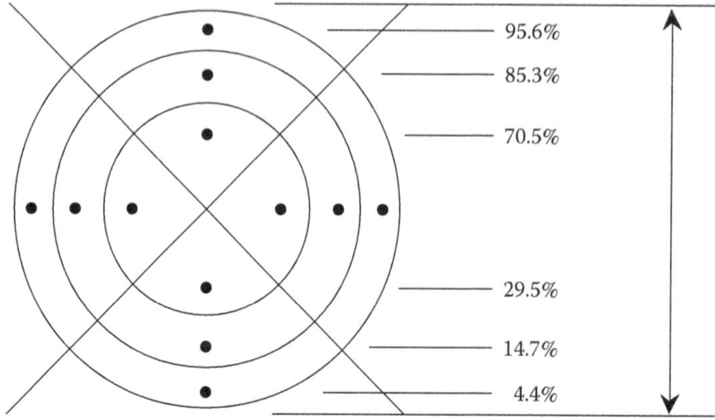

FIGURE 5.2 Traverse point locations for a round duct. The stack cross section is divided into 12 equal areas with the location of traverse points indicated.

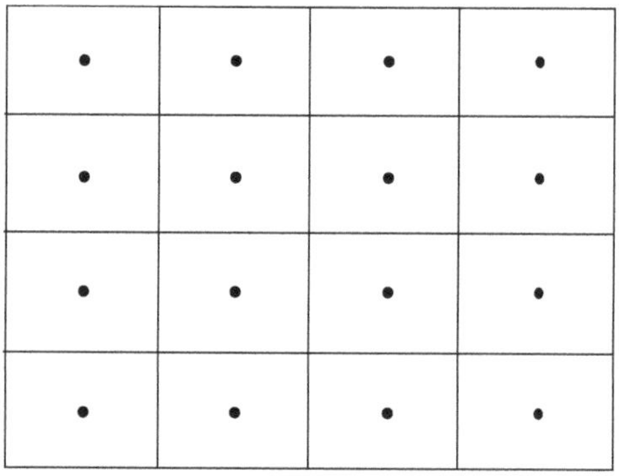

Points	Matrix
9	3 × 3
12	4 × 3
16	4 × 4
20	5 × 4
25	5 × 5
30	6 × 5
36	6 × 6
42	7 × 6
49	7 × 7

FIGURE 5.3 Traverse point locations for a rectangular duct.

duct, and to a lesser extent, the duct size. The minimum number of sample points for a velocity traverse is illustrated in Figure 5.4. The minimum number of samples to be taken for a particulate traverse is illustrated in Figure 5.5.

5.3.3 Isokinetic Sampling

Extracting a particulate sample from a moving gas stream using a probe in a duct requires that the sample be taken at the same velocity as the gas flow, that is, isokinetically. If the velocity of the sample is higher than that of the gas flow, then excess gas moving toward the probe will divert toward the probe and be collected with the sample. Meanwhile, particles with sufficient momentum will tend to continue traveling in a straight line, leaving the gas flow streamlines and will not be carried into

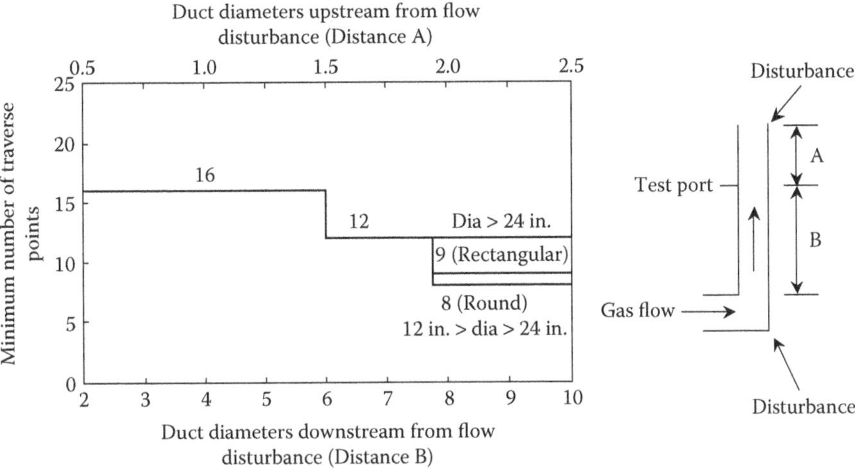

FIGURE 5.4 Minimum number of sample points for a velocity traverse.

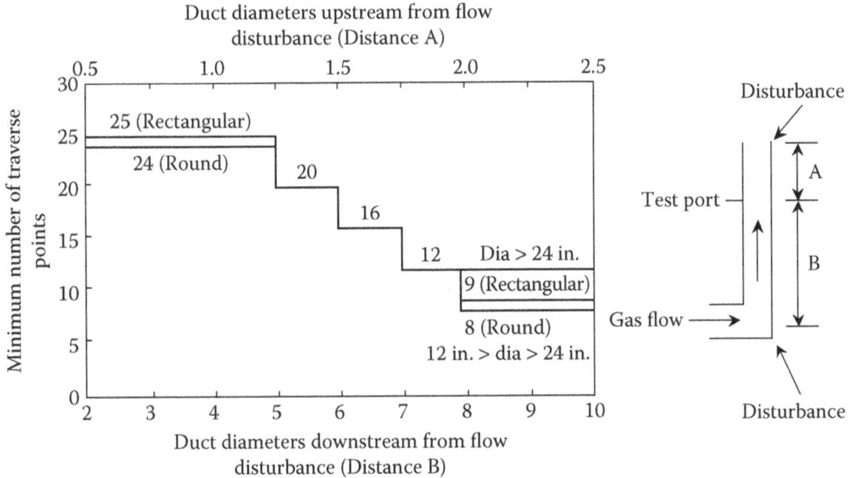

FIGURE 5.5 Minimum number of sample points for a particulate traverse.

the sampling probe, as illustrated in Figure 5.6a. This produces a sample that, after measuring the collected gas volume and weighing the collected particulate filter, has an erroneously low particulate concentration. Similarly, if the sample velocity is too low, excess gas is diverted away from the probe, while particles are carried into the probe, as illustrated in Figure 5.6b, resulting in an erroneously high particulate concentration.

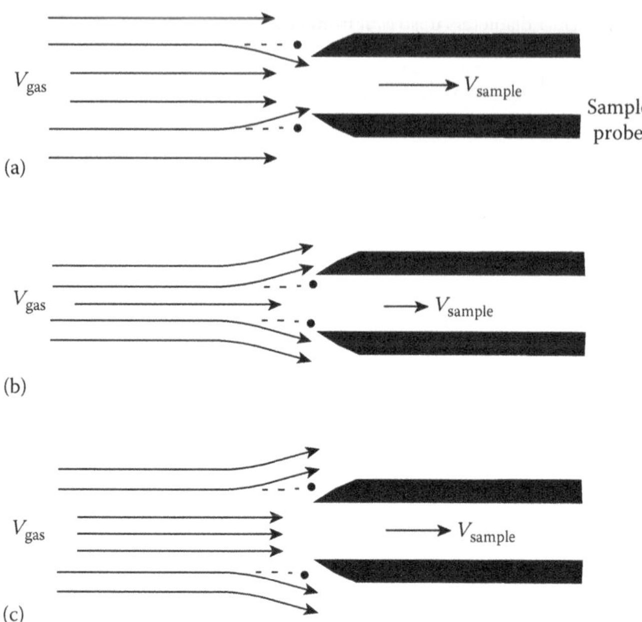

FIGURE 5.6 Reason for isokinetic sampling: (a) $V_{gas} < V_{sample}$; (b) $V_{gas} > V_{sample}$; (c) $V_{gas} = V_{sample}$.

In Figure 5.6c is isokinetic sampling where the sampling flow rate is equal to the gas flow rate. This is the ideal sampling flow rate. Here the parallel flow streams flow into the sample inlet carrying with them particles of all diameters capable of being carried by the stream flow.

The correct isokinetic sample flow rate is determined by conducting a velocity traverse prior to collecting a particulate sample. During the particulate sample, the collected gas volume is measured with a gas meter. After the sample is taken and as data are being evaluated, the sample velocity as a percentage of gas velocity is determined and reported as a quality check on the particulate sample.

6 Ambient Air Quality and Continuous Emissions Monitoring

6.1 AMBIENT AIR-QUALITY SAMPLING PROGRAM

Air pollution monitoring is conducted to determine either emission concentration or ambient air quality. The range of emission concentrations, temperature, and pressures encountered are sometimes many magnitudes greater than those found at an ambient air sampling station. For this reason, sampling and analysis techniques and equipment are different for each case, even though the same general principles may be employed. This chapter deals with both ambient air-quality sampling procedures and monitoring and continuous emissions monitoring.[1]

6.2 OBJECTIVES OF A SAMPLING PROGRAM

Instrumentation for ambient air-quality monitoring is perhaps best described in terms of the types of measurements that would be made in a typical community by the local air pollution control agency. Usually, a sample network would be installed that would blanket the area with a series of similar stations. The object would be to measure the amount of gaseous and particulate matter at enough locations to make the data statistically significant. It is not uncommon to find each station in a network equipped with simple, unsophisticated grab sampling devices. However, quite a few sophisticated monitoring networks have been developed, which contain continuous monitors with telemetry and computer control. Meteorological variables are also monitored and correlated with the concentration data. The information is then used

1. To establish and evaluate control measures.
2. To evaluate atmospheric-diffusion model parameters.
3. To determine areas and time periods when hazardous levels of pollution exist in the atmosphere.
4. For emergency warning systems.

6.3 MONITORING SYSTEMS

Ambient air-quality data may be obtained through the use of mobile or fixed sampling networks and using integrated samplers or continuous monitors. Decisions regarding these monitoring techniques constitute the first important steps in the design of a monitoring network.

6.3.1 Fixed versus Mobile Sampling

Fixed-point sampling entails a network of monitoring stations at selected sites, operating simultaneously throughout the study. Stations are permanent or, at least, long-term installations. In a mobile sampling network, the monitoring/sampling instruments are rotated on schedule among selected locations. The monitoring/sampling is not conducted simultaneously at all locations, and the equipment is generally housed in trailers, automobiles, or other mobile units. An advantage of fixed sampling is that measurements are made concurrently at all sites, providing directly comparable information, which is particularly important in determining relationship of polluting sources to local air quality and in tracing dispersion of pollutants throughout the area. The chief advantage of mobile sampling is that air-quality measurements can be made at many sites—far more than would be feasible in a fixed sampling program. Mobile sampling provides better definition of the geographical variations if the program is long enough to generate meaningful data.

6.3.2 Continuous versus Integrated Sampling

Continuous monitoring is conducted with devices that operate as both sampler and analyzer. Pollutant concentrations are instantaneously displayed on a meter, continuously recorded on a chart, magnetic tape, or disk. Integrated sampling is done with devices that collect a sample over some specified time interval after which the sample is sent to a laboratory for analysis. The result is a single pollutant concentration that has been integrated, or averaged, over the entire sampling period. This is an older technique and currently in limited use.

Continuous or automatic monitoring instruments offer some advantages over integrating samplers; for example, there is a capability for furnishing short-interval data, and there is a rapid availability of data. Moreover, output of the instruments can be electronically sent to a central point. Also, continuous monitors require less laboratory support. They also may be necessary to monitor some pollutants where no integrating method is available or where it is necessary to collect data over short averaging times, for example, 15 min. Automated monitors also have some drawbacks. They require more sophisticated maintenance and calibration, and the operators and maintenance personnel have to be more highly technically trained.

The selection of a monitoring system is influenced by the averaging time for which concentrations are desired, that is, 15-min, 1-h, 3-h, 8-h, 12-h, or 24-h time intervals. It should be consistent with the averaging times specified by air-quality standards. For example, in assessing SO_2 levels, good coverage can be provided by use of integrated samples, widely dispersed over the area, and one or more continuous sampler/analyzers situated in heavily populated areas. The integrated sampler defines SO_2 levels over a broad area, and the continuous devices provide detailed information on diurnal patterns. The short averaging time of interest for CO and ozone dictates the use of continuous monitors for these pollutants.

6.3.3 Selection of Instrumentation and Methods

Choice of instrumentation for an air-monitoring network depends on the following factors:

- Type of pollutants
- Averaging time specified by air-quality criteria or standards
- Expected pollutant levels
- Available resources
- Availability of trained personnel
- Presence in the air of materials that would interfere with the operation of the monitoring instrument

Most pollutants may be monitored by a number of different methods and techniques. The selection of the methodology to be used is an important step in the design of the monitoring portion of the assessment study.

6.4 FEDERAL REFERENCE METHODS AND CONTINUOUS MONITORING

In order to evaluate the current air quality and the effect of air pollution control measures, most larger communities maintain monitoring networks. The Environmental Protection Agency (EPA), as well as most of the states, maintains its own surveillance networks. The ideal objective when installing a monitoring network is to be able to obtain continuous real-time data. Table 6.1 lists standard methods of measurement. Only three of these standard methods employ continuous or semicontinuous monitors.

No satisfactory device exists as yet for determining suspended particulate on a continuous basis. However, Table 6.2 lists continuous methods, which are now commonly employed for the measurement of the five regulated air pollutants.

TABLE 6.1
Federal Reference Methods

Pollutant	Collection Method	Analysis
Sulfur dioxide	Absorption in bubbler	Spectrophotochemically
Particulate matter	Filtration in hi-volume sampler	Gravimetrically
Carbon monoxide	Continuous nondispersive infrared analyzer monitor	Absorption of infrared energy
Nitrogen dioxide	Continuous monitor	Chemiluminescence reaction with ozone
Hydrocarbons	Semicontinuous monitor	Chromatographic separation with flame ionization detector
Ozone	Continuous monitor	Chemiluminescence reaction with ethylene
Lead	Filtration in hi-volume sampler	Extracted by acid, atomic absorption spectrometry

TABLE 6.2
Continuous Monitor Methods for Federally Regulated Pollutants

Pollutant	Continuous Monitor Method
Sulfur dioxide	Conductrometric
	Coulometric
	Flame chemiluminescence
Carbon monoxide	Nondispersive infrared or gas chromatograph, with conversion to methane, flame ionization detection, a semicontinuous method
Nitrogen dioxide	Coulometric
	Chemiluminescent reaction with ozone
Hydrocarbons	Gas chromatograph
	Flame ionization detector
	Semicontinuous
Ozone	Coulometric
	Chemiluminescent reaction with ethylene

Particulate matter emissions can be continuously detected through opacity measurements. Opacity is a function of light transmission through the plume and is defined by the following equation:

$$\mathrm{OP} = \left(1 - \frac{I}{I_0}\right) \times 100 \tag{6.1}$$

where:

OP is the percent opacity
I is the light flux leaving the plume
I_0 is the incident light flux

The following information, Documentation of the Federal Reference Methods for the Determination of the Regulated Air Pollutants, can be found in the *Code of Federal Regulations*, Title 40 (CFR 40)—*Protection of Environment*, Chapter 1, Environmental Protection Agency, Subchapter C—Air Programs, Part 50, National Primary and Secondary Ambient Air Quality Standards. The following is a list of those methods with reference to CFR 40.

Documentation of the Federal Reference Methods for the Determination of Regulated Air Pollutants

Sulfur Dioxide

Appendix A-1

Reference Method Principle and Calibration Procedure for the Measurement of Sulfur Dioxide in the Atmosphere (Ultraviolet Fluorescence Method)

(Continued)

Appendix A-2

Reference Method for the Determination of Sulfur Dioxide in the Atmosphere (Pararosaniline Method)

Particulate Matter

Appendix B

Reference Method for the Determination of Suspended Particulate in the Atmosphere—High Volume Method (Total Suspended Particulate Has Been Replaced by PM_{10} and $PM_{2.5}$)

Carbon Monoxide

Appendix C

Measurement Principle and Calibration Procedure for the Measurement of Carbon Monoxide in the Atmosphere (Nondispersive Infrared Spectrometry)

Ozone

Appendix D

Measurement Principle and Calibration Procedure for the Measurement of Ozone in the Atmosphere

Appendix E

[Previously codified as the Reference Method for Determination of Hydrocarbons Corrected for Methane, but this method has been removed, and the appendix is referenced as *Reserved*]

Nitrogen Dioxide

Appendix F

Measurement Principle and Calibration Procedure for the Measurement of Nitrogen Dioxide in the Atmosphere (Gas Phase Chemiluminescence)

Lead

Appendix G

Reference Method for the Determination of Lead in Total Suspended Particulate Matter

Ozone

Appendix H

Interpretation of the 1-Hour Primary and Secondary National Ambient Air Quality Standards for Ozone

Appendix I

Interpretation of the 8-Hour Primary and Secondary National Ambient Air Quality Standards for Ozone

PM_{10}

Appendix J

Reference Method for the Determination of Particulate Matter as PM_{10} in the Atmosphere

Particulate Matter

Appendix K

Interpretation of the National Ambient Air Quality Standards for Particulate Matter

(*Continued*)

Appendix L
Reference Method for the Determination of Fine Particulate Matter as $PM_{2.5}$ in the Atmosphere

Appendix M
[Reserved]

Appendix N
Interpretation of the National Ambient Air Quality Standards for $PM_{2.5}$

Appendix O
Reference Method for the Determination of Coarse Particulate Matter as $PM_{10-2.5}$ in the Atmosphere

Appendix P
Interpretation of the Primary and Secondary National Ambient Air Quality Standards for Ozone

Appendix Q
Reference Method for the Determination of Lead in Particulate Matter as PM_{10} Collected from Ambient Air

Appendix R
Interpretation of the National Ambient Air Quality Standards for Lead

Appendix S
Interpretation of the Primary National Ambient Air Quality Standards for Oxides of Nitrogen (Nitrogen Dioxide)

Appendix T
Interpretation of the Primary National Ambient Air Quality Standards for Oxides of Sulfur (Sulfur Dioxide)

6.5 *COMPLETE* ENVIRONMENTAL SURVEILLANCE AND CONTROL SYSTEM

An ideal surveillance and control system can be devised employing continuous monitoring, telemetering, and electronic data processing. It is possible to assemble such a system from the hardware components that are now available. The major drawback of this automatic system is the limitations of the computer software; there is little economic information available for formulating the ambient air quality and optimizing models. Figure 6.1 illustrates one conception of a surveillance and control system. This system would demand instruments, which could be calibrated to a known standard, would retain their calibration over long periods of time, would be free of electronic drift over these long time periods, and would possess suitable dynamic response.

Many automated environmental surveillance systems employing continuous monitors exist in the United States and throughout the world. None are quite as sophisticated as would be implied by the system of Figure 6.1.

6.6 TYPICAL AIR SAMPLING TRAIN

A typical air pollution sampling train is applicable to the intermittent collection of an air sample containing either gaseous or particulate pollutants. The sample is retained in the collection equipment, which is then removed for the sample train.

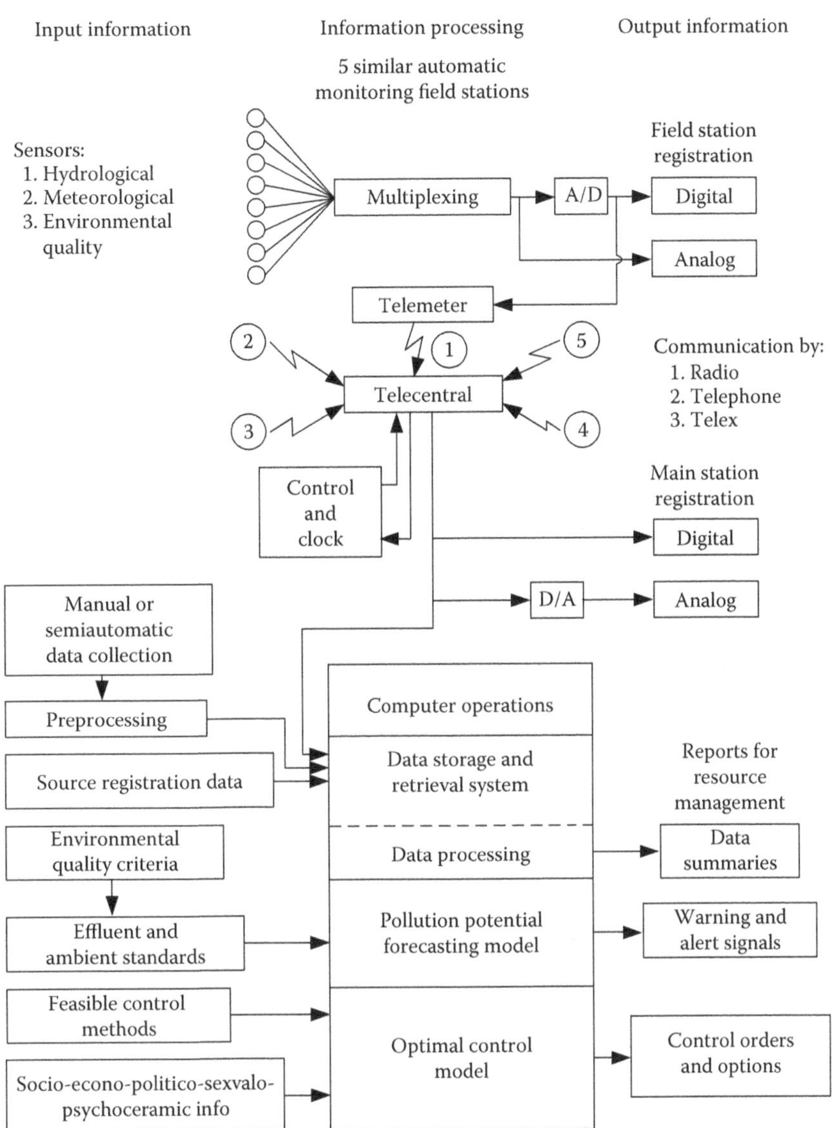

FIGURE 6.1 Conception of a surveillance and control system.

Further processing takes place to prepare the sample for analysis. Most of the analysis techniques are standard procedures involving one or more of the following methods:

1. Gravimetric
2. Volumetric
3. Microscopy
4. Instrumental

a. Spectrophotometric
    i. Ultraviolet
    ii. Visible (colorimetry)
    iii. Infrared
b. Electrical
    i. Conductometric
    ii. Coulometric
    iii. Titrimetric
c. Emission spectroscopy
d. Mass spectroscopy
e. Chromatography

6.7 INTEGRATED SAMPLING DEVICES FOR SUSPENDED PARTICULATE MATTER

Suspended particulate are small particulate that vary in size from less than 1 μm to approximately 100 μm. They remain suspended in the atmosphere for long periods of time and absorb, reflect, and scatter the sunlight, obscuring visibility. When breathed, they penetrate deeply into the lungs. They also cause economic loss because of their soiling and corrosive properties.

The new EPA ambient particulate matter definition includes only the part of the size distribution that could penetrate into the human thorax. This requires a sampling inlet with a 10 μm cutpoint to mimic deposition in the extra thoracic regions. Figure 6.2 is a

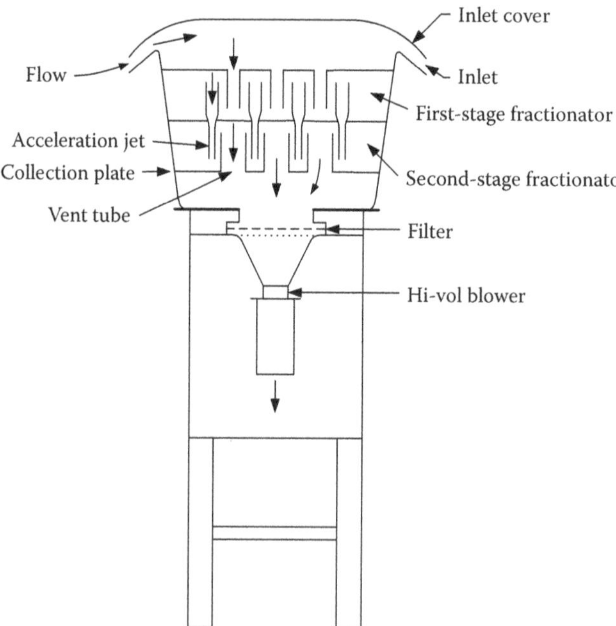

FIGURE 6.2 Two-stage particulate sampler.

schematic of a sampler designed to meet this requirement. There is a two-stage selective inlet. Air is drawn into the inlet and deflected downward into acceleration jets of the first-stage fractionator. Larger, noninhalable particles are removed. Air then flows through the first-stage vent tubes and then through the second fractionation stage. More noninhalable particles are removed, and the remaining aerosol is drawn through the usual 8 × 10 in. sampling filter. The new PM_{10} standard includes only those particles with an aerodynamic diameter less than or equal to 10 μm. This standard went into effect on July 31, 1987.

6.8 CONTINUOUS AIR-QUALITY MONITORS

Continuous emissions monitoring (CEM) is required by the Clean Air Act amendments to monitor SO_2, NO_x, CO, CO_2, opacity, total hydrocarbons, and total reduced sulfur. Title IV, which is to ensure compliance with the Acid Rain Program, sets out provisions for CEM in the two-phase utility power industry control strategy. Phase I utilities were required to install CEM by November 15, 1993, and Phase II utilities by January 1, 1995. Title III focuses on 189 hazardous air pollutants, some of which will possibly require CEM. This should spawn CEM techniques optimized for the chemical compound being monitored. Title V will require CEM for compliance assurance. The collection of real-time emission data will be the first step to attaining the national mandated reduction in SO_2 and NO_x emissions. Furthermore, CEM can be used to track use of allowances in the market-based SO_2 emissions trading program.

CEM is carried out by two general methods—*in situ* and extractive. Each of the methods measures on a volumetric basis, parts per million, for example. Therefore, the measurements require conversion to mass emission rates on a lb/h or lb/10^6 BTU basis. Monitoring instrumentation requires at least 10% relative accuracy. Performance specifications and test procedures can be found in Title 40 CFR, Part 60, Appendix B.

The EPA Emissions Measurement Center at Research Triangle Park and the Midwest Research Institute have created a database on CEM. You may use browse techniques or search the database by hazardous air pollutants or analyzer type. The database is found on the EPA home page at http://www.epa.gov/ttn/emc/index.html. Figure 6.3 is a schematic flow diagram of a general continuous air-quality monitor. The device contains the following:

1. Primary air-moving device, usually a vacuum pump, to pull the air sample through the instrument
2. Flow-control and flow-monitoring device, usually a constant pressure regulator and a rotameter
3. Pollutant detection by various primary sensing techniques
4. Automatic reagent addition where needed
5. Electronic circuitry for transducing the primary signal to a signal suitable for recording and telemetering
6. Provisions for automatic calibration, usually several solenoid valves, which can be operated remotely to connect the inlet gas to a scrubbing train for removal of all pollutants and establishing a chemical zero, or, alternatively, to one or more span gases for setting the chemical range of the instruments

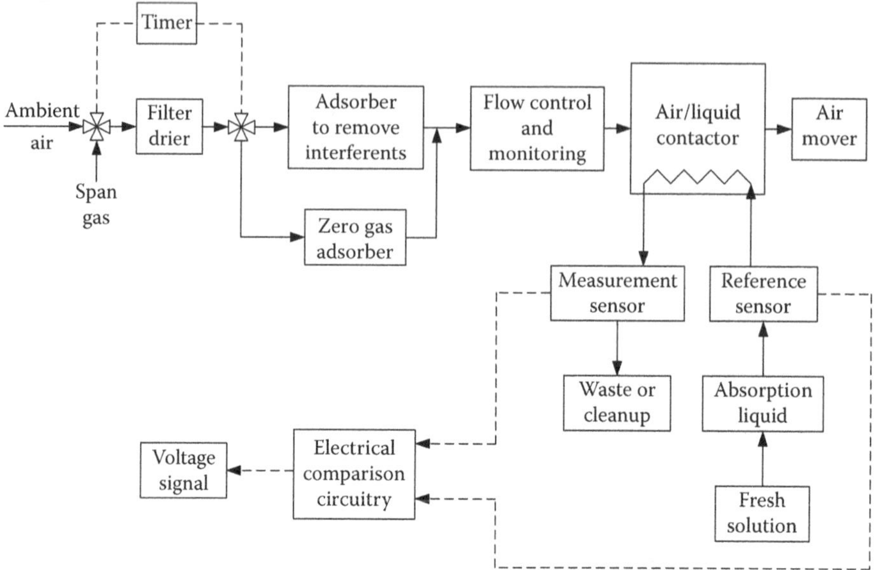

FIGURE 6.3 Generalized automatic continuous air pollution monitor.

Many monitors of the general type described above have been developed for all of the federally regulated gaseous pollutants and others as well. The remainder of this section will provide details of these devices. This list is not exhaustive. Furthermore, although the devices described below are indicated for a particular pollutant; they can be used for other types of pollutants as well.

6.8.1 Electroconductivity Analyzer for SO_2

Electrical conductivity was the basis of the first continuous monitor used to detect and monitor an air pollutant—SO_2. This monitor was built by one Dr. Thomas to monitor SO_2 in a greenhouse during a study of the effects of SO_2 on plants. Later, it was employed by TVA in its original studies of SO_2 from power plant plumes.

In an electroconductivity apparatus, a reagent passes through a reference conductivity cell and then into an absorbing column. Air is drawn by a vacuum pump counter, currently to the reagent flow through the absorbing tube, then through a separator to the exhaust. The SO_2 is absorbed in the reagent, which then passes through a measuring conductivity cell. A stabilized AC voltage is impressed across the conductivity cells, resulting in a current flow that is directly proportional to the conductivity of the solution. The value of this current is measured by connecting a resistor in series with the cell to obtain an AC voltage, which is proportional to the current. This voltage is then rectified to direct current. The DC signals from the rectifiers are connected in opposition, thus resulting in a voltage that induces a current through a meter, which is directly related to the difference in conductivity between the two solutions. To set the zero on the instrument, any SO_2 is removed by passing the air through a soda-lime absorber. The conductivity in both cells should then be the same, and the meter output should be zero.

The principle of operation is as follows:

1. The SO_2 is oxidized in a reagent such as deionized water to form the sulfate ion, which will cause a conductometric change related to amount of SO_2 present. The reaction is as follows:

$$H_2O + SO_2 \rightarrow H_2SO_3$$

$$H_2SO_3 + {}^1\!/_2 O_2 \rightarrow H_2SO_4$$

2. Use of an acidified H_2O_2 solution reduces interference from acid gases such as CO_2.

$$SO_2 + H_2O_2 \rightarrow H_2SO_4$$

3. The method is basically simple. However, since conductivity is temperature dependent, the analyzer section of any instrument must be thermostated.

6.8.2 Coulometric Analyzer for SO_2

Principle of operation of a coulometric analyzer is similar to an electroconductivity apparatus, except three electrodes are required. One is a reference electrode. Figure 6.4 shows the three-electrode circuit. An air sample is drawn through a detector cell, which contains a buffered solution of KI. Then iodine, I_2, is generated at the anode.

$$2I^- \rightarrow I_2 + 2e^-$$

The SO_2 is oxidized by the I_2. Unreacted iodine is reduced to iodide, I^-, at the cathode.

$$I_2 + 2e^- \rightarrow 2I^-$$

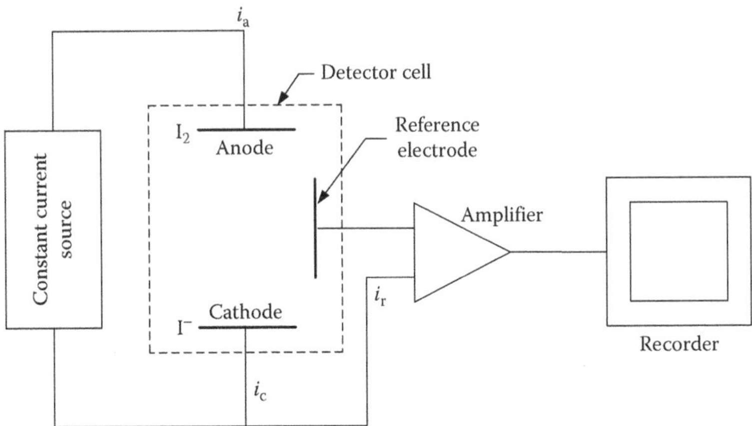

FIGURE 6.4 Electrical circuit for a coulometric continuous monitor.

Due to the loss of I_2 by reaction with SO_2, the cathode current is less than the anode current. The difference is proportional to the SO_2 concentration through a Faradaic expression. The reference electrode serves as a reference to both poles and registers this imbalance in anodic input current and cathodic output current, $i_r = i_a = i_c$. A selective scrubber removes interferents such as O_3, mercaptans, and H_2S, improving specificity of instrument. A zero gas is provided by using an activated carbon filter to remove impurities in air, including SO_2. The gas flow is regulated by pressure control and a capillary tube, which provides a constant pressure and pressure drop, and thus a constant flow. A pulsation damper adds volume to the system to provide stability of flow. The water supply is constantly replenished when evaporation occurs.

6.8.3 Nondispersive Infrared Method for CO

Over the last 40 years, continuous infrared spectrometers have grown to be one of the most satisfactory instruments for determining carbon monoxide concentration in the atmosphere. The nondispersive instruments do not employ spectral separation of the radiation but make use of the specific radiation absorption of heteroatomic gases in the infrared range between 2.5 and 12 μm. The total absorption in the range is measured by an alternating light photometer with two parallel beams and a selective radiation receiver. The sample gas is passed through the sample cell, which is arranged parallel to a reference cell containing a gas, which absorbs no radiation. The radiation emitted by two nickel-chrome filaments reaches the two receiving cells after passing through the sample cell and reference cell, respectively. The receiving cells are filled with the gas component to be measured (CO in this case) and are separated by a metal diaphragm. The incident radiation is absorbed selectively in the specific absorption bands. Every gas has an absorption spectrum consisting of one or two individual absorption bands, which are specific for the gas. Carbon monoxide has a band from 4.5 to 5.0 μm. The absorbed energy is transformed into thermal energy. Any difference in absorbed energy produces a temperature and pressure difference between the two receiving cells. This pressure difference deflects the diaphragm and thus changes the capacitance of the diaphragm capacitor. The diaphragm capacitor is connected to a high-impedance resistor, which generates an alternating millivolt voltage that can be amplified, rectified, and displayed by a recorder.

All infrared analyzers employ the same nondispersive measurement principle, but there are differences in the manufacturer's design. One major difference among the analyzers is the length of the measuring cell. The length of the cell has little effect on the measuring range of the analyzer, and both analyzers are capable of being changed to provide several measuring ranges.

Two infrared sources are used: one for the sample energy beam, and the other for the reference energy beam. The beams are blocked simultaneously 10 times/s by the chopper, a two-segmented blade rotating at five revolutions/s. In the unblocked condition, each beam passes through the associated cell and into the detector. The sample cell is a flow-through tube that receives a continuous stream of sample. The reference cell is a sealed tube filled with a reference gas. The reference gas is selected for minimal absorption of infrared energy of those wavelengths absorbed by the sample component of interest.

The detector consists of two sealed compartments separated by a flexible metal diaphragm. Each compartment has an infrared transmitting window, to permit entry of the corresponding energy beam. Both chambers are filled, to the same subatmospheric pressure, with the vapor of the component of interest. Use of this substance as the gas charge in the detector causes the instrument to respond only to that portion of net difference in energy due to the presence of the measured component.

In operation, the presence of the infrared-absorbing component of interest in the sample stream causes a difference in energy levels between the sample and reference sides of the system. This differential energy increment undergoes the following sequence of transformation:

1. *Radiant energy*: In the sample cell, part of the original energy of the sample beam is absorbed by the component of interest. In the reference cell, however, absorption of energy from the reference beam is negligible.
2. *Temperature*: Inside the detector, each beam heats the gas in the corresponding chamber. However, since energy of the reference beam is greater, gas in the reference chamber is heated more.
3. *Pressure*: Higher temperature of gas in the reference chamber raises the pressure of this compartment above that of the sample chamber.
4. *Mechanical energy*: Gas pressure in the reference chamber distends the diaphragm toward the sample chamber. The energy increment is thus expended in flexing the diaphragm.
5. *Capacitance*: The diaphragm and the adjacent stationary metal button constitute a two-plate variable capacitor. Distention of the diaphragm away from the button decreases the capacitance.

When the chopper blocks the beams, pressures in the two chambers equalize, and the diaphragm returns to the undistended condition. As the chopper alternately blocks and unblocks the beams, therefore, the diaphragm pulses, thus changing detector capacitance cyclically. The detector is part of an amplitude modulation circuit that impresses the 10 Hz information signal on a 10 MHz carrier wave provided by a crystal-controlled radio-frequency oscillator. Additional electronic circuitry in the oscillator unit demodulates and filters the resultant signal, yielding a 10 Hz signal. The 10 Hz signal is routed to the amplifier/control section for amplification and phase inversion, then back into the analyzer section for synchronous rectification. The resulting fullwave-rectified signal is returned to the amplifier/control section for filtering and additional conditioning, as required, to drive the meter and recorder.

6.8.4 Flame Photometric Detection of Total Sulfur and SO_2

Sulfur compounds introduced into a hydrogen-rich flame produce strong luminescent emissions between 300 and 423 nm. A narrow-band optical filter placed between the flame and detector permits transmission at 394 ± 5 nm. This produces a specificity ratio of between 10,000 and 30,000 to 1 for sulfur compounds. A photomultiplier tube detects the emission.

When it is desired to study a mixture of sulfur compounds such as SO_2, H_2S, CS_2, and CH_2SH, the mixture is first passed through a gas chromatograph. A typical column employs 24 ft of 1/8 in. i.d. tubing packed with 30 to 40 mesh Teflon®, which is coated with 10% polyphenol containing phosphoric acid. For the usual ambient conditions, the use of a chromatographic column is not warranted since the sulfur in ambient air is usually in the form of SO_2.

6.8.5 Hydrocarbons by Flame Ionization

A simple hydrocarbon monitor can be built using a flame ionization detector. Carbon atoms produce ions in a hydrogen flame. Thus, the air stream containing hydrocarbons is fed into a hydrogen flame. The ions produced are detected by an electrometer. The hydrocarbon concentration is proportional to the current.

It has been found that CH_4 can be separated by gas chromatographic technique from other hydrocarbons. An employing flame ionization detection has been constructed. Carbon monoxide is detected in the device by catalytically converting the CO to CH_4 over a nickel catalyst after it has been separated from the rest of the gases by the gas chromatographic technique.

6.8.6 Fluorescent SO_2 Monitor

Sulfur dioxide molecules fluoresce when irradiated by ultraviolet light in the 1900 to 3000 nm waveband. The optimum excitation takes place in the narrower range between 2100 and 2300 nm. An ultraviolet light such as a quartz deuterium lamp is used to produce the source of radiation. Selective narrow-band transmission is achieved by optical filters. Subsequently, the collimated narrow-band light beam then passes through a reaction chamber, where the sample also flows. The resultant fluorescence, directly proportional to the number of SO_2 molecules present in the reaction chamber, is then measured by a blue sensitive photomultiplier tube.

6.8.7 Chemilumenescence for Detection of Ozone and Nitrogen Oxides

In 1889, Arrhenius first proposed that an energy of activation was required in order for a chemical reaction to take place. By acquiring this activation energy, a reacting molecule or atom could combine with other molecules or atoms in a chemical reaction. Around 1914 to 1930, the collision theory of reaction prevailed in which the molecules were postulated to acquire the necessary energy of activation by colliding with each other. Molecules were then conceived to be somewhat like hard spheres, and the rate of reaction was determined to be proportional to the number of collisions of these marble-like molecules in a given time. Quantum theory of the 1920s introduced the idea of probability, and the round-sphere idea was replaced by the probability function.

Building on the Arrhenius activation energy idea and the probability theory of quantum mechanics, Eyring and Polyani in 1931 proposed an activated complex theory[2] of chemical reaction. This theory looks at the potential energy of the reacting system as the molecules come closer together. The rate of reaction is then found as the probability of a molecule crossing an energy barrier. In order to cross the barrier, the molecule must become an activated complex. The picture one can write of a chemical reaction is illustrated by the nitric oxide-oxygen reaction.

$$2NO + O_2 \rightarrow (NO - O_2) \quad \text{Activated}$$

$$(NO - O_2) \rightarrow 2NO_2 \quad \text{Complex}$$

Thus, the collision of hard spheres is replaced by a smooth continuous transition of reactants to products.

In a photochemical reaction, the activation energy comes from a quarter of light. This light may be obtained from the ultraviolet or infrared region, as well as from the visible range. Photochemical reactions are primary processes; however, luminescence belongs to a class of reactions in which a secondary photochemical process takes place after the acquisition of an initial energy. Four luminescence processes are recognized.

1. Fluorescence—after excitation by light, there is an emission of light of a different wavelength.
2. Phosphorescence—where an emission of light continues after the excitation source is removed.
3. Thermoluminescence—where, for example, crystals are irradiated with X-rays. Electrons are excited and trapped at a higher energy level. When the crystal is heated, the electrons drop to a lower energy level with the emission of light.
4. Chemiluminescence—which is an emission of light as a result of a chemical reaction at environmental temperature.

An example of chemiluminescence is the reaction between nitric oxide (NO) and ozone (O_3).

1. When NO in lower energy state than O_3 becomes activated as

$$NO \rightarrow (NO)^{\ddagger}$$

2. Then NO crosses over to the O_3 energy level.
3. An activated complex is formed between NO and O_3.

$$(NO)^{\ddagger} \rightarrow (NO - O_3)^{\ddagger}$$

4. Reaction proceeds to nitrogen dioxide, leaving it in a higher energy state.

$$(NO - O_3)^{\ddagger} \rightarrow (NO_2)^{\ddagger} + O_2$$

5. NO_2 drops to a lower energy state with the emission of a quanta of light, $h\nu$.

$$(NO_2)^{\ddagger} \rightarrow NO_2 + h\nu$$

Commercial monitors have been developed employing this process. Ozone is detected in reaction with ethylene. Nitric oxide is detected in reaction with ozone, and nitrogen dioxide is detected by first reducing it to nitric oxide and then using the ozone reaction.

6.8.8 Calibration of Continuous Monitors

6.8.8.1 Specifications for Continuous Air-Quality Monitors

It is unfortunate that terms used in specifying and calibrating air-monitoring instrumentation are sometimes ambiguous. Thus, we find the term *dynamic* dilution being used to mean diluting a flowing gas stream under carefully controlled constant flow conditions to produce an accurately and precisely known concentration. A preferred term would be *steady-state* dilution, which can be used to describe gas in flow but has none of the connotation of transient, which the term *dynamic* might imply. To be most useful, specifications for instrumentation where flowing fluids are involved should provide information about the instrument, which would impart to the user a knowledge about both its steady-state and transient performance.

Steady-state calibrations should be made at several levels of concentration over the expected range of concentrations that will be found. Comparison to the accepted standard method, absorption in a bubbler with West and Gaeke analysis for sulfur dioxide, for example, should be made as well. Replicate runs at the several levels of concentration can be compared through statistical analysis using factors such as the multiple correlation coefficient. In addition, the drift of the zero and span or range of the instrument should be low, so that very little error will be introduced into the accuracy and precision of the instrument.

Dynamic response testing can be used to determine transient and frequency response characteristics of continuous monitors. The most complete dynamic description of a system is its transfer function, which is the ratio of the Laplace transformed output signal to the Laplace transformed input signal. Knowing the transfer function means that the output signal can be determined from any given input signal. Frequency response curves are essentially a graphical record of the transfer function and are thus useful to depict the dynamic response of a system. Transient response curves such as might be provided from a step test are valid only for the particular input function, which produced the curves and are, thus, of limited usefulness. Even more limited in specifying instruments is the rise time and fall time, which provides no information about the shape of the transient curve. The bad news is the difficulty to which the experimenter is put in order to determine the transient characteristics or the transfer function. The good news is that the need for the detailed transient information provided by the transfer function is not usually required.

Additional instrument specifications are concerned with the effect of the ambient conditions imposed on the instrument including interfering substances. A common glossary of terms follows. These terms have been classified according to definition, steady-state specification, dynamic specification, and ambient specification.

6.8.8.2 Steady-State Calibrations

One of the most difficult problems encountered in calibrating air pollution instrumentation is the production of low-level concentrations, which are in the range encountered in the atmosphere. There are several suppliers of gas cylinders with prepared and certified low-level gas concentrations.

Most usually, these gases will have to be diluted before use, which poses a problem in maintaining strictly controlled temperatures, pressures, and flow rates. This writer discovered a paradox when using a cylinder of carbon monoxide prepared to a specified low concentration. A small but significant difference between two low levels of concentration was found when calibrating a nondispersive infrared analyzer. The companies supplying the gases were contacted. One company reported that they were sure of their analysis because they used the best instrument for low-level carbon monoxide concentration determination—a nondispersive infrared instrument. The net result was that the company was comparing nondispersive infrared analyzer instruments and not conducting a calibration with a true standard. From the author's viewpoint, this situation is to be avoided.

The permeation tube is a device that can provide low-level concentrations. These tubes are made of a material such as Teflon and contain a compound whose concentration is desired, sealed into the tube as a liquid. The vapor will then permeate through the walls of the tube at a constant rate dependent on the surface area exposed and the temperatures to which the tube is subjected. The basic determination of permeation rate can be made gravimetrically; thus, a permeation tube can become a true standard. Table 6.3 lists some of the gases for which permeation tubes are available.

Tubes of proper length can be used to calibrate gas analyzers at a steady-state flow condition. An apparatus must be prepared to hold the tube in a pure-air or nitrogen stream of known flow rate and at a constant temperature. A tube holder may be made from a standard vacuum trap with the tube in the center well and the outer section filled with glass beads to promote heat exchange. The tube holder is placed in a constant temperature bath. Concentration can be varied through a manipulation of pure carrier gas flow rate, temperature, or tube length.

The length of the tube required to produce the desired gas concentration can be estimated from the formula given below:

$$L = \frac{CF}{KP_t} \tag{6.2}$$

TABLE 6.3
Available Permeation Tubes

Sulfur dioxide	Propane
Nitrogen dioxide	Propylene
Hydrogen sulfide	Butane
Chlorine	Butene-1
Hydrogen fluoride	*Cis*-butene-2
Dimethyl sulfide	*Trans*-butene-2
Dimethyl disulfide	Ethylene oxide
Methyl mercaptan	Others

where:

L is the length in centimeters

C is the concentration in parts per million (volume)

F is the flowrate of carrier gas in cm^3/min

K is the constant dependent upon gas at 1 atm, 25°C—0.382 for SO_2; 0.532 for NO_2; and 0.556 for propane

P_t is the permeation rate for stated operating temperature

Note that concentration can be varied by changing tube temperature or length or carrier gas flowrate.

7 Cost Estimating

Cost is the basic and crucial decision-making factor in the selection of air-pollution control equipment. Evaluating costs can be straightforward, as long as there is an understanding as to the objective of the analysis and an appreciation of the time value of money. The classic cost trade-off is between capital and operating costs (see Figure 9.1). This chapter is designed to assist the practicing engineer with cost engineering techniques for estimating capital cost of equipment and for determining the annualized cost of operating air-pollution control systems, such as those required for the cost analysis portion of a best available control technology analysis. It is not intended to cover detailed cost accounting methods that include different types of depreciation and taxes.

7.1 TIME VALUE OF MONEY

Because of interest and inflation, a dollar in U.S. currency today will be worth more than a dollar in the future. Right off the bat, we must distinguish between the *real* interest rate, which does not include inflation, and the *nominal* interest rate, which includes inflation. The *real* or *constant* dollar cost analysis considers the return on investment as profit. This type of analysis is frequently used by engineers because it is straightforward, easy to use, and easy to understand. Annual operating and labor costs of $100,000 this year will remain at $100,000/year in 10 years. To be clear, the analysis is stated to be in *year 2014 dollars.*

Conversely, inflation reduces the value or buying power of the currency with time. $100,000 worth of materials and labor this year will require more money in 10 years. The real and nominal interest rates are related by

$$(1+i_n)=(1+i)(1+r) \tag{7.1}$$

where:

i_n is the nominal interest rate
i is the real interest rate
r is the annual inflation rate

Financial decisions that include the effect of income taxes use the nominal interest rate because both depreciation deductions and taxes are figured on the basis of nominal dollars.

A particularly helpful approach is use of a time line that accounts for all revenues and expenses over time. Generally, revenues are shown above the time line and are indicated as positive dollars, and expenses are shown below the time line and are indicated as negative dollars. Figure 7.1 shows this general representation.

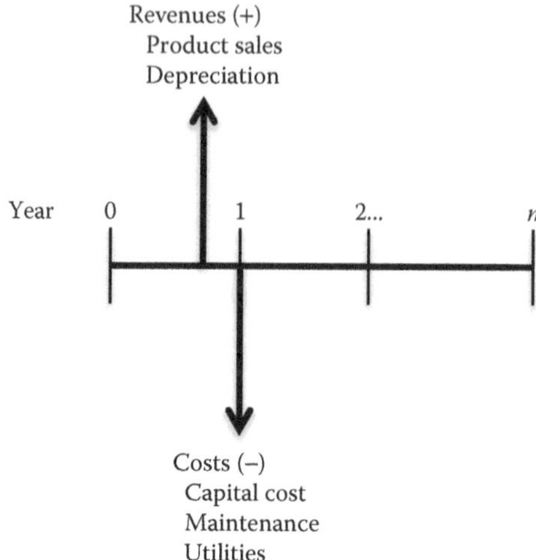

FIGURE 7.1 Timeline illustration for economic analysis.

All revenues and costs are bought to their current value using a preselected interest rate along with the appropriate financial formulas for bringing all future values to a single current value. The summation of all of the current value revenues and costs are then calculated to determine a current net present value. A positive net present value indicates that a project is profitable and a higher positive value indicates a greater level of profitability. This approach is often used to compare multiple process alternatives than can all be employed to solve an air pollution mitigation task.

Consider an example membrane separator whose cumulative membrane cost is \$20,000 and these membranes have to be replaced every four years. In addition, the annual maintenance cost for the membrane separator is \$3000 and the membrane salvage value for recycle after four years of use is \$800. An example timeline showing the economics of this process is shown in Figure 7.2. The values on the timeline

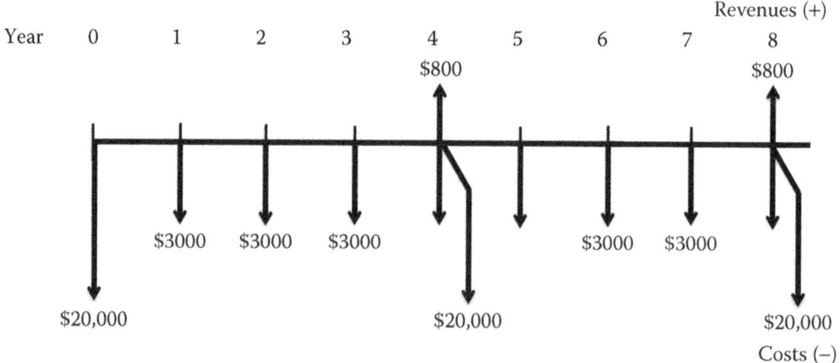

FIGURE 7.2 Economic analysis example using a timeline.

are all brought to their representative current values to show the overall economic impact of this process. Furthermore, adjustments to the membrane replacement costs can be adjusted using escalation factors (cost indexes) that are covered later in Section 7.1.2.

7.1.1 Annualized Capital Cost

The key to annualizing capital cost in real, pretax dollars is to calculate the capital recovery factor.

$$\text{Capital recovery factor} = \frac{i(1+i)^n}{(1+i)^n - 1} \tag{7.2}$$

where:

i is the real pretax marginal rate of return on private investment
n is the economic life of the equipment for the analysis

The product of the capital recovery factor and the capital cost, or investment (P), is the annualized capital recovery cost.

$$\text{Capital recovery cost} = \text{capital recovery factor} \times P \tag{7.3}$$

7.1.2 Escalation Factors

Several cost indexes are available that reflect current costs to a baseline cost and these are also referred to as *escalation factors*. Examples include the Marshall and Swift Equipment Cost Index and the Chemical Engineering Plant Cost Index. Examples of the cost indexes over the period of 1980–2006 are provided in the literature.[1] Specific to the air-pollution control industry is the Vatavuk Air Pollution Control Cost Index.[2] An index is used to convert a known cost at one point in time to an escalated cost at another point in time. The conversion is simply the ratio of the index values for the two points in time.

Indexes are useful for the time periods covered by the index. The Vatavuk Air Pollution Control Cost Index begins in the first quarter of 1994 with a baseline value of 100. Cost data obtained prior to 1994 cannot be escalated using the Vatavuk Index, but the Marshall and Swift Equipment Cost Index can be used from a baseline date of 1926.

Vatavuk Air Pollution Control Cost Indexes are provided for specific air-pollution control devices, including carbon adsorbers, catalytic incinerators, electrostatic precipitators, fabric filters, flares, gas absorbers, mechanical collectors, refrigeration systems, regenerative thermal oxidizers, and wet scrubbers. Updated values of the indexes are published in each biweekly issue of *Chemical Engineering*. Annual values for these indexes are provided in Table 7.1

TABLE 7.1
Vatavuk Annual Average Air-Pollution Control Cost Index

	1994	1995	1996	1997	1998	1999	2000[a]
Carbon adsorbers	101.2	110.7	106.4	104.7	103.6	100.8	107.8
Catalytic incinerators	102.0	107.1	107.0	107.7	106.5	102.9	113.0
Electrostatic precipitators	102.8	108.2	108.0	108.8	109.2	101.2	101.0
Fabric filters	100.5	102.7	104.5	106.2	109.5	111.7	113.2
Flares	100.5	107.5	104.9	105.8	103.6	99.4	104.5
Gas absorbers	100.8	105.6	107.8	107.6	109.7	110.9	112.8
Mechanical collectors	100.3	103.0	103.3	103.9	111.0	119.6	121.8
Refrigeration systems	100.5	103.0	104.4	106.1	107.6	105.7	106.1
Regenerative thermal oxidizers	101.4	104.4	106.3	107.9	108.9	108.1	109.1
Thermal incinerators	101.3	105.9	108.2	109.4	110.5	108.1	108.0
Wet scrubbers	101.3	112.5	109.8	109.0	109.7	108.8	113.7

Source: Seider, W. et al., *Product and Process Design Principles: Synthesis, Analysis and Design*, 3rd ed., 2009.

[a] Preliminary.

7.2 TYPES OF COST ESTIMATES

The accuracy of the project cost estimate and the effort required to calculate it must be appropriate for the decision being made. The American Association of Cost Engineers lists[3] three types of estimates:

- Order of magnitude
- Budget
- Definitive

The *order-of-magnitude* cost estimate uses rules-of-thumb based on prior experience. The only data requirement is the system capacity. Frequently, an exponential factor of 0.6 is used to factor the capacity of large facilities by using the six-tenths factor equation below.[4]

$$\frac{\text{Cost}_2}{\text{Cost}_1} = \left(\frac{\text{Capacity}_2}{\text{Capacity}_1}\right)^{0.6} \tag{7.4}$$

The accuracy of order of magnitude is supposed to be +50% to −30% rather than a factor of 10 as the name implies. However, there is little control of accuracy, and this approach may amount to a best guess with a very high level of inaccuracy.

Budget estimates typically have an accuracy of +30% to −15%. These estimates require knowledge of the site, flow sheet, equipment, and buildings. Also, rough specifications for items such as insulation and instrumentation are needed.

Definitive cost estimates, accurate to within ±20%, require complete plot plans, piping and instrument diagrams, specifications, and site surveys, as well as vendor bids for major equipment.

Perry's Chemical Engineers' Handbook describes five levels of detail and accuracy for cost estimates[5]:

- Order of magnitude
- Study (factored) with an accuracy of ±30%
- Preliminary (budget authorization) with an accuracy of ±20%
- Definitive (project control) with an accuracy of ±10%
- Detailed (firm or contractors) with an accuracy of ±5%

Note that these definitions for level of accuracy are not consistent between the two sources, which can easily lead to confusion when one simply refers to a descriptor such as a *budget* or *definitive* estimate. Even describing an estimate by its desired accuracy can be misleading. To achieve a desired level of accuracy, it is best to define the resources and approach that are to be used to develop the cost estimate.

Air-pollution control equipment cost estimating at the *study* level with ±30% accuracy has received a great deal of attention because of the requirement to determine the *cost-effectiveness* for best available control technology analyses. To make a good study estimate, the following resources[6,7] should be used:

- Location of the source within the plant
- Preliminary process flow sheet
- Preliminary sizing and material specifications for major equipment
- Approximate sizes and types of construction of any buildings required to house the control system
- Rough estimates of utility requirements (e.g., steam, water, and electricity)

7.3 AIR POLLUTION CONTROL EQUIPMENT COST

7.3.1 *OAQPS* CONTROL COST MANUAL

To assist permit applicants and reviewing agencies, and to bring some semblance of consistency to the permit applications, the EPA has published guidelines[5] for estimating the cost of various equipment, including thermal and catalytic incinerators, carbon adsorbers, fabric filters, electrostatic precipitators, flares, refrigerated condensers, gas absorbers, hoods, ductwork, and stacks. The manual provides a good overview of cost-estimating methodology, descriptions of the control equipment and fundamentals of operation, cost correlations based on size and materials, and factors for operating costs.

7.3.2 OTHER COST-ESTIMATING RESOURCES

The *OFFICE OF AIR QUALITY PLANNING AND STANDARDS* (*OAQPS*) *Control Cost Manual* does not discuss all air-pollution control devices, nor does it cover large, complex air-pollution control systems such as flue gas desulfurization systems.

Traditional engineering cost-estimating approaches may be applied in these cases. The design engineer produces a bill of materials including individual equipment items (vessels, pumps, blowers, exchangers, etc.). Using the equipment size and weight and material of construction, a cost estimate can be produced. Common sources of equipment cost estimates include *Perry's Chemical Engineers' Handbook*[5] and Richardson's *Process Plant Construction Estimating Standards.*[7] Additional material has been published by Vatavuk,[8] who is also the primary author and editor of the *OAQPS Control Cost Manual.* A large amount of detailed cost information on utility flue gas desulfurization systems has been sponsored and published by the Electric Power Research Institute.[9,10]

8 Process Design and the Strategy of Process Design

8.1 INTRODUCTION TO PROCESS DESIGN

Process design is the distinguishing feature that differentiates the chemical engineer from all other engineers. As the field of air-pollution control engineering has evolved over the last several decades, it is obvious that the air-pollution control engineer must adopt the ideas and principles of process design. In defining process design, we first define *design* as the development of a plan to accomplish a goal. The engineer must modify that definition to consider *engineering design*. Borrowing from a nineteenth century definition for railway engineering and extending the definition to our time, *engineering design* is "The art of engineering a safe, environmentally sound system for $1.00 which any fool can do for $2.00." The engineer must now accept the responsibility to see that any engineering system is not only economically feasible, but it must be safe and meet environmental regulations. Two books that would be of assistance in developing the ideas of process analysis, synthesis, and design principles are the texts by Seider et al.[1] and by Turton et al.[2] Both these books emphasize the use of computer-aided process design and are up to date in their viewpoint.

8.2 STRATEGY OF PROCESS DESIGN

This book focuses on environmental process design and specifically air-pollution control design. This design proceeds in a sequence of steps from the planning stage to the equipment stage of an *air-pollution control* project. As each decision point is reached, the engineer must evaluate alternatives. The most technically and economically feasible alternatives must be chosen; these alternatives must also be *safe and environmentally sound*. Consider the following process as an example:

It has been determined that NO_2 emissions from the Greasko Limited plant have exceeded emission limitations. The orange cloud continues to be visible every day. The company management is deeply concerned for the community surrounding the plant and has determined that action must be taken to eliminate these emissions. Figure 8.1 illustrates the following likely sequence of events.

- Calculated emission rates are verified by measurement, and sampling and laboratory analysis determine if emission limitations are being exceeded.
- Equipment for the reduction of the NO_2 must be installed.

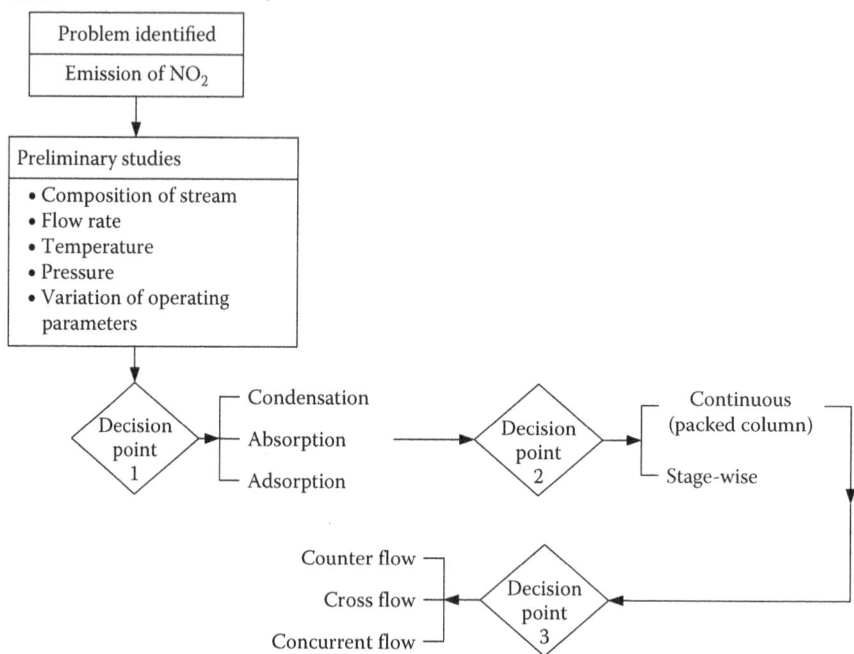

FIGURE 8.1 The process of process design.

- Final problem definition will identify the following factors as part of the design basis:
  - Composition of the stream
  - Flow rate
  - Temperature
  - Pressure
  - Variation of the operating parameters

A series of decisions must now be made.

1. Decision point 1: Alternative control techniques
   a. Condensation
   b. **Absorption**—*chosen*
   c. Adsorption
2. Decision point 2: Absorption—type of absorber
   a. **Continuous**—*packed column—chosen*
   b. Stage-wise
3. Decision point 3: Flow pattern
   a. **Counterflow**—*chosen*
   b. Crossflow
   c. Concurrent flow

4. Decision point 4: *A major impact*—disposing of the collected pollutant
   a. Recycle
   b. Recover
   c. Incinerate
   d. Landfill
   e. React
   f. Neutralize

Emphasizing the decision process, Phillips[3] presents a detailed decision tree for the selection of the right device for removing particulates from a gas stream. He states that his method will help the designer to rapidly identify processes that are suitable for a particular application from 40 or more classes of processes available. In a letter to the editor, Nair[4] takes issue with this detailed decision tree. He points out that a knowledgeable air-pollution control process designer will contact a vendor with the minimum information necessary for the design, and the vendor will do the design. Nair's method leaves the air-pollution control process designer with the question: What is an acceptable and feasible design? The authors present the material in this handbook to assist the air-pollution control engineer in determining and recognizing a safe, environmentally sound, and feasible process design.

8.2.1 Process Flowsheets

Before beginning the process design, a simple basic flowsheet should be made. Even a pencil sketch can help explain better what will be required. Adding the process variables and the flowsheet can be a useful tool at every step. One way to begin is to draw a boundary around the whole process. This diagram can then serve to initiate the basic mass and energy balances, which will be discussed in Section 8.3. As the process design develops, the recycle streams can be added to this input–output sketch. The flowsheet should then be modified and moved toward the final document to be used for the detailed design and cost estimation. Start with the air-pollution control equipment, then add the needed appurtenances such as ducting, piping, pumps, fans, and tanks. Finally, the instrument and automatic controls can be added.

8.3 MASS AND ENERGY BALANCES

The basis for all process design is the mass and energy balances. Thermodynamics serves as the scientific background for these calculations. For a treatment featuring chemical processes see Smith et al.[5] The basic mass and energy balance concept is presented by Felder and Rousseau.[6]

Figure 8.2 presents the general mass and energy balance. Here, the *system* is receiving two input streams from the *surroundings* and is producing one outlet stream to the surroundings. Each stream enters the system at a height z_i above a reference plane. Heat, Q, and shaft work, W_S, are being supplied and considered positive transferred to the system from the surroundings. The overall mass balance in terms of flow rate is

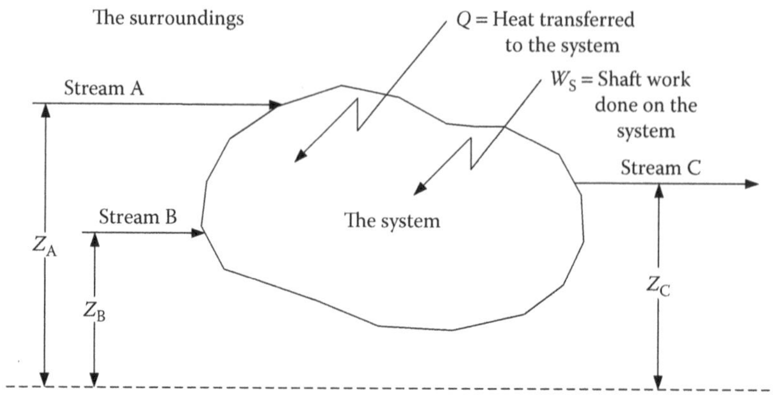

FIGURE 8.2 The general mass and energy balance.

$$M_A + M_B = M_C \tag{8.1}$$

A component mass balance can be made, where the concentration x_{ij} is given in mass or mole fractions depending on the units of the flow rate. Here, i designates the component and j designates the stream.

$$x_{AA}M_A + x_{AB}M_B = x_{AC}M_C \tag{8.2}$$

The energy balance can then be written using the convention of exit quantities minus inlet quantities

$$M_C\left(U_C+\frac{P_C}{\rho_C}+\frac{\bar{U}_C^2}{2\alpha g_c}+\frac{z_C g}{g_c}\right)-M_A\left(U_A+\frac{P_A}{\rho_A}+\frac{\bar{U}_A^2}{2\alpha g_c}+\frac{z_A g}{g_c}\right) -M_B\left(U_B+\frac{P_B}{\rho_B}+\frac{\bar{U}_B^2}{2\alpha g_c}+\frac{z_B g}{g_c}\right)=Q+W_S \tag{8.3}$$

where:

U_j is the internal energy
P_j is the pressure
ρ_j is the density of fluid
$\bar{U}_j$ is the average velocity of stream
α is a correction factor of 1.0 for turbulent conditions and 1/2 for laminar conditions
g is the acceleration due to gravity
g_c is the gravitational constant

Enthalpy is defined as

$$H = U + Pv \tag{8.4}$$

where:

v is the specific volume, which is equal to $1/\rho$

The enthalpy can be calculated from specific heat at constant pressure.

$$C_P \equiv \left(\frac{\partial H}{\partial T}\right)_P \tag{8.5}$$

$$\Delta H = \int_{T_1}^{T_2} C_P dT = C_{P\text{avg}}\left(T_2 - T_1\right) \tag{8.6}$$

Equation 8.3 can be simplified for a single stream in, A, and a single stream out, C. Then,

$$M_A = M_C$$

Using Δ to indicate the output–input difference, we have

$$\Delta H = H_C - H_A$$

$$\Delta \bar{U}^2 = \bar{U}_C^2 - \bar{U}_A^2$$

$$\Delta z = z_C - Z_A$$

Equation 8.3 may now be written as

$$M\left(\Delta H + \frac{\Delta \bar{U}^2}{2\alpha g_c} + \frac{\Delta z g}{g_c}\right) = Q + W_S \tag{8.7}$$

8.3.1 Mass-Balance Example

Two streams enter a separations apparatus as shown in Figure 8.3. The streams mix and are separated. Both streams also leave the apparatus. The following lists the stream compositions:

Stream 1	Stream 2	Stream 3
60% A	100% N	95% A
40% B		5% B

It is also known that the ratio of the flow rate of Stream 2 to Stream 1 is 5/1. The composition of Stream 4 is unknown and must be determined. All of the component A goes into Stream 3. Set the basis for calculation.

Basis: 100 units of Stream 1. The units could be lb mass, kg mass, lb moles, or kg moles. Use kg moles.

Stream 1 = 60 kg moles of A, 40 kg moles of B

Stream 2 = 5 × 100 = 500 kg moles N

A is a *tie substance*, which can be used to determine how much of the total flow goes into Stream 3.

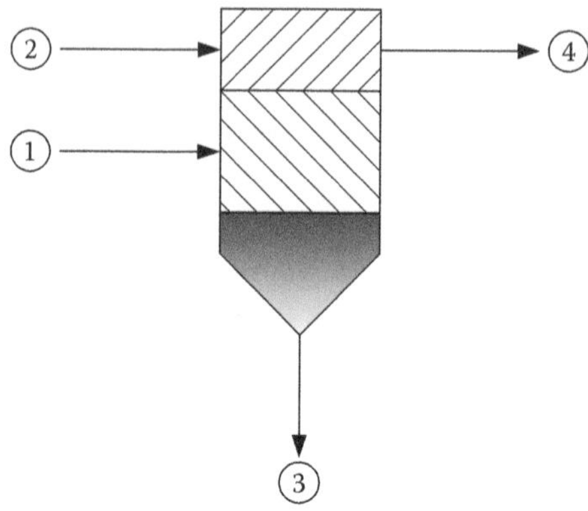

FIGURE 8.3 The mass balance.

Stream 3:

60 kg moles A	= 95% of the stream
60/0.95	= 63.16 kg moles total flow
63.16 – 60	= 3.16 kg moles B
Check: (3.16/63.16) × 100 = 5.00% B—**Ok!**	

Stream 4:

N		= 500 kg moles
B	= 40 – 3.16	= 36.84 kg moles
Total moles	= 500 + 36.84	= 536.84 kg moles
% N	= (500/536.84) × 100	= 93.14%
% B	= (36/536.84) × 100	= 6.86%
Total		= 100.00%

8.3.2 Energy-Balance Example

Hot dirty air at 10,000 acfm, 1.0 atm, and 500°F is blown by a fan into a heat exchanger and then into a waste treatment process. Figure 8.4 illustrates the process. The air enters the blower at 125 ft/s. The blower has a 15 HP motor operating at 80% efficiency. The dirty air is cooled to 130°F at Point 2 and leaves the heat exchanger at 275 ft/s. Water at 80°F is available as a coolant. A 10°F temperature rise is allowed. Determine the amount of coolant required.

Data:

Air density at 500°F: $\rho_{air} = 0.0412$ lbm/ft^3
Air specific heat: $C_{P_{avg}} = 0.241$ BTU/lbm-°F
Water specific heat $C_{P_{avg}} = 1.0$ BTU/lbm-°F

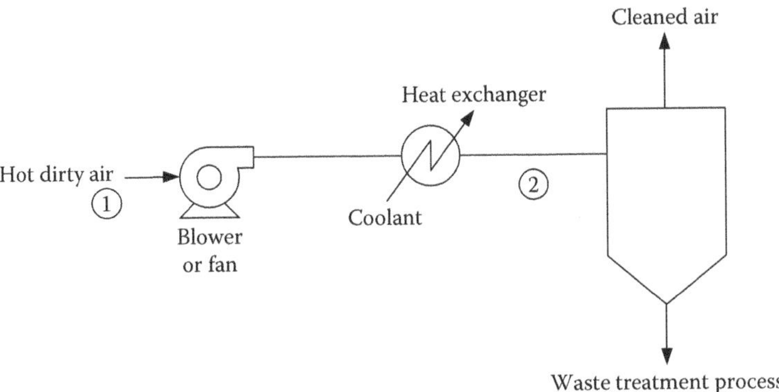

FIGURE 8.4 The energy balance.

Air enthalpy at 1: H_1 = 231.1 BTU/lbm
Air enthalpy at 2: H_2 = 141.1 BTU/lbm

Solve Equation 8.7 for Q.

$$Q = M\left(\Delta H + \frac{\Delta \bar{U}^2}{2\alpha g_c} + \frac{\Delta z g}{gc}\right) - W_S \quad (8.8)$$

Assume $z = z_2 - z_1 = 0.0$ and $\alpha = 1.0$

$$M = 10{,}000\ \text{ft}^3/\text{min} \times 0.0412\ \text{lbm/ft}^3 \times 1.0\ \text{min}/60\ \text{s} = 6.87\ \text{lbm/s}$$

$$\Delta H = C_{\text{Pavg}}(T_2 - T_1) = 0.241(130 - 500) = -89.17\ \text{BTU/lbm}$$

or

$$\Delta H = H_2 - H_1 = 141.1 - 231.1 = -90.0\ \text{BTU/lbm}$$
$$\Delta H = -90\ \text{BTU/lbm} \times 778\ \text{ft-lbf/BTU} \times 6.87\ \text{lbm/s} = -476{,}370\ \text{ft-lbf/s}$$

$$\frac{\Delta \bar{U}^2}{2\alpha g_c} = \left[\frac{(275)^2 - (125)^2}{2 \times 32.174}\right]\frac{\text{ft}^2/\text{s}^2}{\text{ft-lbm/lbf-s}^2} = 932.4\ \text{ft-lbf/lbm}$$

$$\frac{\Delta \bar{U}^2}{2\alpha g_c} = 932.4\ \text{ft-lbf/lbm} \times 6.86\ \text{lbm/s} = 6402.7\ \text{ft-lbf/s}$$

$$W_S = 0.80 \times 15\text{HP} \times 550\ \text{ft-lbf/s/HP} = 6600.0\ \text{ft-lbf/s}$$

$$Q = -476{,}370 + 6403 - 6600 = -476{,}567\ \text{ft-lbf/s}$$

(shows ΔH to be the most significant factor)

$$Q = -476{,}576\ \text{ft-lbf/s} \times \text{BTU}/778\ \text{ft-lbf} = -612.6\ \text{BTU/s}$$

For the coolant $Q_C = -Q$,

$$Q_C = mC_{\text{pavg}}\Delta T,\ \Delta T \text{ limited to } 10°\text{F}$$

$$m = \frac{Q_C}{C_{\text{Pavg}}\Delta T} = \frac{612.6}{1 \times 10} = 61.26\,\text{lbm/s}$$

$$m = 61.26\,\text{lbm/s} \times 60\,\text{s/min} = 3675.3\,\text{lbm/min}$$

$$q_w = 3675.3\,\text{lbm/min} \times 1.0\,\text{gal}/8.34\,\text{lbm} = 440\ \text{gpm}$$

8.4 SYSTEMS-BASED APPROACHES TO DESIGN

Numerous systems-based approaches for the design of air-pollution control systems have been developed over the last two decades. The basis for the development of these design methodologies is that a number of process alternatives (unit operations) can be employed to recover pollutants from air emission streams from a process plant. While the example presented in Section 8.2 for NO_2 emission reduction considered numerous process alternatives, a decision was made in Decision point 1 of Figure 8.1 to select a specific process alternative (absorption). Not only may multiple technologies work for the air pollution control but also numerous separating agents (e.g., a specific absorbent) may be effective in reducing the NO_2 emissions. Systematic techniques extensively reported in the literature and collectively referred to as *process integration* allow the simultaneous evaluation of a number of process alternatives to identify the most cost-effective technology and separating agent(s) to be employed.[7]

The initial motivation to identify the most cost-effective waste recovery system from a large group of process options (multiple technologies and/or separating agents for the recovery task) resulted in the development of mass exchange network *MEN* synthesis.[8] A schematic representation of a single mass exchanger is provided in Figure 8.5. A single mass exchanger is defined as a direct-contact, countercurrent unit that employs a mass separating agent *MSA* to allow the transfer of the pollutant from the gaseous emission stream to the MSA. Examples of mass exchangers would be absorption columns, adsorbers, extraction units, stripping columns, and ion-exchange units; examples of MSAs would be liquid absorbents, adsorbents and extractants such as solvents, activated carbon, and liquid extractants. A mass exchange network *MEN*

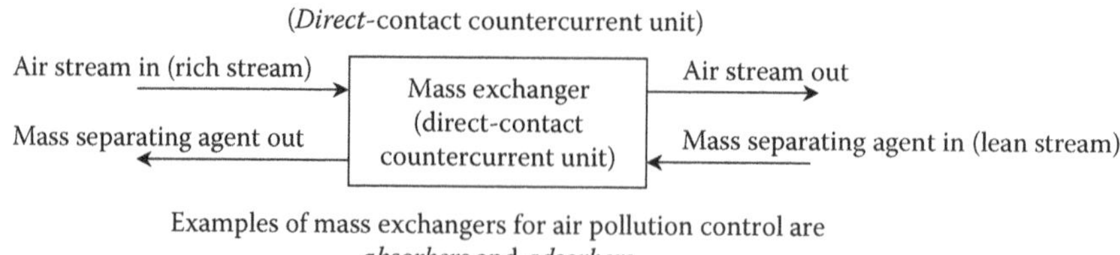

FIGURE 8.5 A schematic of a single mass exchanger for separating pollutants from an air emission stream.

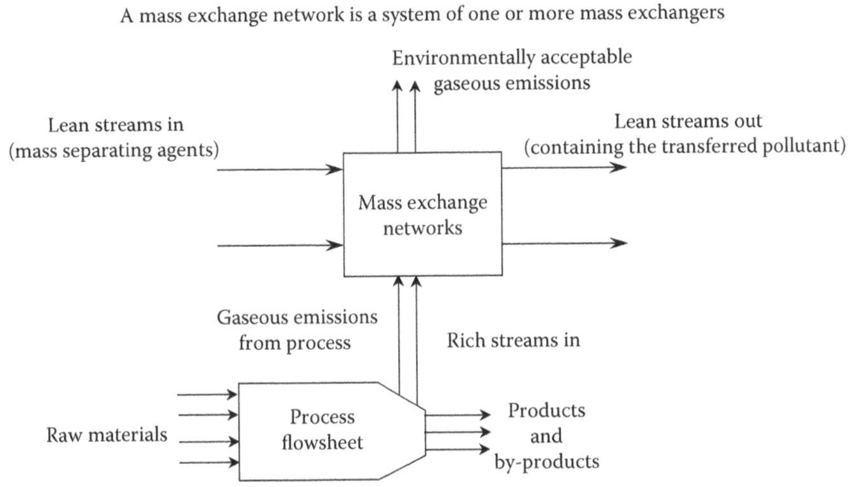

FIGURE 8.6 Mass exchange network synthesis for separating pollutants from air emission streams.

is a network consisting of one or more mass exchangers that collectively satisfy the air pollution mitigation task. The design task of synthesizing an MEN is to systematically identify a network of mass exchangers for the selective transfer of a certain gaseous air pollutant from a set of *rich* (high concentration) streams to a set of *lean* (MSA) streams. Figure 8.6 is included as a general representation of the MEN synthesis task for end-of-the-pipe air-pollution control process design.

Systems-based techniques have also been developed for other separations systems that are traditionally used for air-pollution control process design. For instance, there is a wide class of separation systems that employ energy separating agents *ESAs* (hot and cold process streams and/or utilities such as steam and cooling water) to separate species from an emission stream by using a phase change. These include unit operations such as condensers, evaporators, dryers, and crystallizers, and they are collectively grouped in the category of heat-induced separators. A schematic representation of a heat-induced separator is provided in Figure 8.7. A heat-induced separator is defined as an indirect-contact, countercurrent unit that employs an energy-separating agent *ESA* to allow the removal of the pollutant from the gaseous emission stream by using a phase change. Examples of ESAs would be hot and cold utilities and hot and cold process streams. The design task of synthesizing heat-induced separation networks (HISENs) for air-pollution control process design is to systematically identify a cost-effective system of heat-induced separators and heat exchangers that can achieve a specified pollutant reduction task (single component or multiple component waste streams) by heating/cooling the streams to produce a separation by a phase change of the pollutant.[9] Figure 8.8 is included as a general representation of the HISEN synthesis task for end-of-the-pipe waste minimization process design. HISEN methodologies have been successfully used to identify optimal condensation systems for the recovery of volatile organic compounds (VOCs) from gaseous emission streams.

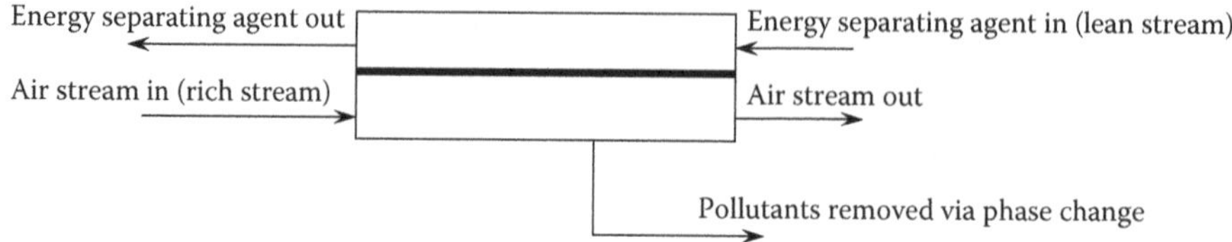

FIGURE 8.7 A schematic of a single heat-induced separator for separating pollutants from an air emission stream.

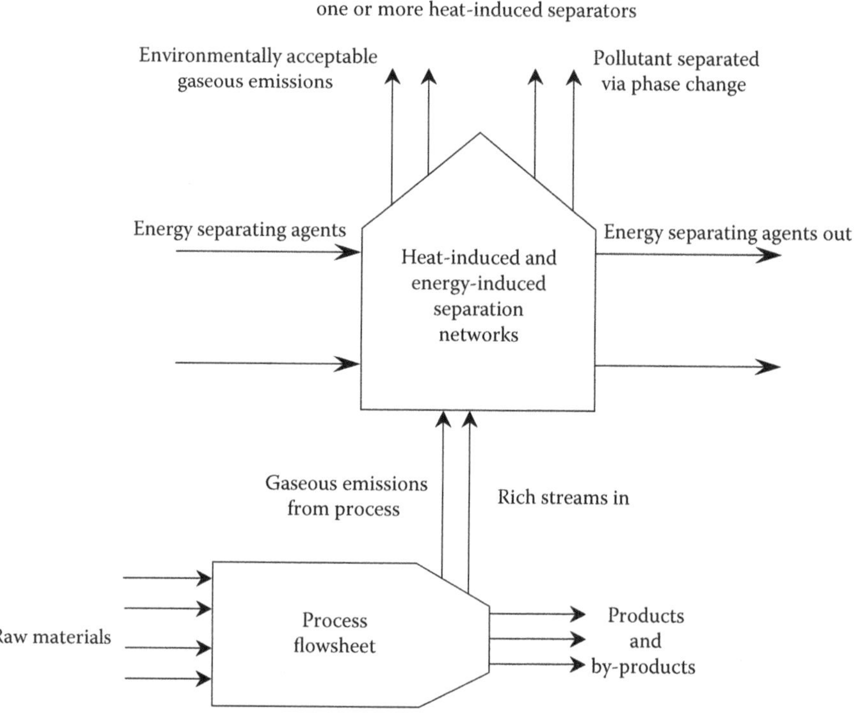

FIGURE 8.8 Heat-induced separation network synthesis for separating pollutants from air emission streams.

Within the last decade, the HISEN synthesis methodology has been extended to include pressurization/depressurization equipment, such as compressors and turbines, in conjunction with heat-induced separation specifically to address air-pollution streams containing VOCs and the resulting recovery system was referred to as an *energy-induced separation network*.[10] The simplest technique for VOC recovery is to use heat-induced separation networks to allow condensation of the VOCs by cooling; however, VOC condensation is, in general, a function of both temperature and pressure. The synergism between pressurization/depressurization and cooling was addressed by the design methodology of Dunn and coworkers, whose objective was to create a cost-effective network of heat-induced separators, heat exchangers, and pressurization/depressurization devices, which can separate one or more VOCs from a set of waste streams via phase change.

9 Profitability and Engineering Economics

9.1 INTRODUCTION—PROFIT GOAL

A major decision point in process design is the selection of the most technically and economically feasible solution alternative. All companies set a profit goal. Most generally, it is to maximize profit, that is, maximize income above the cost of the capital, which must be invested to generate the income. In addition, current governmental regulations require that processes must be safe and ecologically/environmentally sound. Others goals sometimes are used. For example,

- Minimize raw material loss
- Minimize operating cost
- Minimize environmental damage
- Minimize emissions
- Produce a by-product that requires low cost from the company

Furthermore, companies usually have set a required rate of return on the investment. In some cases such as with required environmental control equipment, the process must be built and operated even if it does not meet the rate of return that has been set.

9.2 PROFITABILITY ANALYSIS

An analysis must be made to provide a measure of the attractiveness of a project for comparison to other possible courses of action. In some cases, the capital that would be invested in a process could earn more in some other investment or even in a savings account. For required environmental control processes, there would be no other alternative investment allowed. It should be remembered that a profitability standard can only serve as a guide, and the profitability standard must be weighed against overall judgment and evaluation.

9.2.1 Mathematical Methods for Profitability Evaluation

There are at least five reasonable methods to carry out a profitability evaluation. These methods are as follows:

1. Rate or incremental rate of return on investment
2. Discounted cash flow on full-life performance
3. Net present worth

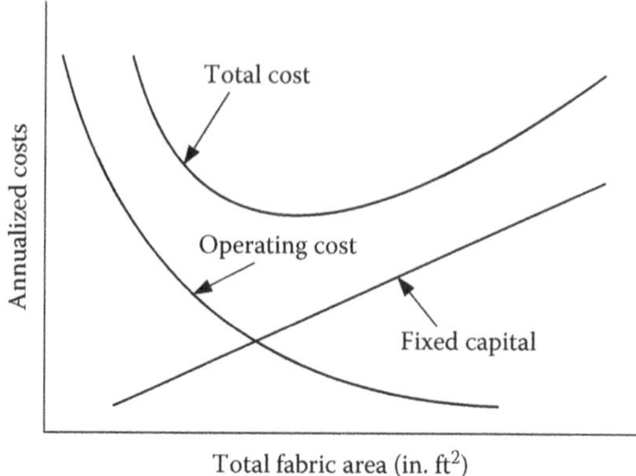

FIGURE 9.1 Optimization of cost for a typical fabric filter or baghouse.

4. Capitalized costs
5. Payback period

Peters and Timmerhaus[1] present an excellent discussion of all these methods. Figure 9.1 illustrates the well-known concept in balancing fixed capital charges for a fabric filter against operating cost to find the minimum of the sum of the two.

9.2.2 Incremental Rate of Return on Investments as a Measure of Profitability

One of the more simple methods of profitability analysis is based on rate of return and the incremental rate of return. It is assumed that there is a *base case* design that will provide the desired pollution control and meet the profitability objective set as the rate of return. As pointed out above, in some cases, a pollution control apparatus must be built even if it does not meet the profitability objective of a company. The base case should represent a system operating under nearly optimum conditions. All other designs are then compared against this base case. Let us define the following terms:

ROI is the percent return on investment
IROI is the incremental percent return on investment
P is the annual profit from investment (income − expenses), or incremental profit or annual savings
I is the total investment or incremental investment

$$\text{ROI or IROI} = \frac{P}{I} \tag{9.1}$$

The process is as follows:

- Set the base case—the alternative case that meets ROI at the least total initial cost.

- Check IROI with the base case compared to the alternative case that meets the ROI and has the second lowest initial cost.
- Accept the second alternative as the base case if IROI is met.
- With this new base case, check ROI compared to the alternative that has the third lowest initial cost and then check the IROI. Set new base case if IROI is met.
- Repeat process with all alternatives and select one that meets the ROI and the IROI by balancing the initial cost the ROI and the IROI.

9.2.2.1 Example of IROI Comparing Two Cases

A company needs to purchase a cyclone to control dust from one of its foundry operations. The company expects to receive 12% ROI before taxes. The company is considering two offers. One manufacturer proposes a carbon steel cyclone to do the job; the other manufacturer proposes a stainless steel cyclone that costs more but has a longer service life and lower maintenance and power cost. Data on the two cyclones are as follows:

Carbon steel cyclone
  Installed cost = \$40,000
  Service life = five years
  Salvage value = 0
Stainless steel cyclone
  Installed cost = \$40,000
  Service life = five years
  Salvage value = 0

Lowered maintenance and power costs = \$1100

Which cyclone should the company purchase?

Straight-line depreciation will be used. Other methods of depreciation will be discussed later in Section 9.3.

The decision will be made based on an IROI calculation. The incremental investment is the difference between the costs of the two cyclones.

Incremental investment = \$60,000 – \$40,000 = \$20,000
Depreciation

$$\text{Carbon steel cyclone} = \frac{\$40,000}{5} = \$8000/\text{year}$$

Depreciation

$$\text{Stainless steel cyclone} = \frac{\$60,000}{10} = \$6000/\text{year}$$

Annual savings with purchase of stainless steel cyclone

Savings = Difference in depreciation + Lowered maintenance and power costs

$$\text{Savings} = \$8000 - \$6000 + \$1100 = \$3100 / \text{year}$$

Incremental return on investment = IROI

$$\text{IROI} = \frac{\text{Amount of savings}}{\text{Cost difference}} = \frac{\$3100}{\$20,000} \times 100 = 15.5\%$$

This is greater than the required 12.0% return.

Therefore, the company should purchase the stainless steel cyclone.

9.2.2.2 Example of IROI with Four Cases

This example requires establishing a base case and comparing the other cases to the base case. The cases are lined up with the least expenses case as the first item. The rest of the cases are then put in order of increasing cost. The least expensive case is taken as the base case and it is determined if it meets the ROI. If it does meet the ROI, then continue on to examine the second case, then the third case and so forth. If any case does not meet the ROI, it must be discarded. After every case is reviewed and all cases that meet the ROI are established, the basic case is then established. The IROI of the next accepted case is then determined, and if it meets the ROI, which was originally set, it is then accepted as the base case and the first case is discarded. The process then continues with the third case and the calculation is continued until all originally accepted cases are reviewed.

As in the following example, if there are only a few number of cases, if the IROI in the first calculation is acceptable, it can then be set as the base case and proceed to compare the third case to it. Then just continue until all cases are reviewed. An example follows.

A chemical plant must install a new control system. Four alternatives A, B, C, and D are being considered. Each alternative will result in a reduction of emissions from the current 2000 T/year. The emissions fee is $35/T, and, therefore, the new equipment will reduce the annual emission fee. Furthermore, two of the processes will result in saving product for which an additional cost saving can be included. The company is demanding a 15% return on any investment. Which of the processes should be selected?

Control System Design	**A**	**B**	**C**	**D**
Installed cost ($)	100,000	150,000	200,000	250,000
Operating cost ($)	5,000	10,000	15,000	10,000
Fixed charges (%)	10	10	10	10
Initial cost ($/year)	10,000	15,000	20,000	25,000
Emission reduction	900 T	1200 T	1200 T	1500 T
($)	31,500	42,000	42,000	52,500
Product saved ($/year)	0	6,000	8,000	21,000

Design A serves as the *base case.*

A. Annual savings $= 0 + 31{,}500 - 5{,}000 - 10{,}000 = \$16{,}500$

$$\text{ROI} = \frac{16{,}500}{100{,}000} \times 100 = 16.5\%$$

Meets ROI criteria.

B. Annual savings $= 6{,}000 + 42{,}000 - 10{,}000 - 15{,}000 = \$23{,}000$

$$\text{ROI} = \frac{23{,}000}{150{,}000} \times 100 = 15.3\%$$

This ROI is acceptable.

$$\text{IROI} = \frac{23{,}000}{50{,}000} \times 100 = 46.0\%$$

This IROI is acceptable, therefore, this becomes the *base case.*

C. Annual savings $= 8{,}000 + 42{,}000 - 15{,}000 - 20{,}000 = \$15{,}000$

$$\text{ROI} = \frac{15{,}000}{200{,}000} \times 100 = 7.5\%$$

This is an unacceptable ROI. Design C must be rejected. Design D is now compared to Design B, which is the current base case.

D. Annual savings $= 21{,}000 + 52{,}500 - 10{,}000 - 25{,}000 = \$38{,}500$

$$\text{ROI} = \frac{38{,}500}{250{,}000} \times 100 = 15.4\%$$

This is an acceptable ROI.

$$\text{IROI} = \frac{38{,}500}{100{,}000} \times 100 = 38.5\%$$

In this case, Design B produces nearly the same ROI and results in a better IROI. Select Design B and get a greater return on the incremental money.

9.3 EFFECT OF DEPRECIATION

Depreciation is used to take into account the value of the equipment when it reaches obsolescence or must be replaced because of new technology. Assets before 1981 could be depreciated using the double declining-balance method. From 1981 to 1986, the accelerated cost recovery system was used. Here, all property was classified in 3-, 5-, 10-, 15-, or 20-year service life groups with no salvage value. The majority was in the five-year group. A half full-year depreciation in the year placed in service was allowed; no deduction was granted in the year when property is disposed. From 1987 onward, the modified accelerated cost recovery system has been used. This is the same as accelerated cost recovery system but also allows a full-year depreciation during the year of disposal. A 200% declining balance (i.e., double declining balance) for class lives of 3, 5, 7, and 10 years and a 150% depreciation for class lives of 15 and 20 years is allowed. A switch to straight-line depreciation at a time appropriate to maximize deduction is permitted.

Straight-line depreciation, although no longer permitted, can be used to illustrate the effect of depreciation. The following definitions apply:

V_R is the initial cost
V_S is the salvage value
V_A is the asset value
n is the useful service life
a is the years in service
d is the annual depreciation

$$d = \frac{V_R - V_S}{n} = \text{annual depreciation}$$

$$V_S = V_R - nd \tag{9.2}$$

$$V_A = V_R - da \text{ or } d = \frac{V_R - V_A}{a}$$

As a part of the example below, the declining balance method will be used. This method is somewhat similar to the accelerated cost recovery system and modified accelerated cost recovery system, but is easier to see in an example. Here, equipment depreciates rapidly over the first few years to minimize taxes during the initial period. The annual depreciation is a fixed percentage at the beginning of the depreciation year.

f is the annual depreciation at a fixed percentage factor

$$V_S = V_R(1 - f)^n$$

$$V_A = V_R(1 - f)^a$$

9.3.1 Example

Companies P and Q bought identical Venturi scrubbers for $50,000. The service life of the scrubbers was eight years with no salvage value. The corporate tax rate is 34% for these companies. Company P uses the straight-line method of depreciation, and company Q uses the double declining-balance method. Compare the tax savings for these two companies for a period of three years.

V_R is the initial cost = $50,000

V_S is the salvage value = 0

N is the useful service life = eight years

A is the years in service = three years

34% tax rate—applied at end of year

- Without depreciation, tax would be 0.34 × 50,000 = \$17,000/year

$$3 \times 17{,}000 = \$51{,}000 \text{ over the three years}$$

- Company P—straight-line depreciation

$$d = \frac{50{,}000 - 0}{8} = \$6{,}250/\text{year}$$

Tax computation

$$\begin{aligned}
&\text{First year:} && 50{,}000 - 6250 = \$43{,}750;\ \text{tax} = \$14{,}875 \\
&\text{Second year:} && 43{,}750 - 6250 = \$37{,}500;\ \text{tax} = \$12{,}750 \\
&\text{Third year:} && 37{,}500 - 6250 = \$31{,}250;\ \text{tax} = \underline{\$10{,}625} \\
& && \text{Total} \qquad \$38{,}250
\end{aligned}$$

$$\text{Savings} = 51{,}000 - 38{,}250 = \$12{,}750$$

- Company Q—double declining-balance depreciation

$$\text{First year factor} = \left(\frac{V_R - V_S}{n}\right)\left(\frac{1}{V_R}\right)$$

$$\text{First year factor} = \left(\frac{50{,}000 - 0}{8}\right)\left(\frac{1}{50{,}000}\right) = 0.1250$$

$$f = 2 \times 0.1250 = 0.2500$$

Tax computation

$$\begin{aligned}
&\text{First year:} && V_R(1-f) = 50{,}000 \times 0.7500 = \$37{,}500;\ \text{tax} = \$12{,}750 \\
&\text{Second year:} && V_R(1-f)^2 = 50{,}000 \times 0.5625 = \$28{,}125;\ \text{tax} = \$9{,}563 \\
&\text{Third year:} && V_R(1-f)^3 = 50{,}000 \times 0.4219 = \$21{,}094;\ \text{tax} = \underline{\$7{,}172} \\
& && \text{Total} \qquad \$29{,}485
\end{aligned}$$

$$\text{Savings} = 51{,}000 - 29{,}485 = \$20{,}515$$

Obviously, a method like the double declining balance is favorable when computing taxes. Again, several methods for depreciation exist and factors such as annual sales, annual operating costs, and other factors should be considered when selecting the method of depreciation. A good summary of these methods are provided in Seider et al.[4]

9.4 CAPITAL INVESTMENT AND TOTAL PRODUCT COST

Numerous costs are required to build and operate a chemical plant other than installed equipment costs and operating costs. However, the usual cost estimation technique is to relate these costs to the installed equipment costs. Guthrie[2] presents a capital cost-estimating technique, which can be useful. Vatavuk and Neveril[3] also discuss a method for estimating capital and operating costs. They present a table of factors, which should be roughly valid at this time.

A key consideration for equipment costing is the economy of scale. In general, the cost of equipment does not double as the size of the equipment doubles. In fact, the general cost relationship for equipment as a function of the equipment capacity is referred to as the *six-tenths factor rule* and is summarized mathematically in Equation 9.3.

$$\frac{\text{Cost}_2}{\text{Cost}_1} = \left(\frac{\text{Capacity}_2}{\text{Capacity}_1}\right)^{0.6} \tag{9.3}$$

This relationship shows that higher equipment capacity results in a lower equipment cost per unit volume of product produced. Of course, this assumes that all of the larger equipment capacity will be utilized and that capital is available for investment into the larger scale equipment.

For example, if the cost of an activated carbon fixed-bed adsorber that is sized for a capacity of 10,000 pounds of activated carbon is known to be \$150,000, then a cost estimate for a larger adsorber with a 20,000 pound capacity can be estimated using Equation 9.4:

$$\text{Cost} = \$150{,}000\left(\frac{20{,}000\,\text{lbs}}{10{,}000\,\text{lbs}}\right)^{0.6} = \$227{,}400 \tag{9.4}$$

Thus, doubling the capacity does not double the cost.

9.4.1 Design Development

In order to employ cost estimation and profitability analysis, it is suggested that the following process be used.

1. Select a process to develop from alternative processes.
2. Prepare the process flowsheet.
3. Optimize the process flowsheet. Generally, a process simulation software package (Aspen Plus, Aspen Hysys, etc.) is used for process optimization studies.
4. Apply process integration technologies to the process flowsheet to minimize utility costs and to minimize waste generation. This step may be included simultaneously with steps 2 and 3.
5. Calculate sizes of all equipment and estimate fixed capital cost.
6. Estimate installed cost.
7. Determine utilities usage and estimate cost.
8. Determine other costs—taxes, buildings, land, and insurance.
9. Undertake a profitability analysis.

A good list of all other costs to be considered can be found in Seider et al.[4]

In carrying out the design, it is recommended to set a budget for pollution control equipment. Then make as accurate a cost estimate as possible. There are four suggested approaches for arriving at a cost estimate. They are as follows:

1. Order-of-magnitude estimate (±50%): Based on bench-scale laboratory data sufficient to determine the type of equipment and its arrangement
2. Study estimate (±35%): Based on preliminary process design
3. Preliminary estimate (±20%): Based on detailed process design, including optimization
4. Definitive estimate (±10%): Based on detailed plant design with equipment drawings and vendor cost estimates

Keep in mind that the designer must maintain a satisfactory profit structure if alternate choices are available. The following factors will affect equipment costs:

- Company policies
- Local and federal government regulations
- Design standards
- Union contracts and agreements
- Agreements with fabricators
- Economy of scale (equipment size)
- Equipment materials of construction
- Operating pressures

10 Introduction to Control of Gaseous Pollutants

Under the auspices of the Environmental Protection Agency (EPA)'s Center for Environmental Research Information in conjunction with the EPA Control Technology Center, a handbook for design of hazardous air pollution control equipment was published in 1986. A revised version of the handbook was published in June 1991.[1] Current information is furnished through the Clean Air Technology Center. This center serves as a resource on all areas of emerging and existing air pollution prevention and control technologies and their cost. The information now may be found on the following website: http//www.usepa.gov/ttn/catc/.

There are six main processes by which a gaseous pollutant may be removed from an air stream. Table 10.1, taken from the *EPA handbook*,[1] lists those processes with the advantages and disadvantages of using each one. The table may be used as a guide to determine which process may provide the best means of cleaning the air stream. Separation processes are used as a means of air pollution control for both particulate matter and gas. These processes essentially remove the pollutant from the carrier gas resulting in a cleaned gas stream. If the pollutant content of the cleaned stream meets the effluent emission standards, the cleaned stream can be discharged to the atmosphere. Absorption and adsorption are both diffusional separation processes that can be used to collect hazardous air pollutants. In the case of absorption, the pollutant is transferred to the solvent, which then may need further treatment, referred to as *regeneration*. Recovery of the solvent might be undertaken by distillation or by stripping the absorbed material from the solvent. The problem of treating the waste material in the stream separated from the solvent remains. If the pollutant material has a value, adsorption may provide the means for the material to be more readily recovered. In the case of particulate matter, wet scrubbing collects the particles primarily through the mechanism of inertial impaction. Gaseous contaminants such as sulfur oxide, nitrogen oxide, and hydrochloric acid, if present along with the particulates, may be collected simultaneously by absorption.

From a systems perspective, as indicated in Chapter 8, adsorption columns and adsorbers used for the removal of gaseous pollutants are classified as mass exchangers, since they recover the gaseous pollutant by use of direct-contact mass transfer processes. The solvents used in absorption columns and the adsorbents employed in adsorbers to remove gaseous air pollutants are collectively referred to as *mass separating agents*. Absorber and adsorber systems may require regeneration of their mass separating to allow the separation and recovery of the gaseous pollutant and to allow reuse of the mass separating agents. Simultaneous synthesis of a mass exchange networks of direct contact exchangers along with its respective regeneration system has been addressed in the literature.[2] Also, from a systems perspective,

TABLE 10.1
Volatile Organic Compounds Control Technologies

Device	Inlet Conc. PPMV	Efficiency (%)	Advantages	Disadvantages
Absorption	250	90	Especially good for inorganic acid gasses	Limited applicability
	1,000	95		
	5,000	98		
Adsorption	200	50	Low capital investment	Selective applicability
	1,000	90–95	Good for solvent recovery	Moisture and temperature constraints
	5,000	98		
Condensation	500	50	Good for product or solvent recovery	Limited applicability
	10,000	95		
Thermal incineration	20	95	High destruction efficiency	No organics can be recovered
	100	99	Wide applicability Can recover heat energy	Capital intensive
Catalytic incineration	50	90	High destruction efficiency	No organics can be recovered
	100	>95	Can be less expensive than thermal incineration	Technical limitations that can poison
Flares		>98	High destruction efficiency	No organics can be recovered Large emissions only

as indicated in Chapter 8, condensers used for the removal of gaseous pollutants are classified as heat-induced separators since they recover the gaseous pollutant by use of indirect-contact cooling of the gaseous stream. The coolants used in condensers to remove gaseous air pollutants are collectively referred to as *energy separating agents*. Condensation systems may require coolant regeneration, such as through a refrigeration cycle, if the coolant employed is a refrigerant. Mass exchangers (absorbers, adsorbers, etc.) and heat-induced separators (condensers) are considered *recovery* technologies because the gaseous pollutant is removed and recovered for sale or reuse. Incinerators and flares (thermal oxidation processes) are considered *destructive* technologies since the gaseous pollutant is eliminated. A more detailed discussion of absorption technology, adorption technology, condensation, and thermal oxidation are provided in Chapters 11 through 14, respectively.

Many organic materials may be removed by condensation, which is essentially a diffusional operation. If a suitable coolant is available and the pollutant concentration is high enough, condensation can be very effective in recovering material that may be used again. For organic pollutants when the concentration is low or recovering the material is not desired, incineration can be used to convert the pollutant to carbon dioxide and water. For large emissions such as those that would be found in petroleum refineries, the pollutant may be flared.

10.1 ABSORPTION AND ADSORPTION

Both absorption and adsorption are diffusional processes employed in the cleanup of effluent gases before the main carrier gas stream is discharged to the atmosphere. Both of these operations are controlled by thermodynamic equilibrium. In pollution control, the concentrations of gases to be treated are relatively low. Thus, the equipment design is one in which it is reasonable to assume that gas is very dilute. In absorption, it is quite likely that the liquid effluent will be dilute as well. Absorption of the contaminant from the dilute gas results in a chemical solution of the contaminating molecule. However, adsorption is a surface phenomenon in which the molecules of the contaminant adhere to the surface of the adsorbent.

In diffusional operations, where mass is to be transferred from one phase to another, it is necessary to bring the two phases into contact to permit the change toward equilibrium to take place. The transfer may take place with both streams flowing in the same direction, in which case the operation is called *concurrent* or *co-current flow*. When the two streams flow in opposite directions, the operation is termed *countercurrent flow*, an operation carried out with gas entering at the bottom and flowing upward, and the liquid entering at the top and flowing down. This process is illustrated in Figure 10.1. Figure 10.2 shows a combined operation in which the contaminated gas is first cleaned in a countercurrent operation, and then the gas is further treated to remove more of the contaminant in a co-current operation.

Countercurrent operation is the most widely used absorption equipment arrangement. As the gas flow increases at constant liquid flow, liquid holdup must increase. The maximum gas flow is limited by the pressure drop and the liquid holdup, which will build up to flooding. Contact time is controlled by the bed depth and the gas velocity. In countercurrent flow, mass transfer driving force is maximum at the gas

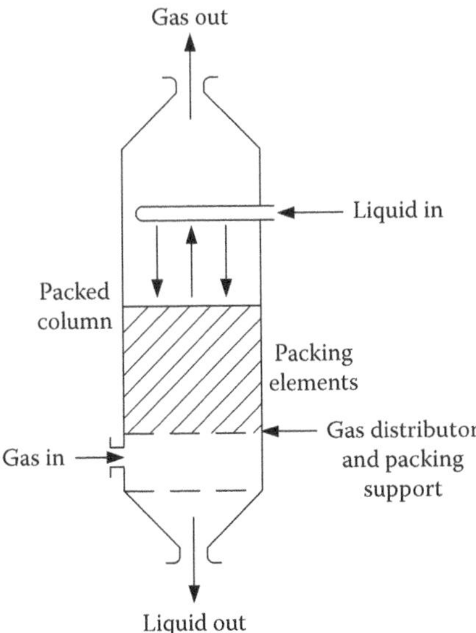

FIGURE 10.1 Countercurrent flow.

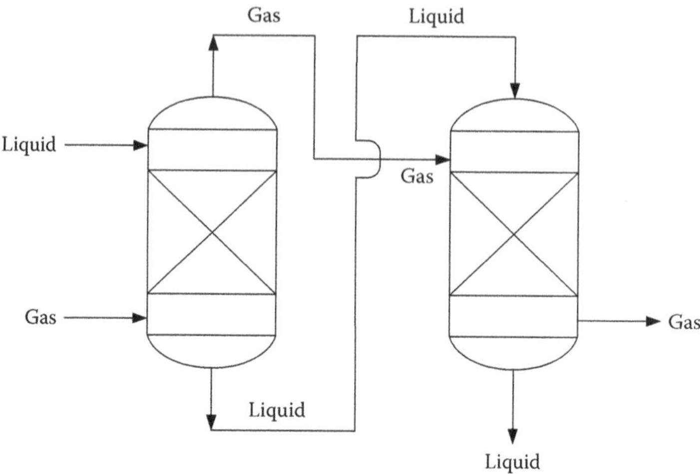

FIGURE 10.2 Combined countercurrent–co-current operation.

entrance and liquid exit. Co-current operation can be carried out at high gas velocities because there is no flooding limit. In fact, liquid holdup decreases as velocity increases. However, the mass transfer driving force is smaller than in countercurrent operation.

Some processes for both absorption and the removal of particulates employ a cross-flow spray chamber operation. Here, water is sprayed down on a bed of packing material. The carrier gas, containing pollutant gas or the particulate, flows horizontally through the packing, where the spray and packing cause the absorbed gas or particles to be forced down to the bottom of the spray chamber, where they can be removed. Figure 10.3 illustrates a cross-flow absorber. The design of cross-flow

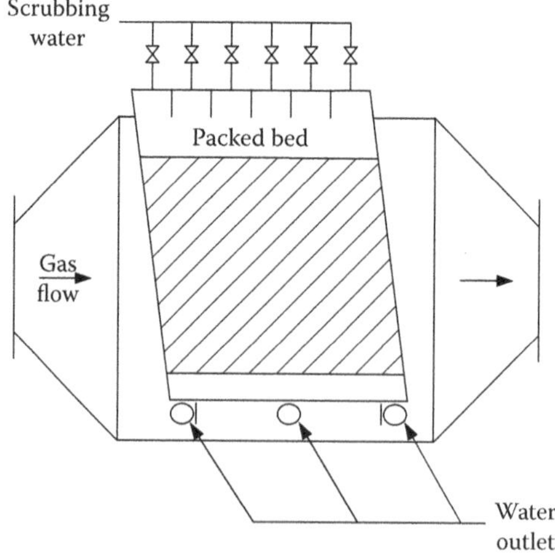

FIGURE 10.3 Cross-flow absorber operation.

absorption equipment is more difficult than vertical towers because the area for mass transfer is different for the gas and liquid phases.

Continuous and steady-state operation is usually most economical. However, when smaller quantities of material are processed, it is often more advantageous to charge the entire batch at once. In fact, in many cases, this is the only way the process can be done. This is called *batch operation* and is a transient operation from startup to shutdown. A batch operation presents a more difficult design problem. Adsorption is a semi-batch operation, in which the contaminant in the carrier gas adheres to the absorbent until the adsorbent is saturated. The process must then be stopped to regenerate or replace the adsorbent.

Absorption takes place in either a staged or continuous contactor. However, in both cases, the flow is continuous. In the ideal equilibrium stage model, two phases are contacted, well mixed, come to equilibrium, and then are separated with no carryover. Real processes are evaluated by expressing an efficiency as a percent of the change that would occur in the ideal stages. Any liquid carryover is removed by mechanical means.

In the continuous absorber, the two immiscible phases are in continuous and tumultuous contact within a vessel, which is usually a tall column. A large surface is made available by packing the column with ceramic, plastic, or metal materials. The packing provides more surface area and a greater degree of turbulence to promote mass transfer. The penalty for using packing is the increased pressure loss in moving the fluids through the column, causing an increased demand for energy. In the usual countercurrent flow column, the lighter phase enters the bottom and passes upward. Transfer of material takes place by molecular and eddy diffusion processes across the interface between the immiscible phases. Contact may be also co-current or cross-flow. Columns for the removal of air contaminants are usually designed for countercurrent or cross-flow operation.

10.1.1 Fluid Mechanics Terminology

Defining velocity through a column packed with porous material is difficult. Even if a good measure of porosity has been made, it is not possible to assure that the same porosity will be found the next time a measurement is made after the packing has been changed. Also, during operation, the bed may expand or in the case of a two-phase, gas-liquid operation, liquid holdup can occur, which varies with the flow. Therefore, determining the unoccupied tower cross-sectional area is difficult, and it becomes advantageous to base the velocity on the total tower cross section, which is the usual way to calculate tower flow, especially in absorption design.

The conservation of mass principle at steady state is

$$m = \rho A \bar{V} \tag{10.1}$$

where:

m is the mass flow rate
ρ is the mass density

A is the cross-sectional area
$\bar{V}$ is the mean velocity in compatible units

The volumetric flow rate Q is given by

$$Q = A\bar{V} \tag{10.2}$$

Therefore,

$$\bar{V} = Q/A \tag{10.3}$$

If G is defined as the mass rate of gas flow, then a superficial mass velocity can be defined as $\bar{G}$, where

$$\bar{G} = G/A \tag{10.4}$$

defines a superficial velocity, which is dependent upon the total tower cross-sectional area.

Note that the mean velocity can be calculated from

$$\bar{V} = \frac{m}{\rho A} \tag{10.5}$$

10.1.2 Removal of Hazardous Air Pollutants and Volatile Organic Compounds by Absorption and Adsorption

Absorption is widely used as a product-recovery method in the chemical and petroleum industry. As an emission control technique, it is more commonly employed for inorganic vapors. Some common absorption processes for inorganic gases are as follows:

- Hydrochloric acid vapor in water
- Mercury vapor in brine and hypochlorite solution
- Hydrogen sulfide vapor in sodium carbonate and water
- Hydrofluoric acid vapor in water
- Chlorine gas in alkali solution

In order for absorption to be a suitable process for emission control, there must be a suitable solvent, which can readily be treated after it leaves the process. Both vapor–liquid equilibrium data and mass-transfer data must be available or capable of being estimated. Absorption may be most effective when combined with other processes such as adsorption, condensation, and incineration.

Adsorption can be used to treat very dilute mixtures of pollutant and air. Activated carbon is the most widely used adsorbent. Silica gel and alumina are also frequently used adsorbents. Removal efficiencies can be as high as 99%. The maximum inlet concentration should be about 10,000 ppmv, with a usual minimum outlet concentration at 50 ppmv. In some cases, it may be advisable to design for minimum outlet concentration of 10–20 ppmv. The maximum concentration entering an adsorption bed is limited by the carbon capacity and in some cases by bed safety. Exothermic reactions can occur when some compounds are mixed in an adsorption bed. Thus, if concentrations are too high, the bed may reach a flammable condition, which could lead to an explosion. It is best to keep the entering concentration to less than 25% of the lower explosive limit. For excessively high concentrations, condensation or dilution could be used to bring the concentration to a more reasonable lower level.

Other limitations for adsorption operation are concerned with the molecular mass of the adsorbate. High molecular weight compounds are characterized by low volatility and are strongly adsorbed. Adsorption technology should be limited to compounds whose boiling points are below 400°F or molecular mass is less than about 130. Strongly adsorbed high molecular mass compounds are difficult to remove when regenerating the adsorbent. With low molecular mass, compounds below a molecular mass of 45 are not readily adsorbed due to their high volatility. On the other hand, lower molecular weight compounds are more readily removed during the regeneration process. Furthermore, gases to be treated may have liquid or solid particles present or have a high humidity. Pretreatment may then be required. Humidity needs to be reduced below 50% in most cases, or the water will selectively adsorb to such a great extent that the desired adsorbate to be removed will be blocked out. Gases to be treated may also be required to be cooled if the temperature is greater than 120°F–130°F and the possibility of exothermic reactions exist.

10.2 PROCESS SYNTHESIS TECHNOLOGY FOR THE DESIGN OF VOLATILE ORGANIC COMPOUNDS RECOVERY SYSTEMS

In the 1970s, the main environmental activity of chemical processes was end-of-the-pipe treatment. This approach is based on installing pollution control units that can reduce the composition of contaminants in waste streams to acceptable levels. Most of these units employ destructive techniques that convert the contaminants into more benign species (e.g., incineration and biotreatment). In the 1980s, the chemical process industries have shown a strong interest in implementing recycle/reuse policies, in which pollutants are recovered from terminal streams (typically using separation processes) and reused or sold. This approach has gained significant momentum due to the realization that *waste streams and pollutants* may be valuable materials that can be recovered in a cost-effective manner. Since separation systems are essential in recycle/reuse policies, the past few years have witnessed considerable progress in the area of designing separation networks for waste recovery. In particular, the powerful concepts of process integration and synthesis were developed to systematically accomplish the following:

1. Provide a global view of all waste streams in the plant (instead of tackling each waste stream independently).
2. Simultaneously screen all potential separation technologies that can be employed in recovering the pollutants.
3. Identify performance targets (such as minimum cost of separating agents and maximum extent of mass integration) ahead of detailed design.

To utilize this design concept, the designer must have the capability of identifying the optimal process configuration without enumerating all design possibilities. In addition, the designer must be aware that a design based on heuristics (what has worked in the past for a similar separation task) will almost certainly identify a nonoptimal design that results in higher cost to the company than should have been necessary. Two major problems exist with a heuristic-based design approach. First, this approach assumes that what was previously employed for a separation task was the most cost-effective approach. Second, new emerging technologies (such as membrane separations) may not possess a wealth of previous process applications. This in no way is an indication that this new technology should not be incorporated in a current optimal process configuration.

To apply this approach to the design of volatile organic compounds (VOC) recovery systems, the designer must be aware of two key points. First, the designer must recognize that several process scenarios exist for satisfying a desired separation task. For instance, for VOC aqueous wastes, as indicated in Chapter 2, air stripping, steam stripping, activated carbon adsorption, reverse osmosis, and/or pervaporation can be employed separately, or in the form of some combined hybrid process, to separate the VOC from the wastewater. Second, the designer must understand that the thermodynamic effectiveness (not necessarily related to the cost-effectiveness) of the system is based on the system equilibrium data. Once the designer understands these two principles, it should become apparent that the number of process designs that are applicable for the separation task are too numerous to envision. In fact, a common design approach has been to envision several process scenarios, perform design calculations for the scenarios to determine process flowrates and equipment sizes, and then determine the economics associated with each design scenario. A common trap is then to designate the least cost scenario as the *optimal* design, when in fact the true optimal design (lowest cost of any process scenario) in many cases may not be one of the originally envisioned scenarios.

Process synthesis is an emerging field of study in engineering designed to answer the above question. Process synthesis is the integration of all process elements to generate the optimal process configuration (overall process design) that meets certain objectives (minimize annualized cost, minimize waste emissions, maximum energy efficiency, etc.). Process synthesis tools provide a systematic framework for tackling numerous process design tasks. A review of process synthesis methodologies has been provided in the literature.[3] Furthermore, El-Halwagi has provided a detailed overview of numerous process integration methodologies in his textbook.[4]

11 Absorption for Hazardous Air Pollutants and Volatile Organic Compounds Control

11.1 INTRODUCTION

Absorption is a diffusional mass-transfer operation by which a soluble gaseous component is removed from a gas stream by dissolution in a solvent liquid.[1] A summary of absorption containing a general description, advantages, and disadvantages for the removal of volatile organic compounds (VOCs) from gaseous emission streams is provided below.[2–4]

11.1.1 Description

- VOC gaseous emissions flow into the bottom of a packed or tray column and is distributed throughout the absorption column.
- A heavy oil/hydrocarbon flows into the top of the column and the VOC is transferred (an amount based on solubility levels) from the gas to the oil via direct contact, thus, the airstream is *scrubbed.*
- The VOC/oil mixture exits from the bottom of the column and is subsequently distilled to allow separation of the VOC and the oil.
- *VOC-free* gas exits from the top of the column.

The driving force for mass transfer is the concentration difference of the solute between the gaseous and liquid phases. In the case of absorption, this driving force can be interpreted as the difference between the partial pressure of the soluble gas in the gas mixture and the vapor pressure of the solute gas in the liquid film in contact with the gas. If the driving force is not positive, no absorption will occur. If it is negative, desorption or stripping will occur and the concentration of the pollutant in the gas being treated will increase.

Absorption systems can be divided into those that use water as the primary absorbing liquid and those that use a low volatility organic liquid. The system can be simple absorption in which the absorbing liquid is used in a single pass and then disposed of while containing the absorbed pollutant. Alternatively, the pollutant can be separated from the absorbing liquid and recovered in a pure, concentrated form by distillation or stripping (desorption). The absorbing liquid is then used in a closed circuit and is continuously regenerated and recycled. Examples of regeneration alternatives to distillation or stripping are pollutant removal through precipitation and settling; chemical destruction through neutralization, oxidation, or reduction; hydrolysis; solvent extraction; and pollutant liquid adsorption.

11.1.2 Advantages

- Can achieve high recovery efficiencies (95%–98%)
- Can be used for a wide range of gas flow rates (2,000–100,000 cfm)
- Can handle a wide range of inlet VOC concentrations (500–5000 ppm)
- Good for high humidity (>50% relative humidity) air streams

11.1.3 Disadvantages

- It may result in the generation of a wastewater stream.
- It may result in column packing plugging or fouling if particulates are present in the gaseous waste stream.
- Some of the liquid absorbent may be transferred to the exit gas stream, thus creating a new pollution concern.

11.2 AQUEOUS SYSTEMS

Absorption is one of the most frequently used methods for removal of water-soluble gases. Acidic gases such as HCl, HF, and SiF_4 can be absorbed in water efficiently and readily, especially if the last contact is made with water that has been made alkaline. Less soluble acidic gases such as SO_2, Cl_2, and H_2S can be absorbed more readily in a dilute caustic solution. The scrubbing liquid may be made alkaline with dissolved soda ash or sodium bicarbonate, or with NaOH, usually no higher a concentration in the scrubbing liquid than 5%–10%. Lime is a cheaper and more plentiful alkali, but its use directly in the absorber may lead to plugging or coating problems if the calcium salts produced have only limited solubility. A technique often used is the two-step flue gas desulfurization process, where the absorbing solution containing NaOH is used inside the absorption tower, and then the tower effluent is treated with lime externally, precipitating the absorbed component as a slightly soluble calcium salt. The precipitate may be removed by thickening, and the regenerated sodium alkali solution is recycled to the absorber. Scrubbing with an ammonium salt solution can also be employed. In such cases, the gas is often first contacted with the more alkaline solution and then with a neutral or slightly acid contact to prevent stripping losses of NH_3 to the atmosphere.

When flue gases containing CO_2 are being scrubbed with an alkaline solution to remove other acidic components, the caustic consumption can be inordinately high if CO_2 is absorbed. However, if the pH of the scrubbing liquid entering the absorber is kept below 9, the amount of CO_2 absorbed can be kept low. Conversely, alkaline gases, such as NH_3, can be removed from the main gas stream with acidic water solutions such as dilute H_2SO_4, H_3PO_4, or HNO_3. Single-pass scrubbing solutions so used can often be disposed of as fertilizer ingredients. Alternatives are to remove the absorbed component by concentration and crystallization. The absorbing gas must have adequate solubility in the scrubbing liquid at the resulting temperature of the gas–liquid system.

For pollutant gases with limited water solubility, such as SO_2 or benzene vapors, the large quantities of water that would be required are generally impractical on single-pass basis, but may be used in unusual circumstances. An early example from the United Kingdom is the removal of SO_2 from flue gas at the Battersea and Bankside electric power stations, described by Rees.[5] Here the normally alkaline water from the Thames tidal estuary is used in large quantity on a one-pass basis.

11.3 NONAQUEOUS SYSTEMS

Although water is the most common liquid used for absorbing acidic gases, amines (monoethanol-, diethanol-, and triethanolamine; methyldietnolamine; and dimethylanaline) have been used for absorbing SO_2 and H_2S from hydrocarbon gas streams. Such absorbents are generally limited to solid-particulate-free systems, because solids can produce difficult to handle sludges, as well as use up valuable organic absorbents. Furthermore, because of absorbent cost, absorbent regeneration must be practiced in almost all cases.

At first glance, an organic liquid appears to be the preferred solvent for absorbing hydrocarbon and organic vapors from a gas stream because of improved solubility and miscibility. The lower heat of vaporization of organic liquids is a benefit for energy conservation when solvent regeneration must occur by stripping. Many heavy oils, No. 2 fuel oil or heavier, and other solvents with low vapor pressure can do extremely well in reducing organic vapor concentrations to low levels. Care must be exercised in picking a solvent that will have sufficiently low vapor pressure that the solvent itself will not become a source of VOC pollution. Obviously, the treated gas will be saturated with the absorbing solvent. An absorber–stripper system for recovery of benzene vapors has been described by Crocker.[6] Other aspects of organic solvent absorption requiring consideration are stability of the solvent in the gas–solvent system, for example, its resistance to oxidation, and possible fire and explosion hazards.

11.4 TYPES AND ARRANGEMENTS OF ABSORPTION EQUIPMENT

Absorption requires intimate contact between a gas and a liquid.[7] Usually, means are provided to break the liquid up into small droplets or thin films, which are constantly renewed through turbulence to provide high liquid surface area for mass transfer, as well as a fresh, unsaturated surface film for high driving force. The most commonly used devices are packed and plate columns, open spray chambers and towers, cyclonic spray chambers, and combinations of sprayed and packed chambers. Some of these devices are illustrated in Figures 10.1 through 10.3. Additional illustrations of absorption equipment are shown in Figures 11.1 through 11.5. Packed towers give excellent gas–liquid contact and efficient mass transfer. For this reason, they can generally be smaller in size than open spray towers. A countercurrent packed tower maximizes driving force, because it brings the least concentrated outlet gas into contact with fresh absorbing liquid. These features make this type of tower design the best choice when the inlet gas is essentially free of solid particulates. However, packed towers plug rapidly when appreciable insoluble particulates are present.

The cross-flow packed scrubber is more plug resistant when properly designed.[8] In plate columns, contact between gas and liquid is obtained by forcing the gas to pass upward through small orifices, bubbling through a liquid layer flowing across a plate. The bubble cap tower is a classical contacting device. In a bubble cap, there is a weir around the hole in the tray. A slotted cap is placed over the weir and the vapor bubbles through the holes and over the weir then out through the cap. This type of plate is very expensive and is not used very much in absorbers today. There is

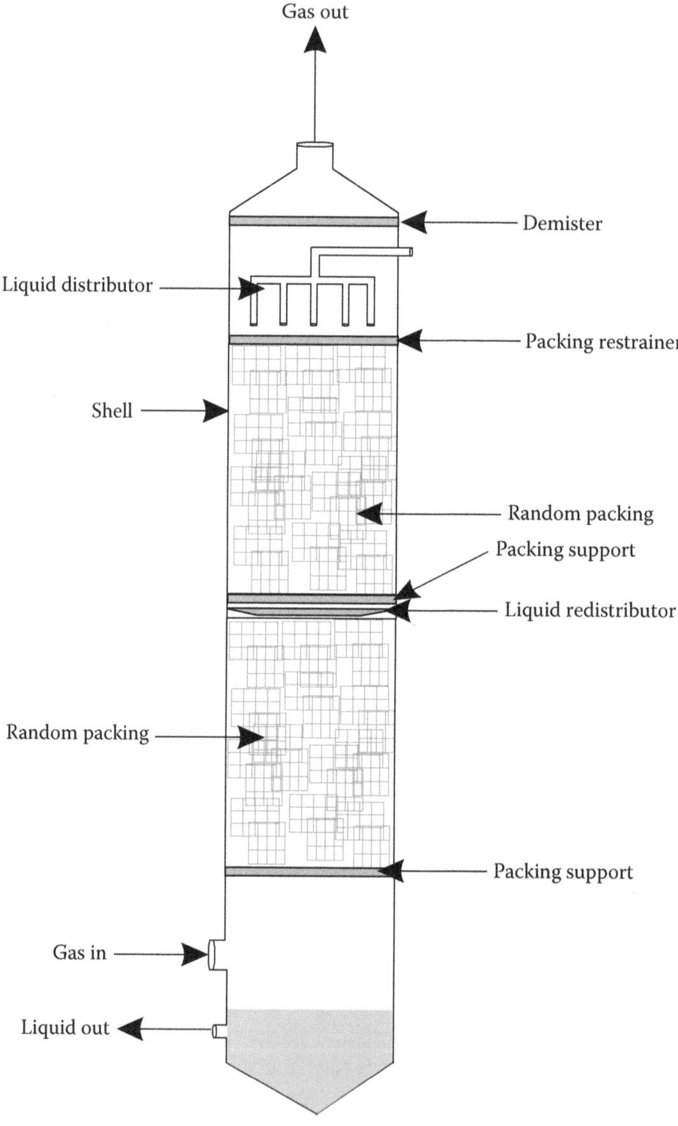

FIGURE 11.1 Packed column with all equipment.

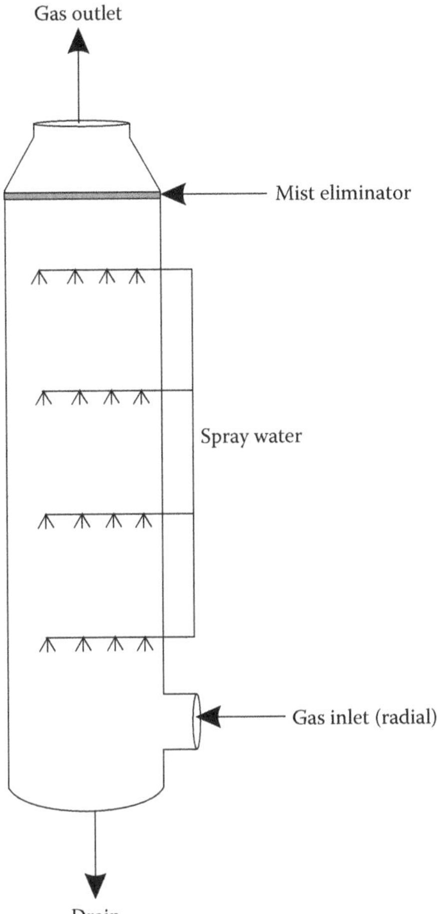

FIGURE 11.2 Spray chamber.

a variation called the *valve tray*, which permits greater variations in gas flow rate without dumping the liquid through the gas passages and is less expensive.

Sieve plates are simple flat plates perforated with small holes. The advantages are low cost and high plate efficiency, but they have narrow gas flow operating ranges. Spray chambers and towers are considerably more resistant to plugging when solid particulates are present in the inlet gas. However, difficulties with plugging in spray towers and erosion can be troublesome when the spray liquid is recycled. Particle settling followed by fine strainers or even coarse filters is beneficial.

Another tower contacting device for absorption is the baffle tower, which has been employed occasionally when plugging and scaling problems are expected to be severe. Gases passing up the tower must pass through sheets downwardly cascading liquid, providing some degree of contact and liquid atomization. Baffle tower design may use alternating segmental baffles (Figure 11.5) or disk and doughnut plates, in

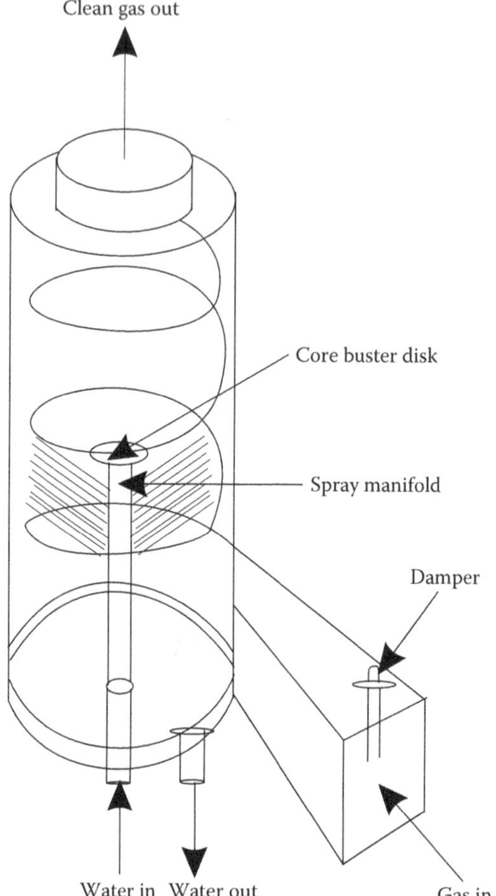

FIGURE 11.3 Cyclonic spray chamber.

which the gas alternately flows upward through central orifices and annuli, traversing through liquid curtains with each change in direction. Mass transfer is generally poor, and information on design parameters is extremely scarce.

11.5 DESIGN TECHNIQUES FOR COUNTERCURRENT ABSORPTION COLUMNS

Although the design of co-current and cross-flow towers are important, countercurrent towers are the most frequently used. Moreover, the design of cross-flow towers involves complicated calculations, which do not enhance the theory of tower design. Therefore, the principles of design of the countercurrent tower will be explained in more detail.

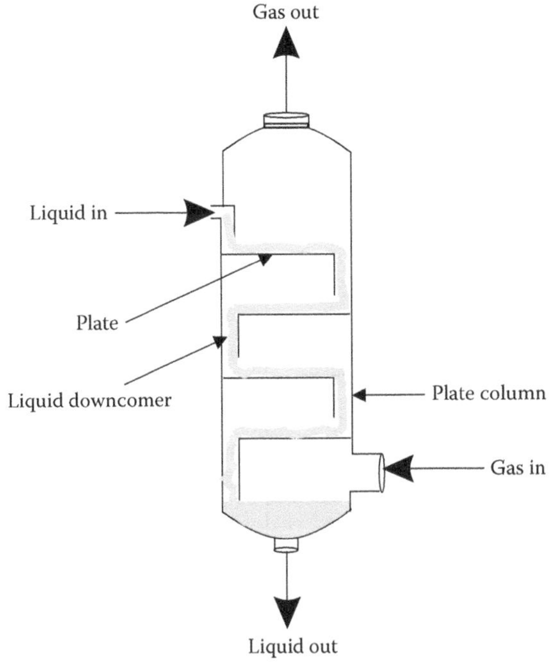

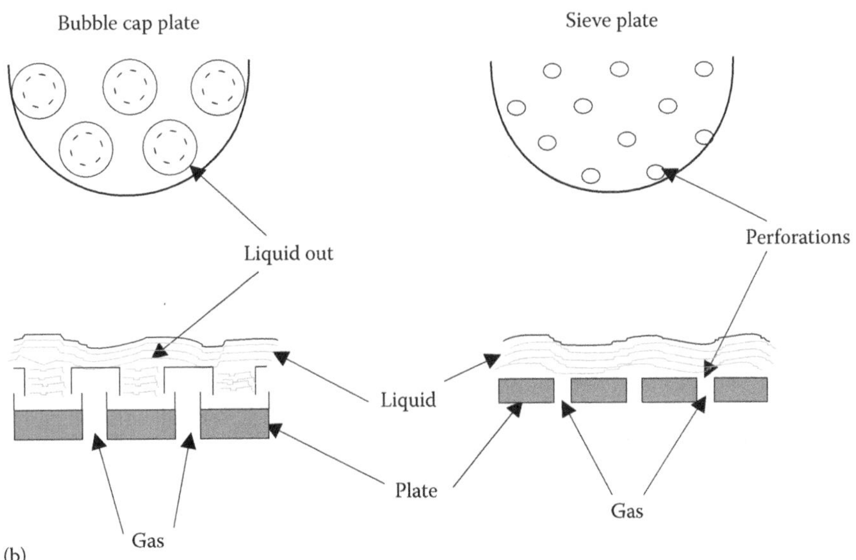

FIGURE 11.4 Plate column: (a) top view and (b) side view.

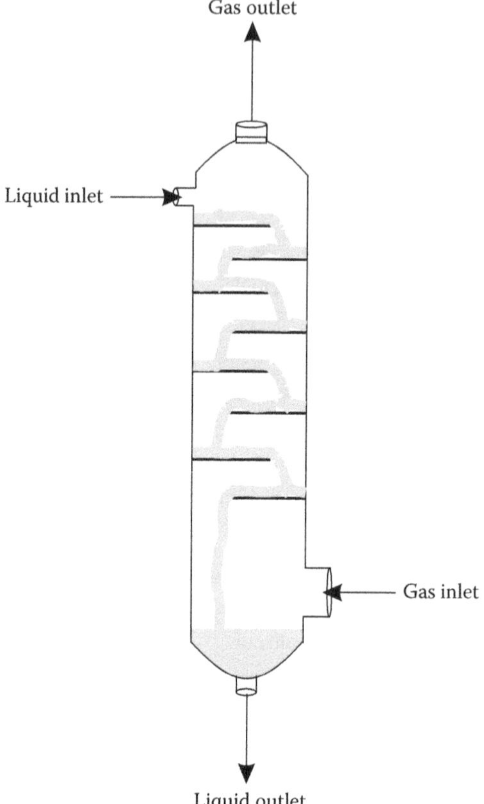

FIGURE 11.5 Baffle tray tower.

The two main factors to be determined in countercurrent absorption column design are the height of packing influenced by the mass-transfer conditions and the tower diameter, influenced by the flow rate of the gas to be treated. Size and type of packing affects both the mass-transfer conditions and the pressure loss through the column.

Height of packing is determined by the rate of mass transfer from one phase to another. The amount of mass transferred is equal to the rate of mass-transfer times the time of contact. The time of contact is dependent upon column size and flow rates. For a very high tower, one might expect that most of the pollutant in the gas stream can be removed.

Tower diameter is determined by the quantity of gas passing up the tower. The upper limit of flow occurs when the tower begins to flood. Most efficient mass transfer occurs at flow rates just short of flooding. Higher flow rates result in higher pressure losses. Thus, an economic optimum column size could be established, if cost data and information regarding mass transfer and pressure loss as a function of flow rate are known.

Because we are dealing with dilute solutions in the case of pollution removal from an effluent, the heat of solution will most usually be negligible. Therefore, we will assume that air contaminant absorption columns operate isothermally, and the design equations presented in this chapter will be for the isothermal, dilute gas case.

11.5.1 Equilibrium Relationships

For a very detailed introductory discussion of phase equilibria see the text by Smith, van Ness, and Abbott.[9] Vapor–liquid equilibria for miscible systems is determined from the equality of fugacity in both phases at the same temperature and pressure:

$$\hat{f}_i^V = \hat{f}_i^L \tag{11.1}$$

For component i in the liquid phase:

$$\hat{f}_i^L = x_i \gamma_i f_i^o \tag{11.2}$$

and in the vapor phase

$$\hat{f}_i^V = y_i \phi_i P \tag{11.3}$$

According to Equation 11.1, therefore,

$$y_i \phi_i P = x_i \gamma_i f_i^o \tag{11.4}$$

This system is structured so that only the fugacity coefficient ϕ_i and the activity coefficient γ_i depend upon the compositions at any given temperature and pressure. The standard state fugacity f_i^o is a property of the pure component only.

The Gibbs–Duhem equation, which involves partial molar solution properties, is extremely useful in dealing with phase equilibria. It can be written as follows for n components:

$$\left(\frac{\partial M}{\partial T}\right)_P dT + \left(\frac{\partial M}{\partial P}\right)_T dP - \sum_i^n x_i d\left(\ln \overline{M_i}\right) = 0 \tag{11.5}$$

Here M is a solution property $\overline{M_i}$ and is the corresponding partial molar property. By using excess Gibbs free energy changes, this equation can be written in terms of the enthalpy and volume changes on mixing ΔH and ΔV, respectively, and the activity coefficient:

$$-\frac{\Delta H}{RT^2} dT + \frac{\Delta V}{RT} dP = \sum_i^n x_i d\left(\ln \gamma_i\right) \tag{11.6}$$

In the case of solutions of nonpolar molecules or for solutions at low concentration, the enthalpy and volume changes on mixing are small and Equation 11.6 for a binary mixture becomes

$$x_i d\left(\ln \gamma_1\right) + x_2 d\left(\ln \gamma_2\right) = 0 \tag{11.7}$$

For a solution at low pressure, where the vapor is an ideal gas, and presuming f_i^o to be equivalent of the pure component fugacity at the temperature and pressure of the solution, Equation 11.7 can be rewritten at constant temperature:

$$x_1 d\ln(y_1 P) + x_2 d\ln(y_2 P) = 0 \tag{11.8}$$

Note that $(y_1 P) = \overline{P_1}$ and $(y_2 P) = \overline{P_2}$, the partial pressures, and $x_2 = 1 - x_1$, therefore,

$$\int_{P_2^\circ}^{\overline{P_2}} d\left(\ln \overline{P}_2\right) = -\int_0^{\overline{P_1}} d\left(\ln \overline{P}_1\right) \tag{11.9}$$

$$\ln\left(\frac{\overline{P}_2}{P_2^\circ}\right) = -\int_0^{\overline{P}_2} \frac{x_1}{(1-x_1)} \frac{d\overline{P}_1}{\overline{P}_1} \tag{11.10}$$

Thus, from partial pressure data of one component of a binary solution, the partial pressure of the second component may be calculated. Use of this technique is recommended when limited experimental data are available.

11.5.2 Ideal Solutions—Henry's Law

A general relationship for fugacity of a component in a liquid mixture in terms of the pure component fugacity is given by the following equation:

$$\ln\left(\frac{\hat{f}_i}{x_i f_i^L}\right) = \frac{1}{RT}\int_0^P \left(\overline{V}_i - V_i\right) dP \tag{11.11}$$

According to Amagat's law, an ideal solution is one in which $\left(\overline{V}_i - V_i = 0\right)$, and the value of the integral in Equation 11.11 is also 0. This makes

$$\hat{f}_i^L = f_i^{id} = x_i f_i^L \tag{11.12}$$

for an ideal solution. This relationship is known as the *Lewis and Randall fugacity rule*. More generally to be consistent with Equation 11.2, the ideal fugacity would be defined as

$$f_i^{id} = x_i f_i^o \tag{11.13}$$

where:

f_i^o is the fugacity of component *i* in the standard state at the same temperature and pressure of the solution

If the standard states adopted were that of the pure substance designated by *i*, then $f_i^o = f_i$.

Equation 11.13 is plotted in Figure 11.6 in such a manner that both broken lines are valid representations of the equation. Each broken line represents one model of the ideal solution while the solid line represents the value of $\hat{f}_i^L$ for a real solution as given by Equation 11.11. From the figure, it can be seen that while $f_i^o(A)$ and $f_i^o(B)$ are both fugacities of pure component i, only $f_i^o(A)$ represents the fugacity of a pure component as it actually exists. Thus, $f_i^o(B)$ represents an imaginary state of the real solution. In either case, the broken line is drawn tangent to the solid curve for the real solution, where it meets the curve. Real solution behavior is represented by $f_i^o(A)$ as $x_i \rightarrow 1.0$, and by $f_i^o(B)$ as $x_i \rightarrow 0.0$. Then,

$$\lim_{x_i \to 1.0} \left(\frac{\hat{f}_i^L}{x_i} \right) = f_i^o(A) = f_i \tag{11.14}$$

Henry's law, which is valid for a dilute solution, may be defined as follows:

$$\lim_{x_i \to 0.0} \left(\frac{\hat{f}_i^L}{x_i} \right) = k = f_i^o(B) \tag{11.15}$$

Equation 11.15 can be written as

$$\hat{f}_i^L = x_i k_i \tag{11.16}$$

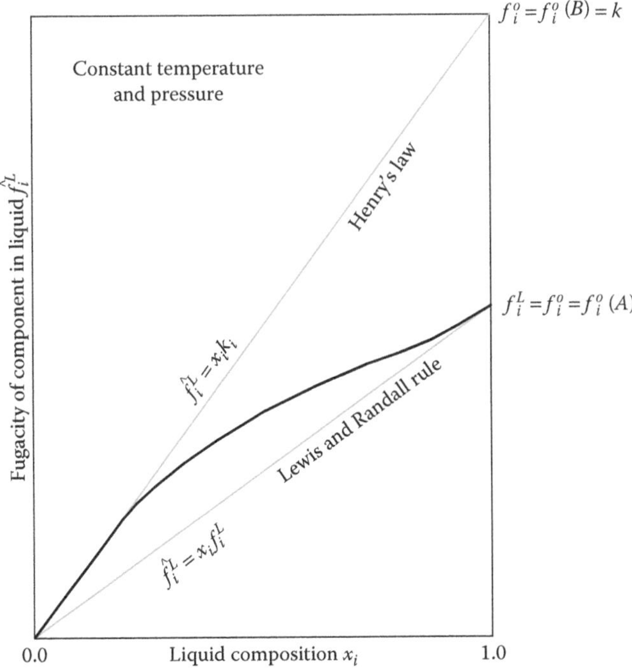

FIGURE 11.6 Defining Henry's law and the Lewis and Randall fugacity rule in relation to the true fugacity.

where:

k_i is Henry's law constant

Equation 11.4 can now be written

$$y_i \hat{\phi}_i P = x_i k_i \tag{11.17}$$

For solutions at low pressure, $\hat{\phi}_i \approx \phi_i \approx 1.0$ and $y_i P = \overline{P}_i$, thus,

$$\overline{P}_i = x_i k_i \tag{11.18}$$

which is a more common form of Henry's law. The composition x_i could be expressed as a weight fraction, as well as a mole fraction. However, k_i must have units to make the equation consistent in units. Noting that $y_i P = \overline{P}_i$ is the partial pressure and putting Equation 11.18 into mole fraction terms on both sides of the equation, results in the following equation:

$$y_i = \left(\frac{k_i}{P}\right) x_i \tag{11.19}$$

Define $m = k_i/P$ and a common definition for Henry's law constant, and Equation 11.19 becomes

$$y_i = m_i x_i \tag{11.20}$$

an equation, which has application to the case of absorption of dilute solutions. Henry's law is an adequate method of representing vapor–liquid phase equilibria for many dilute solutions encountered in absorption work. However, in the case of weak electrolytes, dissociation of the electrolyte may take place upsetting the normal equilibrium relationship. In this case, Henry's law must be altered.

11.5.3 Countercurrent Absorption Tower Design Equations

Figure 11.7 illustrates a packed absorption tower with countercurrent flow. A mass balance is made around a cross section through the tower assuming that the gas and liquid molar flow rates are constant. Making a differential mass balance around the tower in molar units,

$$y_A G + x_A L = (y_A - dy_A)G + (x_A + dx_A)L \tag{11.21}$$

This equation can be reduced to the following:

$$G dy_A = L dx_A \tag{11.22}$$

If it is assumed that the flow rates do not remain constant through the column, Equation 11.22 can be written as

$$d(G y_A) = d(L x_A) \tag{11.23}$$

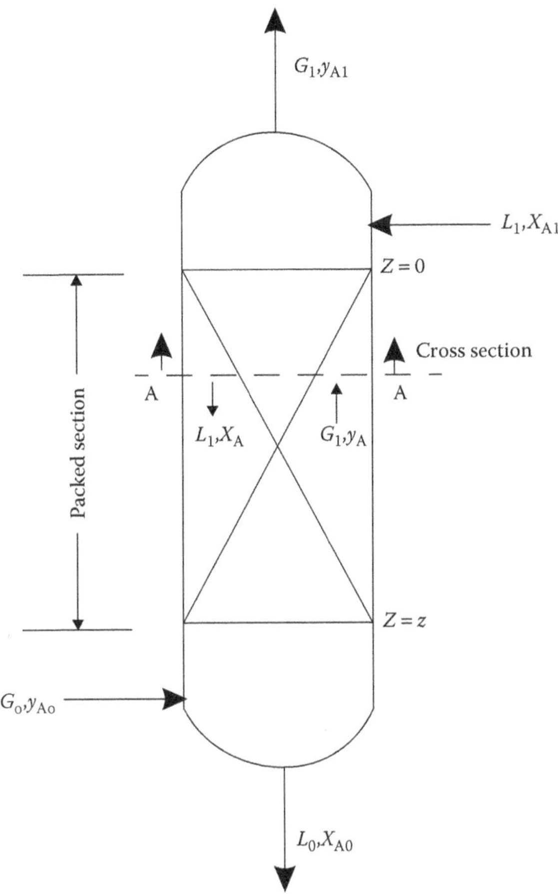

FIGURE 11.7 Packed tower definitions: L and G are the molar flow rates and x and y are the mole fractions.

This equation can now be integrated from the top of the column down to the tower cross section AA.

$$y_A G - y_{A1} G_1 = x_A L - x_{A1} L_1 \tag{11.24}$$

A common practice for solution of the equations for general use, especially when dilute solutions are not expected, is to rewrite the flow rates on a solute-free basis. The following equations present these solute-free flow rates designated by the subscript B.

$$\begin{aligned} G_B &= (1 - y_A)G = (1 - y_{A1})G_{A1} \\ L_B &= (1 - x_A)L = (1 - x_{A1})L_{A1} \end{aligned} \tag{11.25}$$

The material balance of Equation 11.24 can be rewritten in terms of these solute-free flow rates as

$$\left(\frac{y_A}{1-y_A}\right)G_B + \left(\frac{x_{A1}}{1-x_{A1}}\right)L_B = \left(\frac{x_A}{1-x_A}\right)L_B + \left(\frac{y_{A1}}{1-y_{A1}}\right)G_B \tag{11.26}$$

For dilute solutions, such as those found in most air pollution control conditions,

$$(1 - y_A) \approx 1.0 \text{ and } (1 - x_{A1}) \approx 1.0$$

and

$$G_B \approx G \approx G_{A1}$$

$$L_B \approx L \approx L_{A1}$$

Equation 11.26 reduces to Equation 11.24, which is called the *operating line* since it represents concentrations in an operating absorption column. Equation 11.26 is a general equation, which can be used along with a general equilibrium relationship such as Equation 11.4. For this case of a dilute solution, Equation 11.24 will be used along with Henry's law Equation 11.20, $y_A^* = m\,x_A^*$. Both the equilibrium line, 11.20, and the operating line, 11.24, will be plotted as a straight line on an x–y plot.

11.5.4 Origin of Volume-Based Mass-Transfer Coefficients

Interphase mass transfer takes place due to a difference of concentration across the interface between phases. Several important theories have been devised to describe this transport by Higbie[1] and Dankwerts.[10] These theories serve as a realistic description of the dynamic activities at the interface between two fluids. However, the Whitman[11] *two-film* concept remains an excellent basis for defining mass-transfer coefficients and as a picture of interfacial action.

11.5.4.1 Steady-State Molecular Diffusion

Molecular diffusion serves as a basis for an initial definition of mass-transfer coefficients. Molecular diffusion appears as a random motion of molecules in which a net movement of molecules takes place from a region of high concentration to a region of low concentration. The driving force for diffusion across the interface can be considered to be due to the difference in concentration. This driving force continues to act until the concentrations are the same through the vessel containing the molecule. When there is no concentration, temperature, or pressure difference in the areas of the diffusion, it is assumed that complete thermodynamic equilibrium has been achieved.

In molecular diffusion, we can assume that only one component is in motion across the interface. Molecular diffusivity is interpreted in terms of a relative velocity of the diffusing component with respect to an average velocity of the entire stream in a direction normal to the interface. Mass transfer of component A in a mixture with component B is depicted as a movement to or from a fixed surface that is parallel to the interface and stationary in relation to the interface. The molecular motion is accounted for by Fick's law and the total flux is the sum of the molecular diffusion and the convective bulk flow.

Defining $\overline{N}_A$ as the flux of component A with the units of moles per unit time per unit area, and $\overline{N}_B$ similarly for component B, the bulk flux is given by

$$\overline{N}_{bulk} = \overline{N}_A + \overline{N}_B \tag{11.27}$$

Then for component A with mole fraction y_A, the sum of the bulk flux and the flux due to molecular diffusion in the z-direction is given by

$$\overline{N}_A = y_A\left(\overline{N}_A + \overline{N}_B\right) - D_{AB}\rho_m \frac{dy_A}{dz} \tag{11.28}$$

where:

D_{AB} is the molecular diffusivity
ρ_m is the density of the gas

In the case of absorption of a single component A, it is assumed that A is diffusing through a stagnant layer of B, thus $\overline{N}_B = 0.0$. Equation 11.22 then can be rearranged and integrated over a distance L with the following boundary conditions:

$$y_A = y_{AO} \text{ when } z = 0.0$$

$$y_A = y_{AL} \text{ when } z = L$$

The result is

$$\overline{N}_A = \left(\frac{D_{AB}\rho_m}{L}\right)\ln\left(\frac{1 - y_{AL}}{1 - y_{AO}}\right) \tag{11.29}$$

Because

$$y_{AO} + y_{BO} = 1.0 \tag{11.30}$$

and

$$y_{AL} + y_{BL} = 1.0 \tag{11.31}$$

a logarithmic mean concentration can be defined,

$$y_{BM} = \frac{\left(y_{BO} - y_{BL}\right)}{\ln\left(y_{BO}/y_{BL}\right)} = \frac{\left(y_{AO} - y_{AL}\right)}{\ln\left(1 - y_{AL}/1 - y_{AO}\right)} \tag{11.32}$$

Then a mass-transfer coefficient can be defined as

$$k_y = \frac{D_{AB}\rho_m}{L y_{BM}} \tag{11.33}$$

and Equations 11.32 and 11.33 can be substituted into Equation 11.29,

$$\overline{N}_A = k_y\left(y_{AO} - y_{AL}\right) \tag{11.34}$$

11.5.5 Whitman Two-Film Theory

Mass transfer in real absorption equipment resembles a molecular diffusion process only in the basic idea of a concentration difference driving force. However, the Whitman[11] two-film theory can be used to construct a model similar in many respects to the molecular diffusion equations. Figure 11.8 shows a schematic representing the Whitman two-film theory. The theory may be summarized as follows:

1. Visualize two films, the gas film and the liquid film, on either side of the interface.
2. Material is transferred in the bulk of the phases by convection currents. Concentration differences are negligible except in the vicinity of the interface.
3. On each side of the interface, convection currents die out and the two thin films form.
4. Both films offer resistance to mass transfer. Transfer takes place through these films by a mechanism similar to molecular diffusion.
5. The interface is at equilibrium and offers no resistance to mass transfer.
6. There is uniform composition in the main stream.
7. At the interface, y_{Ai} and x_{Ai} are considered to be in equilibrium described by Henry's law, $y_{Ai} = mx_{Ai}$, where m = Henry's law constant.

In a packed absorption column, the fluid is in turbulent motion. Mass transfer through the films is defined by k_y and k_x, which are now turbulent mass-transfer coefficients. An equation similar to Equation 11.34 for molecular diffusion can be used to describe the mass transfer. However, the concentration difference is expressed in terms of mole fractions at the interface.

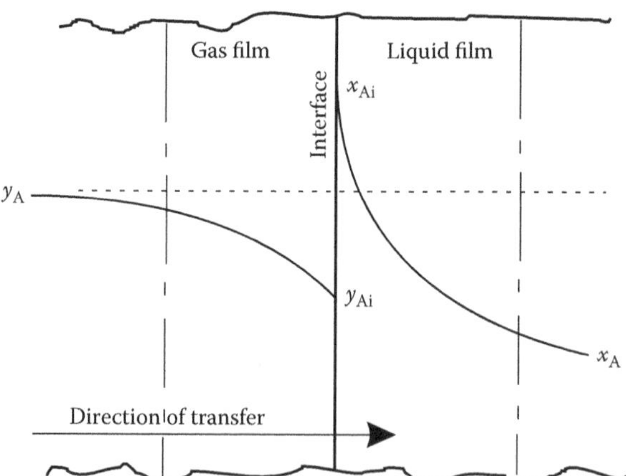

FIGURE 11.8 Whitman two-film theory: y_A is the mole fraction of A in bulk gas stream; x_A is the mole fraction of A in the bulk liquid stream; y_{Ai} is the mole fraction of A in the gas film at the interface; x_{Ai} is the mole fraction of A in the liquid film at the interface; and $y_{Ai} = mx_{Ai}$.

$$\overline{N}_A = k_y(y_A - y_{Ai}) = k_x(x_{Ai} - x_A) \tag{11.35}$$

11.5.6 Overall Mass-Transfer Coefficients

Mass-transfer coefficients defined on the basis of interfacial mole fractions have little practical value since these mole fractions cannot be measured. A new mass-transfer coefficient can be defined, which can be determined from measured data. Two new pseudo-mole fractions are defined as illustrated in the Figure 11.9. Overall mass-transfer coefficients K_y and K_x are defined based on these new pseudo-mole fractions.

y_A^* is the equilibrium mole fraction of the solute in the vapor corresponding to the mole fraction x_A in the liquid

x_A^* is the equilibrium mole fraction of the solute in the liquid corresponding to the mole fraction y_A in the vapor

and the mass transfer may now be written

$$\overline{N}_A = K_y(y_A - y_A^*) = K_x(x_A^* - x_A) \tag{11.36}$$

The mass balance line is termed the *operating line* and this line and the equilibrium curve are both plotted in Figure 11.9. The construction in the figure illustrates how to determine y_A^* for any y_A, and for any x_A. The overall coefficients can be related to the individual phase coefficients making use of Henry's law. Equations 11.35 and 11.36 then can be equated.

$$\overline{N}_A = K_y(y_A - y_A^*) = k_y(y_A - y_{Ai}) = K_x(x_{Ai} - x_A) = \frac{k_x}{m}(y_{Ai} - y_A^*) \tag{11.37}$$

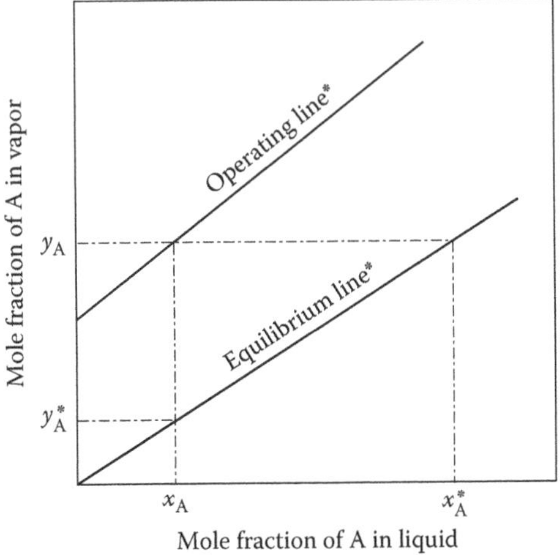

FIGURE 11.9 Illustrating new pseudo-equilibrium mole fractions.

also

$$\overline{N}_A = K_y\left(y_A - y_{Ai} + y_{Ai} - y_A^*\right) \tag{11.38}$$

Then combining Equations 11.37 and 11.38,

$$\overline{N}_A = K_y\left(\frac{\overline{N}_A}{k_y} + \frac{m\overline{N}_A}{k_x}\right) \tag{11.39}$$

Then dividing both sides of the Equation 11.39 by $\overline{N}_A$ and rearranging,

$$\frac{1}{K_y} = \frac{1}{k_y} + \frac{m}{k_x} \tag{11.40}$$

11.5.7 Volume-Based Mass-Transfer Coefficients

Absorption towers are packed with plastic, metal, or ceramic pieces to provide a greater area for contact between the vapor and liquid phases. Each of these packings has a characteristic area per unit volume of packing, which can be denoted by the quantity a, where

$$a = \frac{\text{Interfacial area}}{\text{Unit volume}}$$

Flow rate of both phases, viscosity, density, surface tension, and size and shape of the packing determine the value of a. These same factors affect the value of the mass-transfer coefficients K_y and K_x. Therefore, it is expedient to include a in the mass-transfer equation and define two new quantities K_ya and K_xa. These quantities would then be correlated with the solution parameters as functions of various chemical systems. If A is the absorption tower cross-sectional area and z the packing height, then Az is the tower packing volume. Defining A_i as the total interfacial area:

$$A_i = aAz \tag{11.41}$$

or in differential form

$$dA_i = aAdz \tag{11.42}$$

We can now rewrite Equation 11.36 in terms of a differential rate of mass transfer dN_A, with units (moles/unit time) = (moles/unit time – area) × (area),

$$dN_A = \overline{N}_A dA_i = K_y\left(y_A - y_A^*\right)aAdz \tag{11.43}$$

or

$$dN_A = K_ya\left(y_A - y_A^*\right)Adz = K_xa\left(x_A - x_A^*\right)Adz \tag{11.44}$$

Equation 11.44 defines the volumetric mass-transfer coefficients K_ya and K_xa, which have the typical units moles/(h-m^3-mole fraction). Note also that the term Adz represents the differential packed tower volume.

11.5.8 Determining Height of Packing in the Tower: Height of a Transfer Unit Method

The differential rates of mass transfer given by Equation 11.44 are equal to the differential rate of change of mass Equation 11.23 within a phase. Then, for the gas phase, the differential rate of mass transfer of component A is equal to the differential rate of change of the mass of A in the incoming gas stream.

$$dN_A = d(Gy_A) \tag{11.45}$$

Rewriting Equation 11.25, where G_B is a constant,

$$G_B = (1 - y_A)G \tag{11.46}$$

then

$$d(Gy_A) = d\left[\frac{G_B y_A}{(1-y_A)}\right] \tag{11.47}$$

It can be shown that

$$d\left[\frac{G_B y_A}{(1-y_A)}\right] = \frac{G dy_A}{(1-y_A)} \tag{11.48}$$

Therefore,

$$\frac{G dy_A}{(1-y_A)} = K_y a\left(y_A - y_a^*\right) A dz \tag{11.49}$$

becomes the basic design equation. A similar equation based on the liquid phase can be written, which would be useful in stripping calculations.

We now define a mass-transfer coefficient K_y^o, which is independent of concentration:

$$K_y = \frac{K_y^o}{(1-y_A)_{LM}} \tag{11.50}$$

where $(1-y_A)_{LM}$ is the log mean concentration defined by

$$(1-y_A)_{LM} = \frac{(1-y_A)-(1-y_A^*)}{\ln\left[(1-y_A)/(1-y_A^*)\right]} \tag{11.51}$$

Define set of flow rates based on tower cross section

$$\bar{G} = \frac{G}{A} \text{ and } \bar{L} = \frac{L}{A} \tag{11.52}$$

Then rewrite Equation 11.49,

$$\overline{G}\left(\frac{dy_A}{1-y_A}\right)=\frac{Ky_A^{o}a}{\left(1-y_A\right)_{LM}}\left(y-y_A^{*}\right)dz \tag{11.53}$$

Assuming constant flow rate, $\overline{G}$, through the column, this equation can now be integrated from $z = 0$ to $z = z$ down the column to determine tower height.

$$z=\left(\frac{\overline{G}}{K_y^{o}a}\right)\int_{y_{A1}}^{y_{A0}}\frac{\left(1-y_A\right)_{LM}}{\left(1-y_A\right)\left(y_A-y_A^{*}\right)}dy_A \tag{11.54}$$

The ratio of flow rate to mass transfer has been designated as the height of a transfer unit, or, for the gas phase, H_{OG}.

$$H_{OG}=\left(\frac{\overline{G}}{K_y^{o}a}\right) \tag{11.55}$$

Therefore, H_{OG} has been defined in such a way that it remains constant through the absorption column. The integral portion of Equation 11.54 is designated as the number of overall mass-transfer units, or, for the gas phase, N_{OG}. Thus,

$$N_{OG}=\int_{y_{A1}}^{y_{A0}}\frac{\left(1-y_A\right)_{LM}}{\left(1-y_A\right)\left(y_A-y_A^{*}\right)}dy_A \tag{11.56}$$

The integral may be evaluated by graphical or numerical techniques. The height of the column may now be calculated from

$$z=H_{OG}N_{OG} \tag{11.57}$$

11.5.9 Dilute Solution Case

For dilute solutions, Henry's law is usually a good choice for an equilibrium relationship. In this case, $y_A^{*}=mx_A$, which is associated with the overall mass-transfer coefficient, as defined in Equation 11.36. For a dilute solution,

$$\left(1-y_A\right)\approx\left(1-y_A\right)_{LM}\approx 1.0 \tag{11.58}$$

and Equation 11.56 reduces to

$$N_{OG}=\int_{y_{A1}}^{y_{A0}}\frac{dy_A}{\left(y_A-y_A^{*}\right)} \tag{11.59}$$

The operating line, Equation 11.24, may now be rewritten as

$$y_A=\left(\frac{L}{G}\right)\left(x_A-x_{A1}\right)+y_{A1} \tag{11.60}$$

Rewrite Equation 11.60, multiplying and dividing and multiplying by Henry's law constant, m,

$$y_A = \left(\frac{L}{mG}\right)(mx_A - mx_{A1}) + y_{A1} \tag{11.61}$$

Note that the ratio $L/G = \bar{L}/\bar{G}$. Then an absorption factor, Ab, can be defined as

$$\text{Ab} = \frac{L}{mG} = \frac{\bar{L}}{m\bar{G}} \tag{11.62}$$

Then rewrite Equation 11.61 using Henry's law and the absorption factor and solve for y_A^*

$$y_A^* = \left(\frac{y_A - y_{A1}}{\text{Ab}}\right) mx_{A1} \tag{11.63}$$

Substitute Equation 11.63 in Equation 11.59 and integrate

$$N_{OG} = \frac{\ln\left\{\left[(y_{A0} - mx_{A1})/(y_{A1} - mx_{A1})\right](1 - 1/\text{Ab}) + (1/\text{Ab})\right\}}{(1 - 1/\text{Ab})} \tag{11.64}$$

When a pure solvent such as water is used, $x_{A1} = 0.0$, and Equation 11.64 reduces to

$$N_{OG} = \left(\frac{1}{1 - 1/\text{Ab}}\right) \ln\left[(y_{A0}/y_{A1})(1 - 1/\text{Ab}) + 1/\text{Ab}\right] \tag{11.65}$$

11.5.10 Using Mass Exchange Network Concepts to Simultaneously Evaluate Multiple Mass Separating Agent (Absorbent) Options

As indicated previously in Chapter 8, systems-based methodologies exist that allow the simultaneous evaluation of multiple absorbents that could be used for the removal of contaminants from gaseous emission streams. Examples of contaminants in gaseous emissions that have been removed by liquid absorbents are VOCs (solvents) and sulfur-based contaminants such as hydrogen sulfide. Numerous absorbents can be employed for the separation task. One type of absorbents, physical absorbents, facilitates the direct physical transfer of the contaminants from the gaseous emission stream to the absorbent but the contaminant transferred is not altered. This approach can be typically employed by using heavy oils as adsorbents to remove VOCs from gaseous emission streams. A second type of absorbents, reactive absorbents, facilitates the direct physical transfer of the contaminants from the gaseous emission stream to the absorbent, where a chemical reaction with the absorbent occurs that converts the contaminant transferred to a chemical compound that can be more readily recycled, regenerated, discharged, or sold. A specific example of this approach reported in literature is the use of sodium hydroxide solutions to absorb hydrogen sulfide from gaseous emission streams from pulp and paper plants.

The absorbed hydrogen sulfide reacts with the sodium hydroxide to form sodium sulfide, a raw material that can be recycled for reuse in the pulping manufacturing process.[12] Both physical and reactive absorbents are collectively classified as mass separating agents, as has been previously discussed in Chapter 8. It has previously been shown in the literature that both physical and chemical mass separating agents can be modeled using an operating line and equilibrium line similar to the depiction previously provided in Figure 11.9.[13]

As previously shown, the composition driving force between the operating and equilibrium line drives the transfer of a contaminant from a gaseous emission stream to the absorbent. A visual representation of this concept showing the operating and equilibrium lines along with the composition driving forces is provided as Figure 11.10. The minimum horizontal distance between the operating and the equilibrium line is called the *minimum composition driving force*, ε, as shown in this figure. This minimum composition driving force is used to trade off the capital versus operating costs for an absorption column.[14] As the minimum composition driving force increases, more absorbent is required to achieve the separation while the capital cost of the absorption column decreases and vice versa as the composition driving force decreases. This is analogous to the effect of the minimum temperature driving force on the operating cost and capital cost of a heat exchanger. Ultimately, an optimal minimum composition driving force will result in an absorption column that realizes the minimum total annualized cost as illustrated in Figure 11.11. This concept is particularly helpful when designing a network of multiple absorption columns.

Many industrial design problems involve several gaseous emission streams that have been generated at different locations within a manufacturing process. In addition, numerous liquid absorbents can potentially be used to remove the contaminants from the gaseous emission streams. A systems-based representation of the problem of synthesizing a mass exchange network of absorption columns is provided in

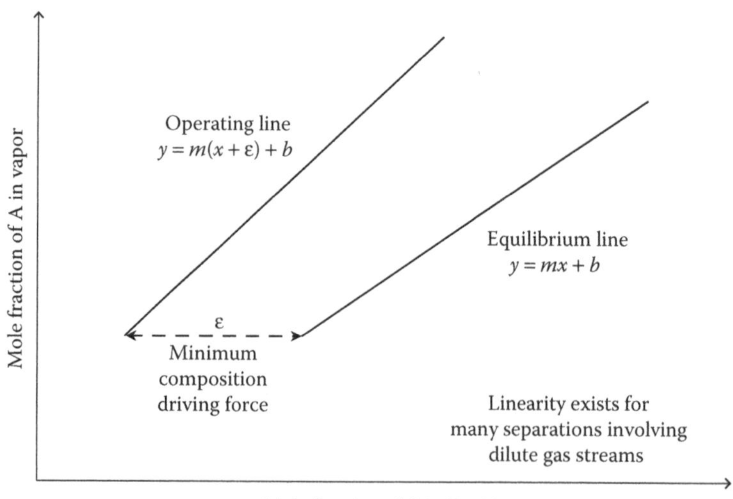

FIGURE 11.10 Minimum composition driving force for absorption mass transfer.

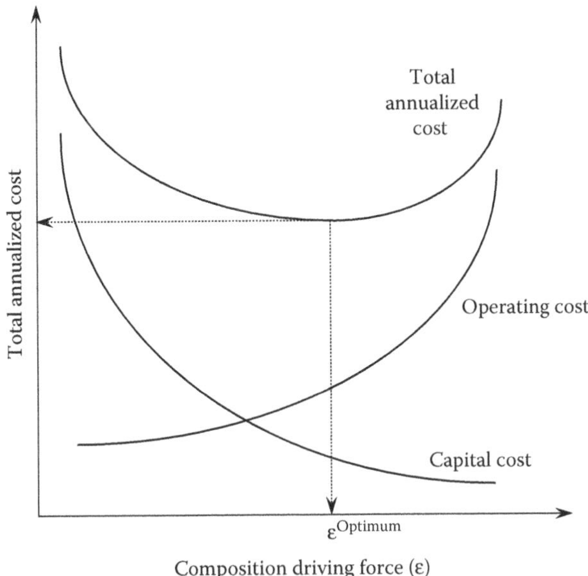

FIGURE 11.11 Identifying the minimum total annualized cost for absorption-based VOC removal from gaseous emission streams.

Figure 11.12. Any number of gaseous emission streams containing a single common contaminant can be simultaneously analyzed using this approach, and El-Halwagi has provided details of along with examples in his recent book.[15]

A useful representation for the simultaneous analysis of multiple absorbents is the composition-interval-diagram.[16] This diagram contains a vertical line for each gaseous emission stream that extends on a rich composition scale from the initial composition of the contaminant in each of the gaseous emission streams to the desired outlet contaminant composition. All of the gaseous emission streams are collectively referred to as *rich streams*. This diagram also has a vertical line for each liquid adsorbent that extends on its composition scale from its initial composition of contaminant to its maximum allowable outlet contaminant composition. All of the liquid absorbent streams are collectively referred to as *lean streams*. Each lean stream has its own composition scale that is based on the thermodynamic equilibrium line separated by a unique composition driving force, ε, as previously shown on Figure 11.10. The rich stream composition scale and the lean stream composition scales can be plotted on the same composition-interval-diagram since each respective lean stream scale can be related to a common rich stream composition scale. An example composition-interval-diagram with nomenclature is provided as Figure 11.13. El-Halwagi has shown that this composition-interval-diagram can be used with a graphical mass pinch diagram or via the use of a two-stage mathematical optimization approach to identify feasibility level designs that represent the most cost-effective network of absorption columns.[15] Details on these design methodologies can be found in El-Halwagi's text.

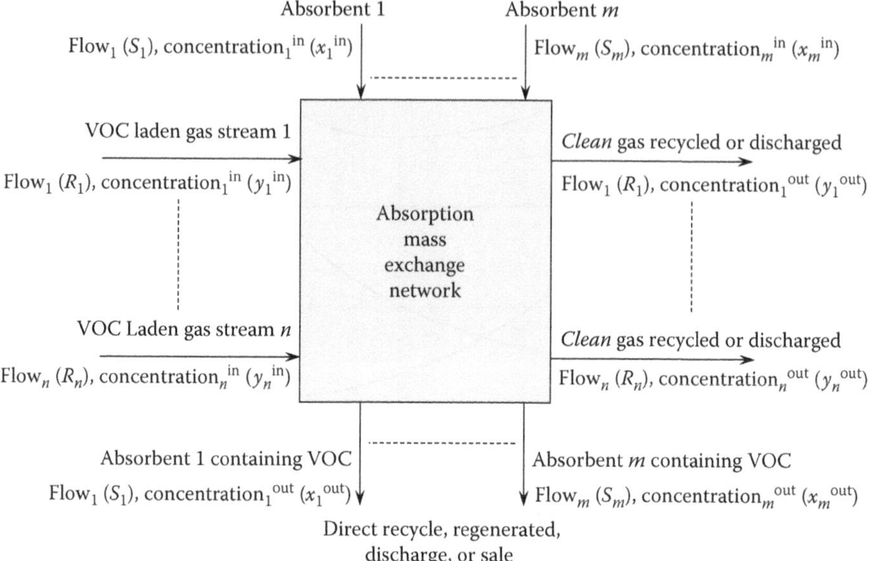

FIGURE 11.12 A schematic representation of the simultaneous analysis of multiple absorbent options for VOC removal from gaseous emission streams.

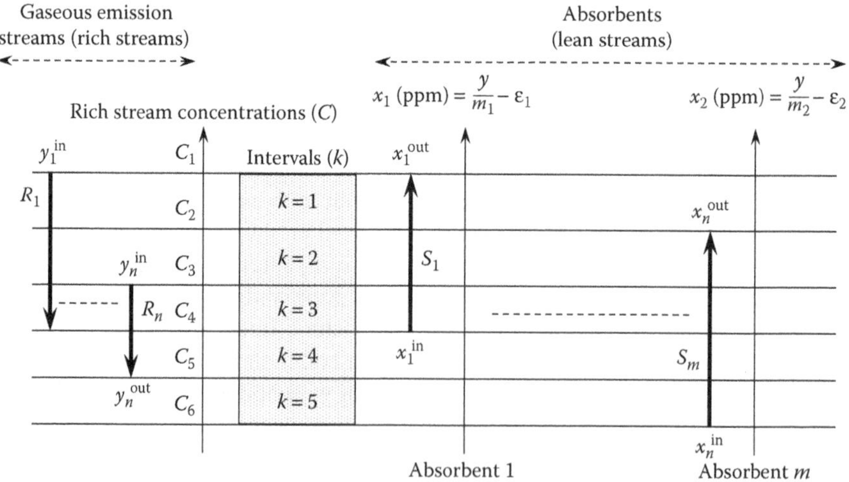

FIGURE 11.13 Composition-interval-diagram for the simultaneous analysis of multiple absorbent options for VOC removal from gaseous emission streams.

11.6 COUNTERCURRENT FLOW PACKED ABSORPTION TOWER DESIGN

11.6.1 General Considerations

Filters, heat exchangers, dryers, bubble cap columns, cyclones, and so on are ordinarily designed and built by process equipment manufacturers. However, units of special design for one-of-a-kind operations such as packed absorption cleanup towers are quite often designed and built under the supervision of plant engineers. Thus, there is a large variety of this type of equipment, none of it essentially standard.

When a packed tower is designed for treating a given quantity of gas/h, the height of packing is determined from mass-transfer considerations, and the diameter or cross-sectional area is determined by the gas velocity in the empty tower cross section. The smaller the diameter of the tower, the higher the gas velocity will be, which will help the gas to overcome the tower pressure drop. This could result in a lower cost of pushing the gas through the packed tower. The chief economic factors to be considered in design are listed in Table 11.1.

11.6.2 Operations of Packed Towers

In the past, absorbers operating as cleanup towers to remove undesirable gaseous effluents were customarily called *scrubbers*. At that time, most of the gases being removed were acid gases being scrubbed with water. The designation of scrubber to scrub the discharge gas and clean it seemed rather natural. Today, we carry out the same kind of operation, but with more stringent regulations imposed by the local air pollution control agency. Since we now also have more stringent regulations imposed for the removal of particulate matter in effluents, the term *scrubber* will be applied to those operations in this book, which may also include the simultaneous removal of gaseous pollutants. Therefore, in this book the term *absorber* will refer to the removal of gaseous contaminants.

Most gas absorption columns are operated as countercurrent contactors with gas entering at the bottom and the liquid at the top. In a packed tower operating in

TABLE 11.1
Economic Factors in Packed Tower Design: Operating and Capital Cost Factors

Operating Costs	Capital Costs
Pumping power for gas and liquid	Tower and shell packing
Labor and maintenance	Packing support
Steam and cooling water	Gas and liquid distributor
Loss of unabsorbed material	Pumps, blowers, and compressors
Disposal of absorbed material	Piping and ducts
Solvent makeup	Heat exchanger
Solvent purification	Solvent recovery system

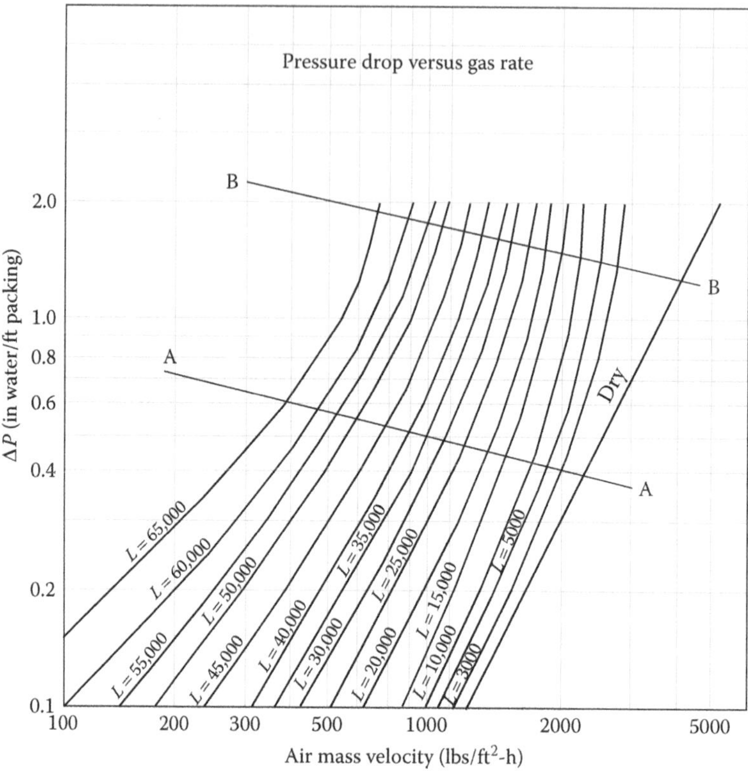

FIGURE 11.14 Pressure drop versus superficial gas mass velocity with superficial liquid mass velocity as the parameter for 3 in. Intalox® ceramic saddles. (Courtesy of Saint-Gobain NorPro Corporation, Akron, OH.)

countercurrent flow at constant liquid rate, the pressure drop varies with the gas mass velocity. With liquid holdup in the packing, there is smaller void space in the column, and a higher pressure drop exists than with dry packing. A typical set of pressure drop curves is shown in Figure 11.14. It can be seen that at low gas flow rates the liquid flow rate lines are nearly parallel to the dry line up to high liquid flow rates. Figure 11.14 shows a continually increasing pressure drop with gas flow at any liquid flow rate until the slope of the curve become infinite. The point at which the rate of change of pressure drop increases more rapidly than a constant value has been called the *upper loading point* approximately indicated by line A on Figure 11.14. The liquid holdup is now increasing with the increasing gas rate.

Above the loading region the column will approach the maximum hydraulic capacity at the point where the slope becomes infinite. This is the upper limit to the gas rate. The velocity corresponding to this upper limit is known as the flooding velocity and indicated by line B on Figure 11.10. Flooding can be observed as a crowning of the packing which begins at the bottom and progresses to the top of the packing where a layer of liquid is visible above the packing. At gas rates greater than the flooding velocity, the column will act as a gas bubbler. The greater the liquid rate, the lower the gas velocity at which flooding occurs. Larger, more open packings flood at high velocities than smaller, more dense packings.

Liquid holdup is a function of liquid flow rate and column pressure drop. Two types of holdup have been defined. Static holdup is the volume of liquid per volume of packing that remains after gas and liquid flows are stopped and the bed has drained. Static holdup depends on packing surface characteristics. The second type is operating holdup, which is the volume of liquid per volume of packing that drains out of the bed after gas and liquid flows have been stopped. Gas flow rate has little effect on holdup below loading.

Packed towers may be operated above 90% of the flooding velocity, when pressure controls are provided to maintain a fixed maximum pressure drop not to exceed the desired flooding velocity. Strigle[17] suggests the following operational characteristics for countercurrent towers.

- Pressure drop between 0.25 in. and 0.60 in. H_2O per foot of packing
- Air velocity between 5.0 and 8.0 ft/s with modern high capacity plastic packing
- Inlet concentrations of contaminant in the gas stream not to exceed 5000 ppm by volume
- Liquid irrigation rates typically between 2 and 8 gpm/ft^2 of column cross-sectional area

11.6.3 Tower Packings

There are three major types of packing: random dumped pieces, structured modular forms, and grids. The structured packing is usually crimped or corrugated sheets, and grids are an open lattice. Packing provides a large interfacial area for mass transfer and should have a low pressure drop. However, it must permit passage of large volumes of fluid without flooding. The pressure drop should be the result of skin friction and not form drag. Thus, flow should be through the packing and not around the packing. The packing should have enough mechanical strength to carry the load and allow easy handling and installation. It should be able to resist thermal shock and possible extreme temperature changes, and it must be chemically resistant to the fluids being processed. Kister[18] provides a very detailed discussion of tower packings in Chapter 8.

11.6.3.1 Random or Dumped Packing

Random packings are dumped into the tower during construction and are allowed to fall at random. With ceramic type packing the tower might be filled with water first to allow a gentler settling and to prevent breakage or the packing may be lowered into the column in buckets. Random dumped tower packing comes in many different shapes. The first generally known commercially made packings were Raschig rings and Berl saddles. These shapes generally prevail with many distinctive alterations of shape. Sizes normally range from 1/4 to 3.5 in. with 1 in. being a very common size. The choice of a packing is most usually dependent upon the service in which the tower will be engaged. Packings are made of ceramic, metal or plastic, dependent upon the service. Ceramic materials will withstand corrosion and are therefore used where the solutions resulting are aqueous and corrosive. Metals are used in noncorrosive and corrosive systems, where stainless steel and other metals such as zirconium are suitable because of their high capacity and better handling characteristics than ceramic packing. Plastic packing may be used in the case of corrosive aqueous solutions and for organic liquids, which are not solvents for the plastic of which the packing is made. Metal packing is more expensive than plastic made of commodity materials such as polypropylene, but provides lower pressure drop, thermal stability, and higher efficiency. When using plastic materials, care must be taken that the temperature is not too high and that oxidizing agents are not present. Ring-type packings are commonly made of metal or plastic except for Raschig rings, which are generally ceramic although available as a metal packing. Ring-type packings lend themselves to distillation because of their good turndown properties and availability in metals of all types that can be press formed. Ring-type packings are used in handling organic solutions when there are no corrosive problems. Saddles are commonly made from ceramic or plastic and give good corrosion resistance. Saddles are best for redistribution of liquid and, thus, serve as a good packing for absorption towers. However, the biggest benefit of saddle-type packing is their greater capacity at lower pressure drop.

11.6.3.2 Types of Random Packing

- First generation
  Original packing was frequently simply stones or essentially gravel. As the use of packed columns became more sophisticated, more sophisticated packing was required.

  The two basic types of first generation packing made artificially were the Raschig ring and Berl saddles (Figure 11.15). These packings were originally ceramic although metal Raschig rings soon became available.
- Second generation
  By cutting windows in a metal Raschig ring and then bending the window tongues inward, the Pall ring, or FLEXIRING™ (Figure 11.16) was developed. This design lowered the friction and improved packing area distribution, wetting, and distribution of the liquid. The result was higher capacity and efficiency and lower pressure drop.

  The Intalox™ saddle (Figure 11.17) was created by modifying the Berl saddle, so that adjacent pieces of packing would not blank off any portion of the liquid wetting the packing. The result was a higher capacity and higher efficiency and lower pressure drop than the Berl saddle.
- Third generation
  The IMTP™ (Figure 11.18) was formed from two circles. In this packing, high void friction is combined with lower aerodynamic drag. It provides greater strength of the rings and arches. However, weight is kept low to prevent greater bed depth. While in a tower there is less entanglement, good, mechanical strength, and improved liquid spread over the surface.

  The plastic Intalox™ Snowflake™ (Figure 11.19) packing was developed with an extremely low height to diameter ratio, which orients its open side facing the vapor flow. Very good liquid drainage and exceptionally reduced friction result.

As can be seen from these examples of third generation packing, a common characteristic is low height to diameter ratios, high void space, and surfaces that are more evenly dispersed in space than the second generation packings.

- Fourth generation
  A fourth generation of random packing has been created that provides improved performance and/or capacity A lower pressure drop results as well. A good packing in this case is the plastic Super Intalox Saddles® (Figure 11.20a). This packing has relatively high holdup, which allows good absorption efficiency with low chemical reaction. The packing also has lower sensitivity to liquid and vapor distribution quality and, therefore, can be used with conventional liquid distributors. A newer packing giving better performance named Intalox™ ULTRA™ is shown in Figure 20b.

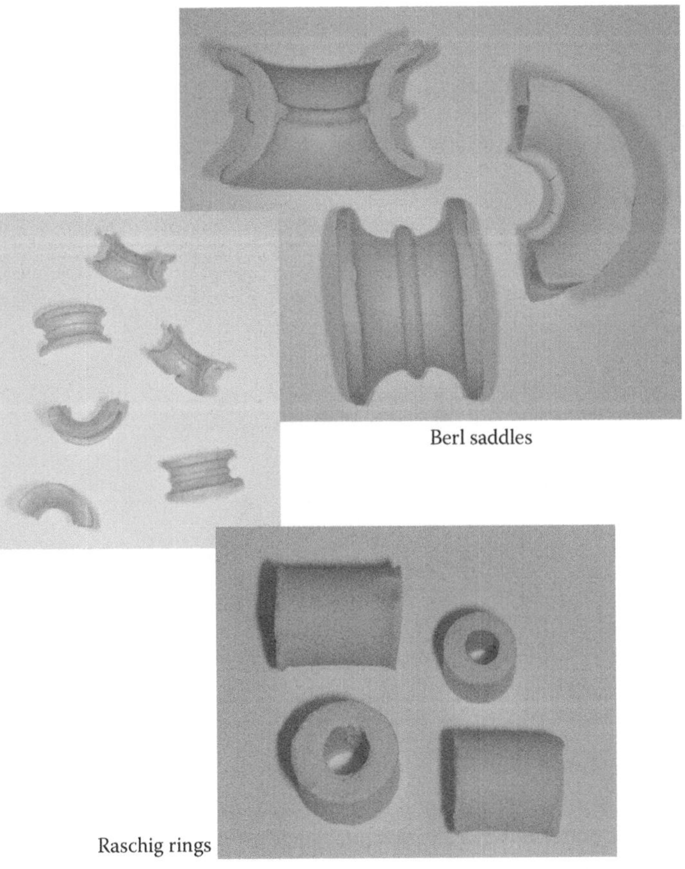

FIGURE 11.15 Raschig ring and Berl saddles random packing. (Courtesy of Carnegie Mellon University, Chemical Engineering Department, Pittsburgh, PA.)

FIGURE 11.16 Pall ring random packing, FLEXIRING™ random packing. (Courtesy of Koch-Glitsch, LP.)

FIGURE 11.17 NorPro®Proware™Ceramic saddles. (Courtesy of Saint-Gobain NorPro.)

11.6.3.3 Structured Packing

Early on after the production of random packings had been used extensively, stacked beds of the conventional random packings such as larger sized Raschig rings were used as ordered packings. Due to the high cost of installation of this type of packing, it was largely discontinued. At that time, multiple layers of corrugated metal lath formed into a honeycomb structure came into use. These early types of structured packing never gained much use. Second generation structured packing was high

FIGURE 11.18 IMTP™ random packing. (Courtesy of Koch-Glitsch, LP.)

FIGURE 11.19 Intalox™ SNOWFLAKE™ Random Packing. (Courtesy of Koch-Glitsch, LP.)

efficiency woven wire-mesh arranged into rows of vertically corrugated elements that came into use, especially in the case of vacuum distillation. Subsequently, other wire-mesh structures have gained favor. These packings are relatively expensive because each wire in the mesh is made of multiple strands. The liquid spreading on these packings depends upon capillary action; therefore, they work best at low liquid rates, so that flowing liquid does not sheet over the packing surface and close off the surface area of each individual wire strand. This feature and the high cost of the

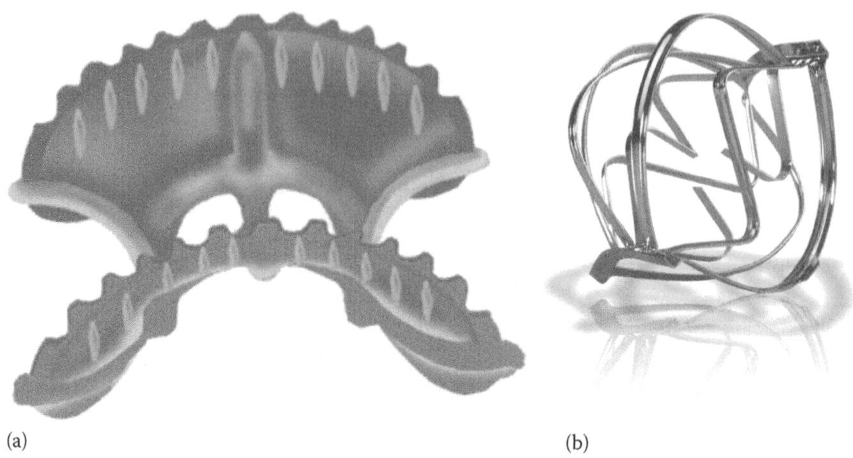

FIGURE 11.20 (a) Super Intalox™ Saddles® random packing. (Courtesy of Koch-Glitsch, LP.) (b) Intalox™ ULTRA™ random packing. (Courtesy of Koch-Glitsch, LP.)

packing limits its use to low liquid rate vacuum distillation applications of high value specialty product.

The third generation type, which is a sheet metal structured packing was developed to reduce the expense of the wire-mesh type mass transfer due to increased surface area, but also has less pressure drop per theoretical stage. Structured type of packing not only promotes mass transfer due to increased surface area but also has greater capacity than random packings with comparable efficiency because of the discrete open gas passages between the sheets of the packing bundle.

11.6.3.4 Types of Structured Packing

- Second generation
  Wire gauze structured packing (Figure 11.21) is an example of wire-mesh structured packing. The packing elements are parallel, perforated, corrugated sheets of wire mesh. Then corrugations are inclined with respect to the tower axis, and the corrugation is reversed on adjacent strips. Packing sections are about 7 in. in height and are stacked in the tower up to the required height. Adjacent sections are rotated by 90°. This packing is available most commonly in 316 stainless steel.
- Third generation
  Intalox™ high-capacity structured packing (Figure 11.22) is an example of the corrugated structured packing type. This packing has relatively flat crimps, somewhat rounded crimp apexes, and deeply embossed surfaces with tiny perforations. It surpasses the best of other metal packings in terms of efficiency and capacity.

FIGURE 11.21 Wire gauze structured packing. (Courtesy of Koch-Glitsch, LP.)

FIGURE 11.22 Intalox™ high capacity structured packing. (Courtesy of Koch-Glitsch, LP.)

FIGURE 11.23 FLEXIPAC® HC™ structured packing. (Courtesy of Koch-Glitsch, LP.)

FIGURE 11.24 FLEXIPAC® structured packing. (Courtesy of Koch-Glitsch, LP.)

- Fourth generation
  FLEXIPAC® HC and FLEXIPAC® are examples of fourth generation structured packing. These packings shown in Figures 11.23 and 11.24 are corrugated packings, in which the corrugation turns vertical at both the top and the bottom of the sheet. FLEXIPAC has grooved and perforated surfaces. Shape increases the capacity of the packing by decreasing the liquid buildup between the layers. The phenomenon of liquid build up between

structured packing layers was first observed by a number of parties through liquid density measurements made under distillation conditions using a gamma scanning technique.

11.6.3.5 Grid-Type Packing

Grid packings are also systematically arranged, which make them like structured packing. However, grid-type uses an open lattice structure and not the wire mesh or corrugated sheets as in the structured packing. Grid packings have high open areas, which promote resistance to fouling and plugging such as in vacuum crude towers. In addition, grids have a high capacity for flow at a low pressure drop. Grids are primarily used in direct contact heat transfer, scrubbing, and de-entrainment operations. Grids are available from several suppliers. Kister illustrates grids and their capabilities in his book[4]. A typical grid packing is high capacity Flexigrid®.

- High capacity Flexigrid
  Flexigrid™ style 2 (Figure 11.25a) is constructed from high open air panels, which are 60 in. long, 16 in. wide, and 2¾ in. high. Welded cross members hold the parallel blades in a fixed relationship to each other. Each successive layer of grid is rotated by 45° to the previous one. The projections are smaller and angled compared to other grids. This results in interference to vapor flow and the area for vapor–liquid contact are reduced, resulting in a higher capacity and lower efficiency.

  PROFLUX™, a new high-performance grid-type packing, is shown in Figure 11.25b.

  PROFLUX is a grid packing that combines the efficiency of structured packing with the robustness and fouling resistance of grid packing. It outperforms both of its predecessors in severe services.

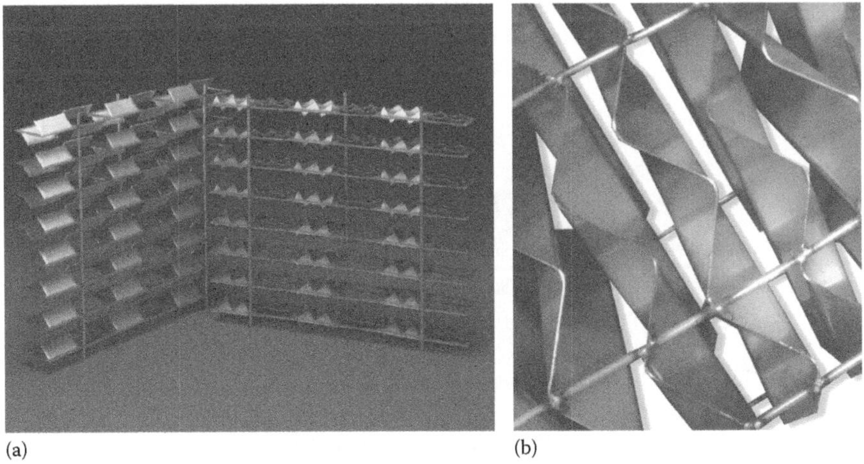

(a) (b)

FIGURE 11.25 Severe service grid packing: (a) Flexigrid® #2 and (b) PROFLUX. (Courtesy of Koch-Glitsch, LP.)

11.6.4 Packed Tower Internals

In addition to the packing, absorption towers must include internal parts to make a successful piece of operating equipment. The first of these internal parts to be discussed will be the packing support plate.

11.6.4.1 Packing Support Plate

The internals begin with a *packing support plate* at the bottom of the tower. The packing support plate is only required for random packing. Structured packing support plates are a very open grid. The plate for random packing must physically support the weight of the packing. It must incorporate a high percentage of free area to permit relatively unrestricted flow of down coming liquid. A flat plate has a disadvantage in that both liquids and gases must pass countercurrently through the same holes. Therefore, a substantial hydrostatic head may develop. Furthermore, the bottom layer of packing partially blocks many of the openings, reducing the free space. Both of these conditions lower tower capacity. A gas-injection plate provides separate passage for gas and liquid and prevents buildup of hydrostatic head. A typical support plate is in the following Figure 11.26.

11.6.4.2 Liquid Distributors

Liquid distributors are used at all locations, where an external liquid stream is introduced. Absorbers and strippers generally require only one distributor, while continuous distillation towers require at least two, one at the feed and one at the reflux inlet. The distributors should be 6 to 12 in. above packing to allow for gas disengagement from the bed. The distributor should provide uniform liquid distribution over the superficial tower area and a large free area for gas flow. The following Figure 11.27 shows a channel distributor.

Good operation of a packed column cannot take place without a careful design of the distributor. It is probably the most important part of the column in order to initiate good distribution of the liquid so that mass transfer can take place under the best of conditions. For a design method refer to the work by Rukovena and Cai.[19]

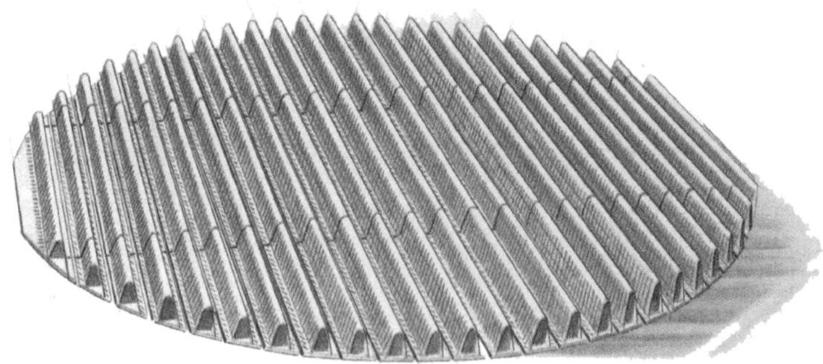

FIGURE 11.26 Support plate. (Courtesy of Koch-Glitsch, LP.)

FIGURE 11.27 Channel distributor. (Courtesy of Koch-Glitsch, LP.)

11.6.4.3 Liquid Redistributors

Liquid redistributors collect down coming liquid and distribute it uniformly to the bed below. Initially, after entering the tower the liquid tends to flow out to the wall. The redistributor makes that portion of the liquid more available again to the gas flow. It also breaks up the coalescence of the down coming liquid, and it will eliminate factors that cause a loss of efficiency in the tower and reestablish a uniform pattern of liquid irrigation. A bed depth of up to 20 ft should be sufficient before redistribution as needed. Figure 11.28 shows a typical redistributor.

11.6.4.4 Bed Limiter

Bed limiter plates are used only with ceramic and carbon tower packings only. They prevent the upper portion of the packed bed from becoming fluidized and breaking up during surges in pressure or at high pressure drop. The plates rest directly on packing

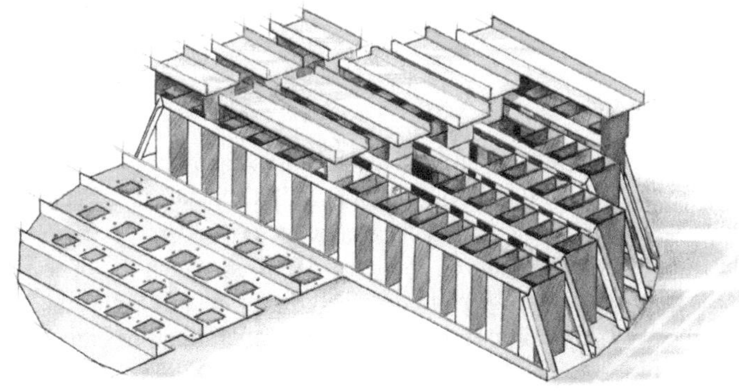

FIGURE 11.28 Deck liquid redistributor. (Courtesy of Koch-Glitsch, LP.)

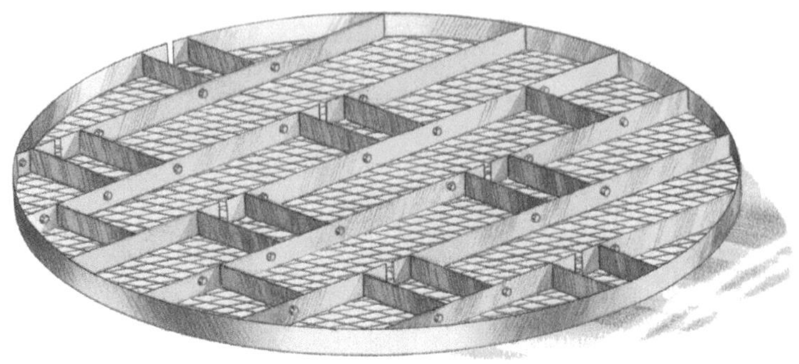

FIGURE 11.29 Bed limiter. (Courtesy of Koch-Glitsch, LP.)

and restrict movement by virtue of the weight of the plate. Retainers or bed limiters prevent bed expansion or fluidization. When operating at high pressure drops, retainers are fastened to the wall. They are designed to prevent individual packing pieces from passing through the plate openings. Figure 11.29 shows a typical hold down plate.

11.6.5 Choosing a Liquid–Gas Flow Ratio

When designing an absorption tower for the clean-up of an off gas, most usually the following variables will be known:

- G is the actual gas flow rate
- y_{A0} is the mole fraction of A, gas at inlet coming from process
- y_{A1} is the mole fraction of A, gas at outlet (specified by control agency regulation if discharged to the atmosphere)
- x_{A1} is the mole fraction of A, liquid into tower, quite frequently is zero if solvent is used only once

Based on Equation 11.24, the material balance case around the entire column can be solved for the exit liquid concentration x_{A0}.

$$x_{A0} = \left(\frac{G}{L}\right)\left(y_{A0} - y_{A1}\right) + x_{A1}$$

Thus, if L or G/L can be found, the entire material balance will be solved. In absorption, on a plot of y_A versus x_A, the equilibrium curve can be concave up or nearly a straight line, or the curve can be *S*-shaped, having a concave downward portion. These situations are illustrated in the two curves of Figure 11.30, in which the operating line as specified by the material balance is plotted as well. In the case of Figure 11.30, one (left panel) with a concave equilibrium line, three possible operating lines are shown.

Note, on these plots, L/G is the slope of the operating line. As can be seen, the line furthest to the right with the smallest slope represents the limit to which the slope of the operating line can be drawn. Thus, this slope represents a minimum L/G ratio in which x_{A0} and y_{A0} are actually in equilibrium, and the line could not be drawn with

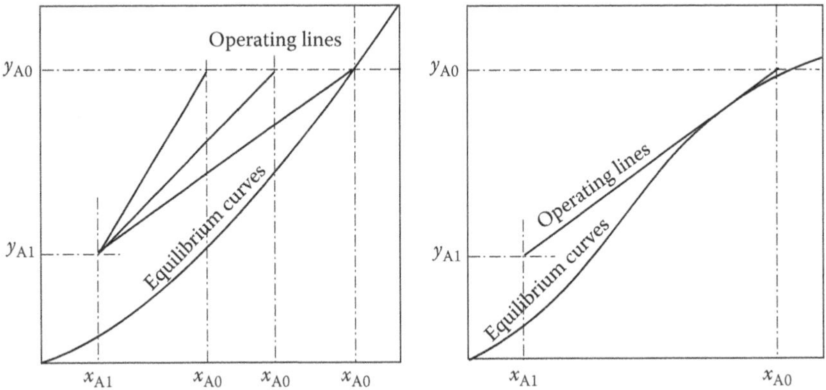

FIGURE 11.30 Graphical absorption tower relations.

a smaller slope because it would go through equilibrium. Since true thermodynamic equilibrium can never be reached practically, this point represents a theoretical minimum *L*/*G*, which never can be attained in practice.

Real *L*/*G* ratios are usually set at 1.1 to 1.7 times this minimum rate. Thus, if *G* is known, *L*/*G* is determined and x_{A0} can be found. In the case of Figure 11.30 (right panel), with *S*-shaped equilibrium curve, the minimum value of *L*/*G* is reached at some point within the column where the operating and equilibrium lines are tangent and thermodynamic equilibrium would be *exceeded* if the *L*/*G* ratio were any smaller.

11.6.6 Determining Tower Diameter—Random Dumped Packing

Knowing the *L*/*G* ratio and the overall material balance, having selected the type of packing and the pressure drop per foot of packing, the tower diameter can now be determined. Current accepted practice is to use a modified Sherwood[20] flooding correlation to determine tower diameter. Such a correlation is presented in Figure 11.31 by Strigle.[17]

However, this correlation no longer presents the flooding line. Flooding is now best determined by the method presented by Kister et al.[21] The flooding line has been replaced by the following simple correlation.

$$\Delta P_{Fl} = 0.12 F_P^{0.7} \tag{11.66}$$

The pressure drop in inches of H_2O per foot of packing is at the flood point is expressed by only the packing factor F_P.

This correlation in Figure 11.31 uses a linear scale for the ordinate, which is expressed in terms of a capacity factor, C_S. Table 11.2 presents the definition of terms used in Figure 11.28. It should be noted that the pressure drop correlation ordinate also contains a term defined as the *packing factor*. The packing factor is a characteristic of a particular packing size and shape. Originally, it was defined as $a/\in^3$, where *a* is the interfacial area per packed volume and $\in$ is the void fraction. However, this definition for F_P did not adequately predict the packing hydraulic performance.

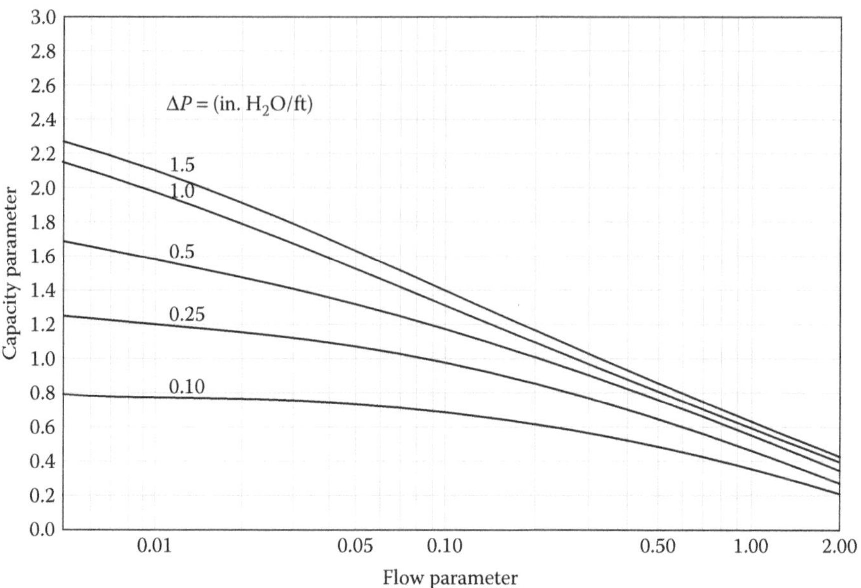

FIGURE 11.31 Generalized pressure drop correlation. (GCPD)[17] for random packing only. (Reprinted from Strigle, R. F., *Packed Tower Design and Applications, Random and Structured Packings,* 2nd ed., Gulf Publishing Co., Houston, TX, 1994.)

TABLE 11.2
Variable Definitions and Units

Flow Rates Are Now in lb-mass Not lb-moles per Unit Time

$\bar{L}$ is the superficial liquid mass velocity in lb/ft²/h
L is the liquid mass flow rate in lb/h
$\bar{G}$ is the superficial gas mass velocity in lb/ft²/h
G is the gas mass flow rate in lb/h
G^* is the superficial gas mass velocity in lb/ft²/s
ρ_G is the gas density in lb/ft³
ρ_L is the liquid density in lb/ft³
ν is the kinematic liquid viscosity in cst
F is the packing factor
$V = G^*/\rho_G$, which is the superficial velocity in ft/s

$$C_s = V\sqrt{\frac{\rho_G}{\rho_L - \rho_G}} = \text{capacity factor in ft/s}$$

Note on Viscosity

Absolute viscosity = μ	Kinematic viscosity = $\nu = \mu/p$
poise = 1 gm/cm-s	stoke = 1 cm²/s
cp = centipoise	cst = centistoke
cp = 0.01 poise	cst = 0.01 stoke

$$\nu = \frac{0.01\,\text{gm/cm-s}}{\text{gm/cm}^3} = 0.01\,\text{cm}^2/\text{s} = \text{cst}$$

TABLE 11.3
Packing Factors, *F*, for Random Dumped Packings

	Normal Packing Size (in.)							
	1/2	**5/8**	**3/4**	**1**	**1¼**	**1½**	**2**	**3 or 3½**
IMTP packing (metal)		51		41		24	18	12
Hy-Pak® packing (metal)				45		29	26	16
Super Intalox® saddles (ceramic)				60			30	
Super Intalox saddles (plastic)				40			28	18
Pall rings (plastic)		95		55		40	26	17
Pall rings (metal)		81		56		40	27	18
Intalox saddles (ceramic)	200		145	92		52	40	22
Raschig rings (ceramic)	580	380	255	179	125	93	65	37
Raschig rings (1/32 in. metal)	300	170	155	115				
Raschig rings (1/16 in. metal)	410	300	220	144	110	83	57	32
Berl saddles (ceramic)	240		170	110		65	45	

Source: Strigle, R. F., *Packed Tower Design and Applications, Random and Structured Packings*, 2nd ed., Gulf Publishing Co., Houston, TX, 1994.

Therefore, Table 11.3 from Strigle[17] presents packing factors, which have been calculated from experimentally determined pressure drop.

Since *L*/*G* is known, we can calculate the value of the abscissa in Figure 11.31, if the gas density ρ_G and the liquid density ρ_L are known. Although these densities may vary through the column, the variations will be small, especially in the dilute solution case, and an average of the top and bottom values from the column should be sufficiently accurate. If the entering gas is very dilute, the entering values of density could be used.

From the abscissa, the packing factor, and the required pressure drop per foot of packing, the ordinate of the curve can be read. Then C_S can be calculated. Note that C_S is in units of ft/s and velocity *V* will be calculated in ft/s. Note that as suggested below the unit of time will be seconds; however, any time unit could be used as long as other variables such as *L* and *G* are in the same time units. In this case the value of G^* in lb/ft²-s is calculated using seconds, which is consistent with the units of velocity. With *G* converted to lb/s the column cross-sectional area A_s is then:

$$A_s\left(\text{ft}^2\right) = \frac{G\left(\text{lb/s}\right)}{G^*\left(\text{lb/s/ft}^2\right)} \tag{11.67}$$

The column diameter can be found from

$$D = \sqrt{\frac{4G}{\pi G^*}} \tag{11.68}$$

11.6.7 Determining Tower Diameter—Structured Packing

Parkinson and Ondrey[22] report that structured packing is becoming more frequently used in air separations. They also report that structured packings are being favored because they have a higher capacity than dumped packing and scaleup is more predictable. For structured packing, Kister[18] has presented a correlation in Figure 11.32 similar to Figure 11.31.

In Figure 11.31, the abscissa is defined as the *flow parameter X* and is given by Equation 11.69.

$$X = \frac{L}{G}\sqrt{\frac{\rho_G}{\rho_L}} \tag{11.69}$$

The ordinate is defined as the *capacity parameter Y* and is defined by Equation 11.70.

$$Y = C_S F^{0.5} \nu^{0.05} \tag{11.70}$$

A generalized method of predicting pressure drop of structured packing has been developed by Bravo et al.[23] A review of the methods for prediction of pressure drop in structured packing is presented by Fair and Bravo.[24] Bravo et al.[25] present a new more comprehensive pressure drop model developed from distillation data. Because the

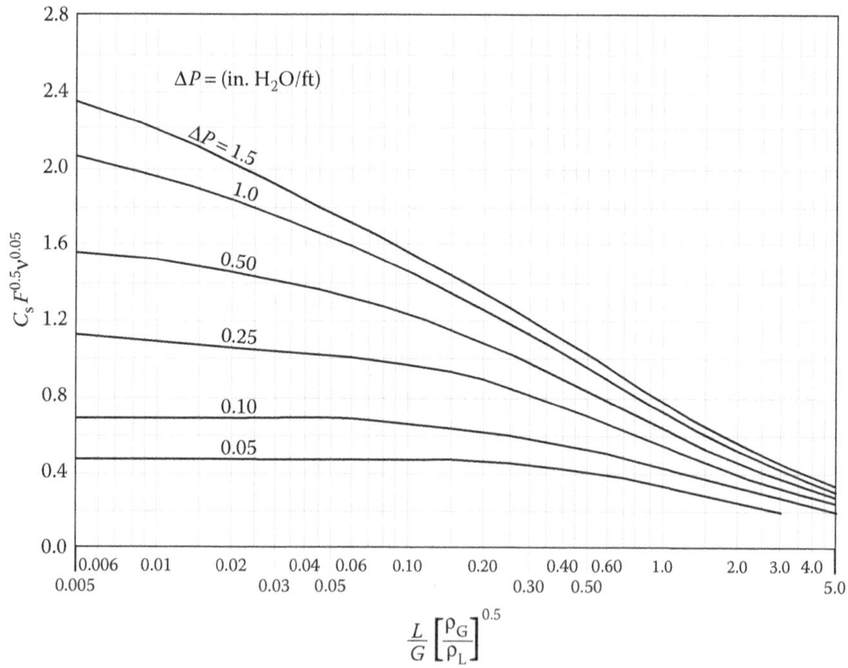

FIGURE 11.32 The Kister and Gill generalized pressure drop correlation[18] for structured packings only. (Reprinted from Kister, H. Z., *Distillation Design*, McGraw-Hill, 1992. Reproduced with permission of McGraw-Hill Education.)

TABLE 11.4
Packing Factors, *F*, for Intalox Structured Packing

	Packing Factor	
Packing Size	**Below 10 psia**	**Above 10 psia**
1T	23.5	28
2T	15.5	20
3T	12.5	15
4T	11	13.5
5T	10	12.5

Source: Strigle, R. F., *Packed Tower Design and Applications, Random and Structured Packings*, 2nd ed., Gulf Publishing Co., Houston, TX, 1994.

pressure drop curves for random and structured packings are similar, Strigle[17] reports that pressure drop for structured packing can be estimated from the generalized correlation of Figure 11.28. However, in this case the packing factor is affected by the operating pressure. Useful packing factors from Strigle are reported in Table 11.4 for five different sizes of Intalox® structured packing. The smaller packing factor should be used for pressures below 10 psia or at liquid rates below 4 gpm/ft^2.

11.6.8 Controlling Film Concept

The Whitman two-film theory states that mass transfer can be determined by the gas–film resistance operating in parallel with the liquid film resistance. Equation 11.40 describes this action. The film that offers the predominant mass-transfer resistance is called the *controlling film*. Equation 11.40 can be solved for K_y.

$$K_y = \frac{1}{\left[\left(1/k_y\right)+\left(m/k_x\right)\right]} = \frac{k_y k_x}{k_x + m k_y} \tag{11.71}$$

Gases with a limited solubility in the liquid phase have a high value of m. Therefore, Equation 11.71 reduces $K_y \approx k_x/m$ to implying that the liquid film resistance is controlling. Gases with a high solubility in the liquid phase have a low value of m. Therefore, Equation 11.71 reduces to $K_y \approx k_y$, implying that the gas film resistance is controlling.

11.6.9 Correlation for the Effect of *L*/*G* Ratio on the Packing Height

Many texts, including Strigle,[17] present a graphical correlation for determining the number of transfer units based on an essentially straight equilibrium and operating lines. This correlation is based on Equation 11.65. Henry's law is a straight equilibrium line, and the correlation presented in this section assumes Henry's law for the equilibrium relationship. At the minimum L/G ratio $y_{A0} = mx_{ao}$. Equation 11.61 can be solved for L/G and rearranged as follows:

$$\left(\frac{L}{G}\right)_{min} = \frac{y_{A0} - y_{A1}}{x_{A0}} = \frac{1 - y_{A1}/y_{A0}}{x_{A0}/y_{A0}} \tag{11.72}$$

and then the equilibrium ratio substituted in the equation to give

$$\left(\frac{L}{G}\right)_{min} = \left(1 - y_{A1}/y_{A0}\right)m \tag{11.73}$$

For any y_A, there is an x_A determined by the operating line of Equation 11.60. There is also a y_A^* in equilibrium with the x_A such that y_A^*/m. Assuming that fresh solvent is being used $x_{A1} = 0.0$ and Equation 11.60 for the operating line may be rewritten.

$$y_A = \frac{y_A^*}{m}\left(\frac{L}{G}\right) + y_{A1} \tag{11.74}$$

and then solved for y_A^*,

$$y_A^* = \frac{m\left(y_A - y_{A1}\right)}{(L/G)} \tag{11.75}$$

Define

$$\text{Ab} = \frac{(L/G)}{m} \tag{11.76}$$

which is the same as Ab in Equation 11.62. Then, define B as a multiple of $(L/G)_{min}$ to set the actual (L/G). Thus,

$$\frac{L}{G} = B\left(\frac{L}{G}\right)_{min} \tag{11.77}$$

and substituting Equation 11.73:

$$\text{Ab} = \frac{B(L/G)_{min}}{m} = B\left(1 - y_{A1}/y_{A0}\right) \tag{11.78}$$

Substitute Equation 11.76 into Equation 11.75 and the following equation results:

$$y_A^* = \left(\frac{1}{\text{Ab}}\right)\left(y_A - y_{A1}\right) \tag{11.79}$$

For the dilute case, Equation 11.79 is substituted into Equation 11.59. Equation 11.59 is now integrated with the following equation resulting:

$$N_{OG} = \left(\frac{1}{1 - 1/\text{Ab}}\right)\text{In}\left[\left(y_{A0}/y_{A1}\right)\left(1 - 1/\text{Ab}\right) + 1/\text{Ab}\right] \tag{11.80}$$

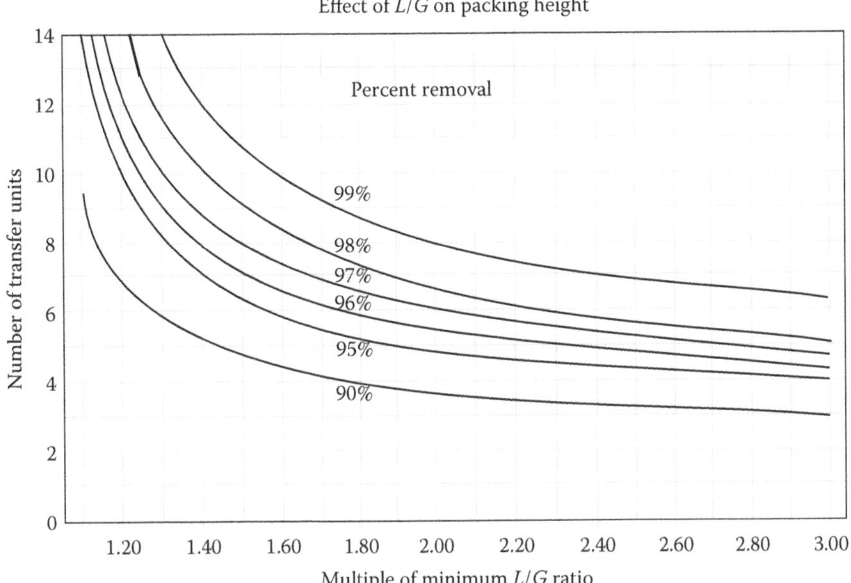

FIGURE 11.33 Effect of *L/G* on packing height.

This equation is the same as Equation 11.65. Colburn[26] put Equation 11.65 into graphical form, which makes it easier to use. See Perry and Green[27] for an example of this plot.

Recall Ab is defined by Equation 11.78 as a function of *B*, the multiple of $(L/G)_{min}$ that sets the actual (L/G). If the value of (y_{A1}/y_{A0}) is fixed at 0.80, 0.85, 0.90, 0.95, 0.96, 0.97, 0.98, and 0.99, N_{OG} can be plotted versus *B* with (y_{A1}/y_{A0}) as a parameter, which is somewhat different from the Colburn plot. Figure 11.33, effect of *L/G* on packing height, is a plot of Equation 11.80.

For any given *B* and percent removal, knowing $\bar{G}$ and $K_y^o a$, H_{OG} can be calculated, and the height of the column can be calculated from Equation 11.57. Since Equation 11.80 is derived for the dilute solution, examination of Figure 11.29 shows that the number of transfer units, N_{OG}, cannot be greatly reduced by increasing *B*, the multiple of the minimum *L/G* ratio, above 1.70.

11.6.10 Henry's Law Constants and Mass-Transfer Information

Henry's law constants can be found in the *Chemical Engineers Handbook*.[27] A listing of these constants for 362 organic compounds in water is given by Yaws et al.[28] It is also possible to calculate Henry's constant for solubility data, seen in the example in Table 11.5.

The measurement of mass-transfer coefficients has occupied many experimenters of the past 70 years. It is surprising that there is not much data available in the open literature. The problems of making satisfactory and consistent measurements are threefold:

TABLE 11.5
Calculation of Henry's Law Constant from Solubility Data

Solubility of ClO_2 in Water

With a mole fraction in the vapor of $y_i = 0.03$

Pressure = 1.00 atm

Solubility = 2.70 g/l of solution

mol wgt ClO2 = 67.5

$$x_i = \frac{2.7/67.5}{997.3/18+2.7/67.5} = 0.0007216$$

$$H = \frac{0.030}{0.0007216} = 41.57\,\text{atm/mole fraction}$$

Then Henry's law constant is

$$m = \frac{41.57}{1.00} = 41.57$$

Source: Perry, R. H. and Green, D. W., *Chemical Engineer's Handbook*, 7th ed., McGraw-Hill Book Co., New York, 1997.

1. Difficulty in determining the interfacial area
2. Difficulty in determining individual coefficients from measurements of overall coefficients
3. Entrance effects

The rate of mass transfer depends upon the available contact area between the phases, that is, on the degree of subdivision of the phases. Tower packing provides the skeleton over which the liquid flows in thin turbulent films. The packing area becomes effective only if it is wetted. Thus, $K_y a$ must include a hidden factor, which reflects the proportion of the packing area that is effective. This effective area is proportional to the following factors:

1. Liquid rate
2. Type of packing
3. Method of packing the tower
4. Nature of the liquid distribution
5. Height and diameter of the tower

Data is available in the literature for many systems. A good summary of available data is presented by Perry and Green, *Chemical Engineers Handbook*.[27] Data is given in graphical form or analytical correlations as either height of transfer units (H_{OG}, etc.) or mass-transfer coefficients ($K_y a$, etc.). Table 11.6 presents mass-transfer coefficient data from Strigle.[17] Mass-transfer coefficients for structured packing are generally greater than dumped packing, indicating improved mass transfer. Manufacturers of structured packing should be consulted for available mass-transfer data.

TABLE 11.6
Mass-Transfer Coefficient Data

$$K_Y a = K_G a P$$

$K_G a$ Values for a Liquid–Film Controlled System

Gas Contaminant	Scrubbing Liquid	Overall $K_G a$ Value[a] (lb-mol/h-ft^2-atm)
Carbon dioxide	4% NaOH	1.5
Hydrogen sulfide	4% NaOH	4.4
Sulfur dioxide	Water	2.2
Hydrogen cyanide	Water	4.4
Formaldehyde	Water	4.4
Chlorine	Water	3.4
Bromine	5% NaOH	3.7
Chlorine dioxide	Water	4.4

$K_G a$ Values for Gas–Film Controlled Systems

Gas Contaminant	Scrubbing Liquid	Overall $K_G a$ Value[a] (lb-mol/h-ft^2-atm)
Hydrogen chloride	Water	14
Hydrogen fluoride	Water	6.0
Ammonia	Dilute acid	13
Chlorine	8% NaOH	10.8
Sulfur dioxide	11% Na_2CO_3	8.9

Relative $K_G a$ Values for Tower Packings

	Size (in.)			
Type of Packing	**1**	**1½**	**2**	**3 or 3½**
Plastic Super Intalox® saddles	1.49	–	1.00	0.63
Metal Hy-Pak® packing	1.52	1.28	1.12	0.76
Metal pall rings	1.62	1.32	1.13	0.65
Plastic pall rings	1.38	1.19	1.05	0.63
Ceramic Intalox® saddles	1.42	1.19	0.99	0.57
Ceramic Raschig® rings	1.21	1.02	0.84	0.51
Plastic Tellerette® packing	1.52	–	1.38	–
Plastic Maspac® packing	–	–	1.00	0.62

Effect of Liquid Rate on Scrubber Efficiency

Liquid Rate (gpm/ft^2)	Relative $K_G a$ Value
2.00	0.80
3.00	0.91
4.00	1.00
5.00	1.07
6.00	1.14
8.00	1.25
10.0	1.34

Source: Strigle, R. F., *Packed Tower Design and Applications, Random and Structured Packings*, 2nd ed., Gulf Publishing Co., Houston, TX, 1994.

[a] $K_G a$ values apply to 2 in. plastic Super Intalox® packing at a gas velocity of 3.5 fps and liquid rate of 4 gpm/ft^2.

11.6.11 Using Henry's Law for Multicomponent Solutions

In a work by Carroll,[29] the use of Henry's law for multicomponent solutions is considered. In this work, Carroll discusses the following cases:

1. A nonvolatile component dissolved in a solvent including the effect of electrolytes
2. The case of mixed solvents
3. The case of mixed solutes
4. Chemically reacting systems

The last two cases are more applicable to the dilute case situation that is discussed in this book. For a binary solvent composed of *j* and *k* pure components, Carroll suggests

$$\ln(H)_{i,\text{mix}} = x_j \ln(H)_{ij} + x_k \ln(H)_{ik} + a_{jk} x_j x_k \tag{11.81}$$

where:

$H_{i,\text{mix}}$ is the Henry's constant for solute i in a mixture of j and k
H_{ij} and H_{ik} are the Henry's constants for the pure solvents
x_j and x_k are the mole fractions of the solvents
a_{jk} is an interaction parameter that must be obtained from experimental data

In the absence of any other information, it may be assumed that a_{jk} is independent of i or set equal to zero.

For mixed solutes at low pressure, it may be assumed that each gas is independent of the other. A pair of equations can then be written for two solutes i and j.

$$x_i H_{ik} = y_i P \tag{11.82}$$

$$x_j H_{jk} = y_j P \tag{11.83}$$

Thus, the solubility of the individual components is determined by its own partial pressure as if the other component were not present. Carroll also presents several examples of the use of these equations.

11.7 SAMPLE DESIGN CALCULATION

11.7.1 Dumped Packing

This sample design calculation is presented to illustrate the use of the design methodology presented in this chapter for an absorption column acting to remove a dilute concentration VOC from an air emission. Therefore, the absorbing solutions will also be dilute.

PROBLEM 11.1

An air stream from a electronic circuitry manufacturer contains 50 ppm of a VOC, which must be treated. The air stream is flowing at 10,000 acfm at about 1.00 atm pressure and a temperature of 20°C. The emission standard of the local control agency requires 80% removal of the VOC in this case. Therefore, the outlet concentration must be reduced to 10.0 ppm. Water, free of any of the VOC, is to be used as the absorbing liquid. It has been determined for this organic that the Henry's law constant is 2.00. Calculations will be made to compare two packings. To be compatible with the two fluids, metal IMPT has been chosen in size #25 and #40. The molar *L*/*G* is to be $1.5 \times (L/G)_{min}$, and the pressure drop has been selected to be 0.50 in of H_2O/ft of packing. Overall mass-transfer coefficients are available in Strigle's book[17]. These mass-transfer coefficients are for the absorption of CO_2/NaOH system in water. They are for a liquid flow rate of 10 gpm/ft^2 and a vapor capacity factor of F_s = 0. 9413 lb$^{0.5}$/ft$^{0.5}$. They are suitable for use in this case to show how they are needed and used in the following calculation. However, for a real design system, values of mass-transfer coefficients for the system and flows used should be found. They may be found in a search of the literature, but more likely they may have to be determined through an experiment for that kind of a study. In this case, the following values were found in Strigle.[17]

$$\text{For \#25 IMPT } K_G a = 3.42 \text{ lb-mol/h-ft}^3\text{-atm}$$

and

$$\text{For \#40 IMTP } K_G a = 2.86 \text{ lb-mol/h-ft}^3\text{-atm}$$

The packing factor *F* (Table 11.3) will be 41 for IMTP #25 and 24 for IMTP #40. At this low VOC concentration, it may be assumed that the molecular mass of the gas stream is essentially that of air, that is, 29.0. The physical properties of the water are as follows:

- Molecular mass = 18
- Temperature = 20°C = 68°F
- Pressure = 1.0 atm
- Density = 62.3 lbm/ft^3
- Viscosity = 1.0 cp
- Kinematic viscosity = 1.0 cst

The density of the air is 0.0753 lbs/ft^3.

Equilibrium is described by Henry's law with m = 2.00:

$$y_1 = 2.00\, x_1$$

The operating line based on Equation 11.24 becomes

$$y_{A0} = \left(\frac{L}{G}\right)\left(x_{A0} - x_{A1}\right) + y_{A1}$$

Minimum *L*/*G* ratio can be calculated by rearranging Equation 11.24:

$$\left(\frac{L}{G}\right) = \frac{\left(y_{A0} - y_{A1}\right)}{\left(x_{A0} - x_{A1}\right)}$$

$$y_{A0} = 0.000050 \quad y_{A1} = 0.000010 \quad x_{A1} = 0.000000$$

Then minimum reflux occurs with x_{A0} in equilibrium with y_{A0}, thus,

$$x_{A0} = x^*_{A0} = \frac{y_{A0}}{m} = \frac{0.000050}{2.00} = 0.000025$$

Then

$$\left(\frac{L}{G}\right)_{min} = \frac{0.000050 - 0.000010}{0.000025} = 1.60$$

The minimum slope operating line is

$$y_A = 1.60x_A + 0.000010$$

$$\text{Operating}\left(\frac{L}{G}\right) = 1.5\times\left(\frac{L}{G}\right)_{min} = 1.5\times1.60 = 2.40$$

Equation 11.24 can be solved for the operating x_{A0} with $x_{A1} = 0.000000$.

$$x_{A0} = \frac{y_{A0} - y_{A1}}{(L/G)}$$

$$= \frac{0.000050 - 0.000010}{2.40} = 0.0000167$$

The operating line then becomes

$$y_a = 2.40x_A + 0.000010$$

The tower diameter can now be calculated. First the *L*/*G* ratio must be converted from a molar ratio to a mass ratio.

$$\frac{L}{G} = 2.40\frac{\text{lb moles H}_2\text{O}}{\text{lb moles air}} \times \frac{\left[(18\text{ lbs H}_2\text{O})/(\text{lb moles H}_2\text{O})\right]}{\left[(29\text{ lbs air})/(\text{lb moles air})\right]}$$

$$= 1.4897\frac{\text{lbs H}_2\text{O}}{\text{lbs air}}$$

Calculate the superficial velocity from Figure 11.27. The abscissa is

$$\left(\frac{L}{G}\right)\left(\frac{\rho_G}{\rho_L}\right)^{0.5} = 1.4897\left(\frac{0.0753}{62.3}\right)^{0.5} = 0.052$$

Read the ordinate value = 1.37.

$$C_s F^{0.5} \nu^{0.05} = 1.37$$

The following calculations will now be done on each of the packings and summarized below:

	IMTP 25 $F = 41$	IMTP 40 $F = 24$
$C_s = \dfrac{1.37}{F^{0.5} \times 1.0}$	$C_s = \dfrac{1.37}{41^{0.5} \times 1.0}$ $C_s = 0.2140$ ft/s	$C_s = \dfrac{1.37}{24^{0.5} \times 1.0}$ $C_s = 0.2797$ ft/s

Calculate the superficial velocity V.

$$V = \frac{C_s}{\left[\rho_G/(\rho_L - \rho_G)\right]^{0.5}} \quad V = \frac{0.2140}{\left[0.0753/(62.3 - 0.0753)\right]^{0.5}} \quad V = \frac{0.2797}{\left[0.0753/(62.3 - 0.0753)\right]^{0.5}}$$

$$= 6.15\,\text{ft/s} \qquad = 8.04\,\text{ft/s}$$

With $Q = 10{,}000$ acfm, calculate the mass flow rate.

$$G = \rho_G Q$$

$$G = 10{,}000 \times 0.0753 = 753\,\text{lbs/min}$$

Calculate the superficial gas mass velocity.

$$\bar{G} = V\rho_G \times 60\,\text{min/s} \quad \bar{G} = 6.15 \times 0.0753 \times 60 \quad \bar{G} = 8.04 \times 0.0753 \times 60$$

$$= 27.79 \frac{\text{lbs}}{\text{min - ft}^2} \qquad = 36.32 \frac{\text{lbs}}{\text{min - ft}^2}$$

Calculate the tower cross-sectional area for a circular tower.

$$A_s = \frac{G}{\bar{G}} \quad A_s = \frac{753\,\text{lbs/min}}{27.79\,\text{lbs / min - ft}^2} \quad A_s = \frac{753\,\text{lbs/min}}{36.32\,\text{lbs/min - ft}^2}$$

$$= 27.10\,\text{ft}^2 \qquad = 20.72\,\text{ft}^2$$

Column diameter is given by

$$D = \sqrt{\frac{4A_s}{\pi}} \quad D = \sqrt{\frac{4 \times 27.10}{\pi}} \quad D = \sqrt{\frac{4 \times 20.72}{\pi}}$$

$$= 5.97\,\text{ft} \qquad = 5.13\,\text{ft}$$

Calculate the height of the tower z from Equation 11.57.

$$z = H_{OG} N_{OG}$$

Calculate the height of a transfer unit (H_{OG}).

The units of $\bar{G}$ must be changed to correspond to the units of the mass-transfer coefficient.

For IMTP 25 $$\bar{G} = 27.79 \frac{\text{lbm}}{\text{min - ft}^2} \times 60\,\text{min/h} \times 1.0\,\text{lb - mole/29 lbm}$$

$$= 57.50 \frac{\text{lb - moles}}{\text{h - ft}^2}$$

For IMTP 40 $$\bar{G} = 36.32 \frac{\text{lbm}}{\text{min - ft}^2} \times 60\,\text{min/h} \times 1.0\,\text{lb - mole/29 lbm}$$

$$= 75.14 \frac{\text{lb - moles}}{\text{h - ft}^2}$$

Calculate the height of the transfer unit H_{OG}.

$$H_{OG} = \frac{\bar{G}}{K_y a} \quad H_{OG} = \frac{57.50\,\text{lb - moles/h - ft}^2}{3.42\,\text{lb - moles/h - ft}^3} \quad H_{OG} = \frac{75.14\,\text{lb - moles/h - ft}^2}{2.86\,\text{lb - moles/h - ft}^3}$$

$$= 16.81\,\text{ft} \qquad\qquad = 26.27\,\text{ft}$$

Calculate the number of transfer units (N_{OG}).

$$y_{A0} = 0.000050, \quad y_{A1} = 0.000010, \quad x_{A1} = 0.000000$$

Because $x_{A1} = 0.000000$, Equation 11.65 applies to this case.

$$N_{OG} = \left(\frac{1}{1 - 1/\text{Ab}}\right) \ln\left[(y_{A0}/y_{A1})(1 - 1/\text{Ab}) + 1/\text{Ab}\right]$$

Calculate the absorption factor Ab.

$$\text{Ab} = \frac{\bar{L}}{m\bar{G}} = 2.40/2.00 = 1.20$$

$$N_{OG} = \left(\frac{1}{1 - 1/1.20}\right) \ln\left[(0.000050/0.000010)(1 - 1/1.20) + 1/1.20\right]$$

$$= 3.065$$

For IMTP 25 calculate the height of the tower, z, from Equation 11.57.

$$z = H_{OG} N_{OG} \quad z = 16.81 \times 3.065 = 51.52$$

Note also the value of N_{OG} can be found from Figure 11.29 with $B = 1.5$ and $(y_{A0} - y_{A1})/(y_{A0}) = 0.80$ or 80% removal, $N_{OG} = 3.06$. Thus,

$$z = 16.81 \times 3.06 = 51.44\,\text{ft}$$

which agrees with the calculation above.

The pressure drop through the column is calculated from the original height of packing, 44.04 ft to be

$$\Delta P = 0.50\,\text{in. of } H_2O\text{/ft of packing} \times 51.52\,\text{ft.} = 25.76\,\text{in. of } H_2O$$

The liquid flow rate Q_L is

$$\bar{L} = (L/G)\frac{\text{lb}\,H_2O}{\text{lb air}} \times \bar{G}\frac{\text{lbs air}}{\text{min - ft}^2}$$

$$\bar{L} = 1.1.4897 \times 27.79 = 41.35\frac{\text{lbm}}{\text{min - ft}^2}$$

$$Q_L = \frac{41.35\text{lbm / min - ft}^2 \times 27.10\,\text{ft}^2}{8.32\text{ lb } H_2O/\text{gal}} = 134.7\text{gpm}$$

The irrigation rate is 134.7/27.10 = 4.97 gpm/ft^2 of tower area. Suggested rates are between 2 and 8 gpm/ft^2 of tower area. Thus the rate is in the suggested range in this case.

For IMTP 40, calculate the height of the tower, z, from Equation 11.57.

$$z = H_{OG}N_{OG} \quad z = 26.27 \times 3.065 = 80.52\,\text{ft}$$

Again, note also the value of N_{OG} can be found From Figure 11.14 with $B = 1.5$ and $(y_{A0} - y_{A1})/(y_{A0}) = 0.80$ or 80% removal, $N_{OG} = 3.06$. Thus,

$$z = 26.27 \times 3.06 = 80.39\,\text{ft}$$

which agrees with the calculation above.

The pressure drop through the column is calculated from the original height of packing, 80.52 ft to be

$$\Delta P = 0.50\,\text{in. of } H_2O\text{/ft of packing} \times 80.52\,\text{ft} = 40.26\,\text{in. of } H_2O$$

The liquid flow rate Q_L is

$$\bar{L} = (L/G)\frac{\text{lb}\,H_2O}{\text{lb air}} \times \bar{G}\frac{\text{lbs air}}{\text{min - ft}^2}$$

$$\bar{L} = 1.1.4897 \times 36.32 = 54.11\frac{\text{lbm}}{\text{min - ft}^2}$$

$$Q_L = \frac{54.11\text{lbm / min - ft}^2 \times 20.72\,\text{ft}^2}{8.32\text{ lb } H_2O/\text{gal}} = 134.75\text{gpm}$$

The irrigation rate is 134.75/20.72 = 6.50 gpm/ft^2 of tower area. Suggested rates are between 2 and 8 gpm/ft^2 of tower area. Thus the rate is in the suggested range in this case also.

11.7.2 Flooding

The capacity factor at flooding can be determined at the same L/G ratio as the problem above; therefore, in Figure 11.27, the pressure drop curve is the same as well. The abscissa is 0.052. The pressure drop at flooding is determined from Equation 11.66.

Then the value of the ordinate of Figure 11.27 can be found. From this ordinate the capacity factor can be calculated and then the flooding velocity calculated.

Flooding Pressure Drop	IMTP 25	IMTP 40
$\Delta P_f = 0.12 F^{0.7}$	$\Delta P_f = 0.12 \times 41^{0.7}$	$\Delta P_f = 0.12 \times 24^{0.7}$
	$\Delta P_f = 1.62 \text{ in} H_2O/\text{ft}$	$\Delta P_f = 1.06 \text{ in } H_2O/\text{ft}$
	Ordinate value	
$C_s F^{0.5} \nu^{0.05}$	1.80	1.65
Capacity factor	$C_s = \dfrac{1.80}{41^{0.5} \times 1.0}$	$C_s = \dfrac{1.65}{24^{0.5} \times 1.0}$
	$C_s = 0.2811 \text{ ft/s}$	$C_s = 3368 \text{ ft/s}$
	Superficial velocity	
$V = \dfrac{C_s}{\left[\rho_G/(\rho_L - \rho_G)\right]^{0.5}}$	$V = \dfrac{0.2811}{\left[0.0753/(62.3 - 0.0753)\right]^{0.5}}$	$V = \dfrac{0.3368}{\left[0.0753/(62.3 - 0.0753)\right]^{0.5}}$
Flooding velocity	$V = 8.08$ ft/s	$V = 9.68$ ft/s

Calculate the gas flow rate.

$$Q = VA_s$$

For IMTP 25:

$$Q = 8.08 \text{ ft/s} \times 27.10 \text{ ft}^2 \times 60 \text{ s/min} = 13{,}139 \text{ acfm}$$

Actual operation is at 10,000 acfm, therefore percent of flooding at

$$\left(\frac{10{,}000}{13{,}139}\right) \times 100\% = 76\%$$

For IMTP 40:

$$Q = 9.68 \text{ ft/s} \times 20.72 \text{ ft}^2 \times 60 \text{ s/min} = 12{,}034 \text{ acfm}$$

Actual operation is at 10,000 acfm, therefore percent of flooding is at

$$\left(\frac{10{,}000}{12{,}034}\right) \times 100\% = 83\%$$

11.7.3 Structured Packing

To treat additional sources or to meet higher electronic circuit production demand, consider the possibility of increasing the air flow to 20,000 acfm by switching to structured packing. The pressure is at 1.0 atm, and the temperature of the steam will be 20°C. Thus, the vapor-liquid data will be the same, and consequently the molar L/G ratio will remain the same, and operation will take place at $1.5 \times (L/G)_{min}$. We will choose Intalox Structured Packing 2T.[17] The flow parameter of Equation 11.69 has already been calculated to be 0.052. The pressure drop at operation will also still be 0.5 in H_2O/ft of packing.

To begin our calculation, we will refer to Kister's book[18] CHART 10.6404 for Norton Intalox 2T packing. Here, Kister indicates that the packing factor F_P is 17. With the flow parameter of 0.052 and with pressure drop to be 0.5 in H_2O/ft of packing, we can read from the chart that the capacity parameter is 1.30. From Strigle[17] in Table 3.6, the overall mass-transfer coefficient for Intalox Structured Packing 2T is listed as 3.80 lb-mol/h-ft³-atm. We can now begin our calculation to determine tower diameter and packing height.

$$C_s F^{0.5} \nu^{0.05} = 1.30$$

Calculate capacity factor C_s.

$$C_s = \frac{1.30}{17^{0.5} \times 1.0^{0.05}} = 0.3153\,\text{ft/s}$$

Calculate superficial velocity V.

$$V = \frac{C_s}{\sqrt{\left[\rho_G/(\rho_L - \rho_L)\right]}}$$

$$V = \frac{0.3153}{\sqrt{\left[0.0753/(62.3 - 0.0753)\right]}} = 9.06\text{ ft/s}$$

Calculate superficial gas velocity $\overline{G}$.

$$\overline{G} = V\rho_G$$

$$= 9.06\,\text{ft/s}\ 0.0753\,\text{lb/ft}^3\ 60\text{ s/min} = 40.93\,\text{lbs/min-ft}^2$$

$$= 40.93\left(\text{lbs/min-ft}^2\right)\left(60\,\text{min/h}\right)\left(\frac{1\,\text{lb mole}}{29\,\text{lbs}}\right) = 84.68\,\text{lb moles/h-ft}^2$$

Calculate column area A and diameter D.

$$A_S = \frac{G}{\overline{G}}$$

$$G = 20000\ \text{ft}^3/\text{min} \times 0.0753\,\text{lbs/ft}^3 = 1506\,\text{lbs/min}$$

$$A_S = \frac{1506\,\text{lbs/min}}{40.93\,\text{lbs/min-ft}^2} = 36.79\,\text{ft}^2$$

$$D = \sqrt{\frac{4 \times A_S}{\pi}}$$

$$= \sqrt{\frac{4 \times 36.79\,\text{ft}^2}{\pi}}$$

$$= 6.84\,\text{ft}$$

Calculate packing height.

The mass-transfer coefficient K_ya = 3.80 lb-mol/h-ft^3-atm. The pressure is 1.0 atm.

The number of transfer units is N_{OG} = 3.065, which the same as for the dumped packing since the vapor liquid equilibria is the same and the L/G ratio = 1.5 $(L/G)_{min}$, which is also the same.

The packing height is

$$H_{OG} = \frac{\bar{G}}{K_ya}$$

$$= \frac{84.68\,\text{lb moles/h-ft}^2}{3.80\,\text{lb moles/h-ft}^3} = 22.28\,\text{ft}$$

Calculate the height of the packing z.

$$z = H_{OG}N_{OG}$$

$$= 22.28\text{ft} \times 3.065$$

$$= 68.30\text{ ft}$$

11.7.3.1 Flooding

As noted above we will refer to Kister's book[18] CHART 10.6404 for Norton INTALOX 2T packing to find the parameters for determining flooding. Here, Kister indicates that the packing factor F_P is 17. That chart indicates that flooding occur at a pressure drop of about 1.51 in of water per foot pf packing. Therefore, we will use the *capacity parameter* from the CHART equal to 1.60 found at the *flow parameter* of 0.052 under which we are operating.

The *capacity parameter* is as noted:

$$C_s F^{0.5} \nu^{0.05} = 1.60$$

Calculate capacity factor C_S.

$$C_s = \frac{1.60}{17^{0.5} \times 1.0^{0.05}} = 0.3881\,\text{ft/s}$$

Calculate superficial velocity V at flooding.

$$V = \frac{C_s}{\sqrt{[\rho_G/(\rho_L - \rho_L)]}}$$

$$V = \frac{0.3881}{\sqrt{[0.0753/(62.3 - 0.0753)]}} = 11.16 \text{ ft/s}$$

Actual operation for the stacked packing inches situation is 20,000 acfm. Calculate the gas flow rate at flooding.

$$Q = VA_S$$

$$Q = 11.16 \text{ ft/s} \times 36.79 \text{ ft}^2 \times 60 \text{ s/min} = 24{,}635 \text{ acfm}$$

Therefore, percent of flooding is at

$$\left(\frac{20{,}000}{24{,}635}\right) \times 100\% = 81\%$$

Now let us consider using Kister's Equation 11.66 for pressure drop at flooding. This is what was done for the two calculations above under random packing.

Since we are using the same L/G ratio as the other problems The capacity factor at flooding can be determined at the same L/G ratio as the problem above the flow parameter will be the same of 0.052. The pressure drop at flooding is determined from Equation 11.66. Then the value of the ordinate of Figure 11.27 can be found. From this ordinate the capacity factor can be calculated and then the flooding velocity calculated.

Flooding pressure drop by Kister's method:

$$\Delta P_f = 0.12 F^{0.7} \quad \Delta P_f = 0.12 \times 17^{0.7}$$

$$= 0.872 \text{ in of } H_2O/\text{ft}$$

The *capacity parameter* is also read from Kister's chart and is as follows:

$$C_s F^{0.5} \nu^{0.05} = 1.45$$

Calculate capacity factor C_s

$$C_s = \frac{1.45}{17^{0.5} \times 1.0}$$

$$C_s = 0.3517 \text{ ft/s}$$

Calculate superficial velocity at flooding.

$$V = \frac{C_s}{[\rho_G/(\rho_L - \rho_G)]^{0.5}}$$

$$V = \frac{0.3517}{\left[0.0753/(62.3-0.0753)\right]^{0.5}} = 10.11\,\text{ft/s}$$

Calculate the gas flow rate at flooding

$$Q = VA_s$$

$$Q = 10.17\,\text{ft/s} \times 36.79\ \text{ft}^2 \times 60\,\text{s/min} = 22{,}317\ \text{acfm}$$

Actual operation is at 20,000 acfm, therefore percent of flooding at

$$\left(\frac{20{,}000}{22{,}317}\right) \times 100\% = 89.6\%$$

Compared to the original calculation above, this is within the range of the data used to determine Kister's equation.

12 Adsorption for Hazardous Air Pollutants and Volatile Organic Compounds Control

12.1 INTRODUCTION TO ADSORPTION OPERATIONS

In adsorption operations, solids, usually in granular form, are brought in contact with gaseous or liquid mixtures. The solids must have the ability to preferentially concentrate or adsorb on their surfaces specific components from the mixture. This phenomenon is possible because the attractive forces that exist between the atoms, molecules, and ions holding the solids together are unsatisfied at the surface and are thus available for holding the components in the mixtures to be adsorbed. Consequently, these components can be separated from each other and the carrier fluid.

Adsorption is used broadly in the chemical process industries. Early on, one of its most significant applications was the drying of wet air over beds of solid desiccants for pneumatic control instruments. Original process applications range from solvent reclamation in dry cleaning to the recovery of ethyl acetate and toluene from cellophane drying operations. A broad area for adsorption application is in component recovery. Processes of this sort have been patented for fractionated petroleum products by oil and chemical companies. A typical early application was that of the hypersorption process developed by the Union Oil Company of California in the mid-1940s, in which a moving bed of carbon separated light hydrocarbons.[1] Another widespread application of adsorption is the recovery of valuable solvents from air streams.

Of most relevance in this book are adsorption processes, which are used to eliminate impurities from emissions to the ambient air. Solvent recovery, removal of other organic compounds, and odor removal are all important in producing clean effluents. The best applications of adsorption are in handling large volumes of air flow with dilute pollution levels and removal of the contaminants, especially volatile organic compounds (VOCs), down to trace levels such as 1.0 ppmv. Table 12.1 summarizes some of these operations and the type of adsorbent used.

A brief summary of adsorption containing a general description, advantages, and disadvantages for the removal of VOCs from gaseous emission streams is provided below.[2–5]

12.1.1 DESCRIPTION

- VOC gaseous emissions flow into the top or bottom of an adsorption column, filled with porous activated carbon, and is distributed throughout the carbon bed.

TABLE 12.1
Adsorption Processes and Type of Adsorbent

	Adsorbent			
Substance to Be Removed	**Activated Carbon**	**Activated Alumna Gel**	**Silica Gel**	**Molecular Sieves**
Odors	x			
Oil	x	x	x	x
Hydrocarbons	x	x	x	
Fluorocarbons	x		x	
Chlorinated hydrocarbons	x		x	
Organic sulfur				
Compounds	x	x		x
Solvents	x			
Moisture	x	x	x	

- Two adsorption processes exist: temperature-swing adsorption and pressure-swing adsorption. Temperature-swing adsorption is the approach commonly used for VOC recovery and the process description, advantages, and disadvantages listed in this section correspond to the temperature-swing adsorption process.
- Carbon adsorption beds can be fixed or moving, with respect to the carbon. For moving beds, the flow of activated carbon is countercurrent to the flow of the gas; however, fixed beds are more common in industry.
- The VOC is adsorbed onto the surface of the activated carbon and onto the surface of the pores. At some point, the carbon becomes saturated with VOC and loses its capacity for additional adsorption. This results in the concept of *breakthrough*, where significant quantities of VOC become apparent in the gas stream exiting the adsorption process. When this occurs, the carbon must be regenerated for reuse or replaced with virgin carbon.
- Multiple fixed beds are generally employed, so that as one or more beds are adsorbing, at least one bed can be regenerating. Regenerating a bed of activated carbon typically involves the direct injection of steam, hot nitrogen or hot air to the bed, which causes the VOC to release from the carbon and exit the bed via a vapor or condensate stream. The regenerated stream, containing a higher concentration of the VOC than the original wastewater stream, is subsequently condensed. If the VOC is immiscible in water, the condensate will form an aqueous layer and a solvent layer that can be separated using a decanter. If the VOC is miscible in water, additional distillation can be used to further separate the VOC and water.
- *VOC-free* gas exits the adsorber after contacting the activated carbon.

12.1.2 Advantages

- A widely used technology with well-established performance levels.
- Can achieve high recovery efficiencies (90%–98%).
- Can be used for a wide range of gas flow rates (100–60,000 cfm).
- Can handle a wide range of inlet VOC concentrations (20–5000 ppm).
- Can efficiently handle fluctuations in gas flow rates and VOC concentrations.

12.1.3 Disadvantages

- VOCs having high heats of adsorption (typically ketones) can cause carbon bed fires.
- Carbon attrition properties (permanent bonding of small quantities of VOC through each adsorption cycle) require the periodic replacement of carbon with virgin or reactivated carbon. Spent carbon may need to be disposed of as a hazardous waste depending on the VOC(s) adsorbed.
- Carbon fines generated from the continual regeneration of the carbon may carry over to downstream equipment if not adequately filtered.
- Carbon efficiency decreases for high humidity (>50% r.h.) air streams.

12.2 ADSORPTION PHENOMENON

Adsorption is based on the capability of porous solids with large surfaces such as silicon gel and activated carbon to selectively retain and release compounds on the surface of the solid. Two general phenomena are recognized in adsorption. Physical adsorption is a low-temperature process similar to condensation. Chemisorption, which occurs at high temperatures, is a process in which forces are very strong in the nature of an actual chemical bond. Essentially, only a monomoleculayer can be formed.

In gas separation, physical adsorption is of primary importance. The adsorbate molecules diffuse from the gas phase across the boundary layer to the surface of the adsorbent, where they are held by fairly weak van der Waals forces. Heat is liberated approximately equivalently to the heat required for condensation. At saturation, these two adsorption processes lead to a complete covering of the solid surface with the adsorbate substance. In physical adsorption, more than a monomolecular layer can build up. The packing density of molecules at saturation will reach approximately the density of the molecules in liquid form.

Another important phenomenon that occurs is capillary condensation. At higher pressures, the gases and vapors begin to condense in the pores of the adsorbates. The adsorption forces reduce the vapor pressure in the capillaries, so that it is possible to condense vapors at temperature well above the condensation temperature.

12.3 ADSORPTION PROCESSES

The techniques used in adsorption processes include both stagewise and continuous-contacting methods applied to both continuous and semicontinuous operations. These operations are analogous to absorption when only one component of a gas is strongly absorbed. When more than one component of the gas is strongly adsorbed, the operation is one that is analogous to fractionation, and in particular it becomes much like extraction.

12.3.1 Stagewise Process

For example, in the stagewise drying of air with silica gel, the silica gel is contacted countercurrently in the upper part of the tower with the air to be dried. The contact takes place on perforated trays in relatively shallow beds, the gel moving from tray to tray through down spouts. In the lower part of the tower, the gel is dried by similar contact with a hot gas, which desorbs and carries off the moisture. The dried gel is recirculated to the top of the absorber by an air lift. In cases where the adsorbed component is to be recovered, for example, an organic solvent, the regeneration might include steam stripping of the adsorbent with distillation or decantation of the organic solvent from the water. The adsorbent would then be air dried and returned to the tower.

12.3.2 Continuous Contact, Steady-State, Moving-Bed Adsorbers

Countercurrent, continuous-contact, steady-state, moving-bed adsorbers in which uniform solid flow is obtained without channeling or localized flow irregularities have been developed. One such device for the fractionation of light hydrocarbon gases is the hypersorber built for the Union Oil Company process previously mentioned.[1] This device uses very hard active coconut shell or fruit pit activated carbon. The feed is introduced centrally, and the solids flow downward from the top. In the upper section, the more readily adsorbed components are picked up by the descending solid. The top gas product contains the poorly adsorbed constituents. Solids passing the feed point contain all the feed components. In the lower section, a rising stream of gas displaces the most volatile constituents, which pass upward. The adsorbent then leaves the column rich in readily adsorbed components. In the lowest section, the adsorbent is removed from the solid by heating and by steam stripping. Part of the desorbed gas is removed as product, while a portion continues up the column as reflux. The solid is recycled to the top by a gas lift.

12.3.3 Unsteady-State, Fixed-Bed Adsorbers

Due to the higher cost of transporting solids in a moving bed as required in steady-state continuous operations, frequently a stationary bed of adsorbent is used. Such a bed adsorbs increasing amounts of solute in an unsteady-state process, which continues until the bed is saturated. One of the most important applications of this type of adsorber is the recovery of solvent vapors. Figure 12.1 is a typical arrangement of this type of adsorption vessel. Recovery of 99%–99.8% of solvent is possible from mixtures containing as little as 0.05% by volume of the solvent. Thus, air-vapor mixtures well below the explosive limit may be handled. In most cases, the pressure drop through the bed is kept small to reduce power costs. Thus, granular rather

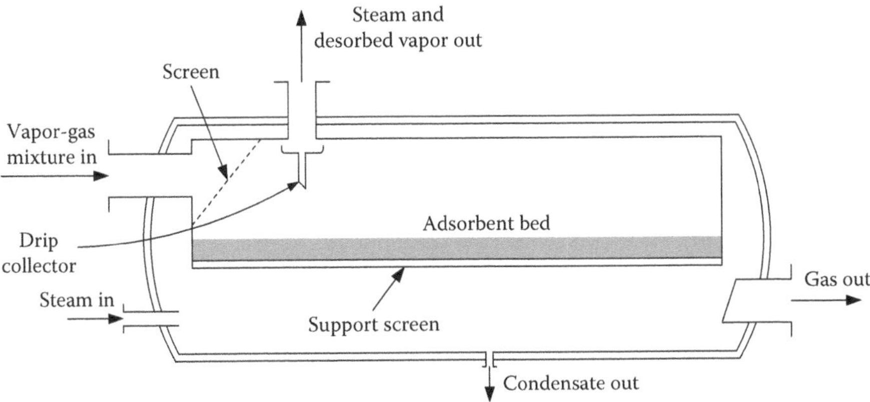

FIGURE 12.1 Fixed-bed adsorber.

than powdered adsorbents are used, and bed depths run between 0.30 m (12 in.) and 1.50 m (60 in.). The superficial gas velocity may be in the range of 0.23–0.56 m/s (0.75–1.83 ft/s).

After the adsorber becomes saturated, the gas flow is diverted to a second similar vessel. The adsorbent is regenerated by low pressure steam or hot clean gases. If steam is used, the steam condenses and provides the heat of desorption, as well as lowering the pressure of the vapor in contact with the solid. The steam vapor is condensed and the condensed solvent recovered by decantation if it is insoluble in water or by distillation if it is water soluble. The water-saturated adsorbent is readily dried when fresh gas is admitted to the vessel. If moisture is very undesirable in the gas to be treated, the bed can be first air dried then cooled by unheated air prior to reuse for solvent recovery. Our design considerations will be limited to these fixed-bed types of adsorbers.

12.3.4 Newer Technologies

12.3.4.1 Rotary Wheel Adsorber

In this newer configuration, the rotary wheel adsorber, a circular medium is coated with carbon or hydrophobic molecular sieves.[6] These adsorbents remove the VOC from the air as the device rotates. One part of the wheel is adsorbing, while the other part is regenerating. The device is most effective for high flow rates with concentrations below 1000 ppmv and required efficiencies below 97%.

12.3.4.2 Chromatographic Adsorption

In chromatographic adsorption, a cloud of solid adsorbent is sprayed into the effluent gas stream.[7] The adsorbent and effluent gas travel concurrently through the containing vessel. Adsorption takes place on the adsorbent, which is then removed from the gas stream in a conventional bag filter. Adsorption also takes place on the adsorbent particles trapped on the filter bags.

12.3.4.3 Pressure Swing Adsorption

In pressure swing adsorption, the adsorbent bed is subjected to short pulses of high pressure gas containing the VOC to be adsorbed.[7] The higher pressure results in better adsorption. The pressure is then reduced, and the adsorbed material will vaporize, regenerating adsorbent. By controlling the pressure and cycle time, the pollutant is transferred from the effluent stream to the low-pressure gas regeneration stream.

12.4 NATURE OF ADSORBENTS

All solids possess an adsorptive ability. However, only certain solids exhibit sufficient specificity and capacity to make an industrially useful material. Furthermore, unlike solvents for absorption, the adsorptive characteristics of solids of similar chemical composition depend mostly on their method of manufacture. All carbon is adsorptive, but only *activated* carbon is useful in industrial processes. Carbon can be activated in two ways: one is by use of a gas to create a pore structure by burning of the carbon at 700°C to 1000°C followed by treatment with steam at 700°C to 900°C; the other is by removing water from the pores of uncarbonized raw materials, such as sawdust, using a solution of zinc chloride, phosphoric acid, or sulfuric acid.

The adsorbent must possess appropriate engineering properties, dependent on applications. If used in a fixed bed, it must not offer too great a pressure drop, nor must it be easily carried away in the flowing stream. It must have adequate strength so as not to be crushed in beds, nor by being moved about in moving-bed adsorbers. If it is to be frequently transported, it must be free flowing. Table 12.2 is a summary

TABLE 12.2
Properties of Representative Adsorbents

	Particle Form[a]	Mesh Size[b]	Bulk Density (lbm/ft³)	Effective Diameter (ft)	Internal Void Fraction	External Surface (ft²/ft³)	Reactivation Temperature (°F)
Activated carbon	P	4/6	30	0.0128	0.34	310	200–1000
	P	6/8	30	0.0092	0.34	446	200–1000
	P	8/10	30	0.0064	0.34	645	200–1000
	G	4/10	30	0.0110	0.40	460	200–1000
	G	6/16	30	0.0062	0.40	720	200–1000
	G	4/10	28	0.0105	0.44	450	200–1000
Silica gel	G	3/8	45	0.0127	0.35	230	250–450
	G	6/16	45	0.0062	0.35	720	250–450
	S	4/8	50	0.0130	0.36	300	300–450

(Continued)

TABLE 12.2 (*Continued*)
Properties of Representative Adsorbents

	Particle Form[a]	Mesh Size[b]	Bulk Density (lbm/ft^3)	Effective Diameter (ft)	Internal Void Fraction	External Surface (ft^2/ft^3)	Reactivation Temperature (°F)
Activated alumina	G	4/8	52	0.0130	0.25	380	350–600
	G	8/14	52	0.0058	0.25	480	350–600
	G	14/28	54	0.0027	0.25	970	350–600
	S	(1/4)	52	0.0208	0.30	200	350–1000
	S	(1/8)	54	0.0104	0.30	400	350–1000
Molecular sieves	G	14/28	30	0.0027	0.25	970	300–600
	P	(1/16)	45	0.0060	0.34	650	300–600
	P	(1/8)	45	0.0104	0.34	400	300–600
	S	4/8	45	0.0109	0.37	347	300–600
	S	8/12	45	0.0067	0.37	565	300–600

Source: Stenzel, M. H., *Chem. Eng. Prog.*, 89(4), 36–43, 1993.

[a] G, granules; P, pellets; S, spheroids.

[b] Mesh refers to the number of openings per inch in a mesh screen.

of some of the common properties of adsorbents. A description of these common adsorbents for air-pollution control follows.

1. *Activated carbon:* This is made by carbonization of coconut shells, fruit pits, coal, and wood. It must be activated, essentially a partial oxidation process, by treatment with hot air or steam. It is available in granular or pelleted form and is used for recovery of solvent vapors from gas mixtures, in gas masks, and for the fractionation of hydrocarbon gases. It is revivified for reuse by evaporation of the adsorbed gas.
2. *Silica gel:* This is a hard, granular, very porous product made from the gel precipitated by acid treatment of sodium silicate solution. Its moisture content prior to use varies from roughly 4% to 7%, and it is used principally for dehydration of air and other gases, in gas masks, and for fractionation of hydrocarbons. It is revivified for reuse by evaporation of the adsorbed matter.
3. *Activated alumina:* This is a hard, hydrated aluminum oxide, which is activated by heating to drive off moisture. The porous product is available as granules or powders, and is used chiefly as a desiccant for gases and liquids. It may be reactivated for reuse.
4. *Molecular sieves:* These are porous, synthetic zeolite crystals, metal aluminosilicates. The *cages* of the crystal cell can entrap adsorbed matter, and the diameter of the passages, controlled by the crystal composition, regulates the sizes of the molecules, which may enter or be excluded.

The sieves can thus separate according to molecular size, but they also separate by adsorption according to molecular polarity and degree of unsaturation. They are used for dehydration of gases and liquids, separation of gas and liquid hydrocarbon mixtures, and in a great variety of processes. They are regenerated by heating or elution.

12.4.1 Adsorption Design with Activated Carbon

12.4.1.1 Pore Structure

The pore structure determines how well an adsorbent will perform in a particular VOC recovery process. Coconut-shell-activated carbon pore diameters average less than about 20 Å. A very high surface volume results and produces a high retentivity for small organic molecules. Thus, coconut-shell-activated carbon is an ideal adsorbent for VOCs. A smaller portion of the porosity of coal-based activated carbon is in the lower micropore diameter size. Coal-based activated carbons are typically used to remove both low molecular weight hydrocarbons, such as chlorinated organics, and high molecular weight materials, such as pesticides.

12.4.1.2 Effect of Relative Humidity

The relative humidity severely reduces the effectiveness of activated carbon at values greater than 50% relative humidity. At this point, capillary condensation of the water becomes very pronounced, and the pores tend to fill up selectively with water molecules. To reduce relative humidity, the air stream can be cooled first to drop out the moisture, or if the relative humidity is not too high, the air stream can simply be heated 20°F or 30°F, or it can be cooled first, then heated.

12.5 THEORIES OF ADSORPTION

When a gas is brought into contact with an evacuated solid, a part of the gas is taken up by the solid. The molecules that are taken up and enter the solid are said to be adsorbed, similar to the case of a liquid absorbing molecules. The molecules that remain on the surface of the solid are said to be adsorbed. The two processes can occur simultaneously and are spoken of as *sorption*. If the process occurs at constant volume, the gas pressure drops; if it occurs at constant pressure, the volume decreases. To study adsorption, the temperature, pressure, and composition must be such that very little absorption takes place. If a gas remains on a surface of a solid, two things may happen: there may be a weak interaction between solid and gas similar to condensation, or there may be a strong interaction similar to chemical reaction. The first interaction is termed *physical adsorption* or its synonym, *van der Waals adsorption*. The name van der Waals implies that the same forces that are active in condensation, that is, the van der Waals forces, are also active in physical adsorption. The second interaction, in which the forces involved are strong as in chemical bonding, is termed *chemical adsorption or chemisorption*, or another synonym, *activated adsorption*. The implication here is that this type of adsorption requires an energy of activation, just as in chemical reactions.

The differences between physical adsorption and chemisorption may be briefly summarized in six points.

1. The most fundamental difference between the two types of adsorption is in the forces involved.
   Physical adsorption ≈ van der Waals forces = condensation
   Chemisorption = chemical reactions
2. The differences manifest themselves in the strength of the binding between adsorbate and adsorbent.
   Physical adsorption ≈ van der Waals forces = heat of condensation chemisorption = heat of reaction
3. The difference also manifests itself in the specificity of the process. At sufficiently low temperature, physical adsorption takes place between any surface and any gas, but chemisorption demands a chemical affinity between adsorbate and adsorbent.
4. In physical adsorption, the rate of adsorption is rapid, while in chemisorption, the energy of activation must be supplied before the adsorbent-adsorbate complex can form.
5. The adsorption isotherm in chemisorption always indicates unimolecular adsorption, while in van der Waals adsorption, the process may be multi-molecular.
6. The adsorption isobar of gases that can be adsorbed by the two processes in the same adsorbent shows both van der Waals and chemisorption regions in which the adsorption decreases with temperature.

Essentially, physical desorption may be called *surface condensation* and chemisorption may be called *surface reaction*. Since the processes are so different, the fundamental laws that deal with the mechanisms are different. On the other hand, laws that deal with equilibrium states only, such as the Clausuis-Clapeyron equation, may be used to calculate the heat released for both physical and chemisorption. Similarly, equations such as the Freundlich equation, which merely describes the shape of the isotherm without implying any mechanisms, may be applied to both types of adsorption.

One other factor distinguishes the two types of adsorption and that is the ability to readily reverse the physical adsorption process, while removal of chemisorbed gases is more difficult. Simple evacuation combined with heating, or even a simple heating, will remove physically adsorbed gases, leaving the chemisorbed material behind.

An additional mechanism, which adsorbed gases may undergo, is capillary condensation. Most adsorbents are full of capillaries, and the gases make their way into these pores adsorbing on the sides of the pore. If a liquid wets the walls of a capillary, the vapor pressure will be lower than the bulk vapor pressure. Thus, it has been assumed that adsorption in capillaries takes place at a pressure considerably lower than the vapor pressure. The capillaries with the smallest diameters fill first at the lowest pressures. As the pressure is increased, larger capillaries fill until at saturation pressure all pores are filled with liquid.

It is apparent that capillary condensation plays a role in physical adsorption. Multimolecular adsorption and capillary condensation are necessarily preceded by unimolecular adsorption. One complete theory must be applicable to all of this range, from capillary condensation to multimolecular adsorption. The theory credited to Brunauer, Emmett and Teller, called the BET theory, covers this entire range of adsorption. The theory is based on the assumption that the same forces that produce condensation are chiefly responsible for the binding energy of multimolecular adsorption.

12.6 DATA OF ADSORPTION

When a gas or vapor is admitted to a thoroughly evacuated adsorbent, its molecules are distributed between the gas phase and the adsorbed phase. The rate of adsorption is so fast in some cases that it is most difficult to measure. In other instances, the rate is more moderate and can be readily measured. After a time, the process stops, and a state of stable equilibrium is reached. The amount of gas adsorbed per gram of adsorbent at equilibrium is a function of temperatures and pressure, and the nature of the adsorbent and the adsorbate.

$$a = f(P,T) \tag{12.1}$$

where:

a is the amount absorbed per gram of adsorbent
P is the equilibrium pressure
T is the absolute temperature

When the temperature is held constant and the pressure varied, the plot produced is known as the *adsorption isotherm*:

$$a = f(P) \tag{12.2}$$

If the adsorption occurs under constant pressure and variable temperature, the plot is called the *adsorption isobar*:

$$a = f(T) \tag{12.3}$$

An adsorption isotere occurs when the variation of the equilibrium pressure with respect to temperature for a definite amount adsorbed is measured.

$$P = f(T) \tag{12.4}$$

The amount adsorbed is usually expressed as the volume of gas taken up per gram of adsorbent at 0°C and 760 mm of pressure (STP) or as the weight of gas adsorbed per gram of adsorbent. The amount of adsorption decreases as temperature increases, as shown in Figure 12.2, the adsorption of *n*-hexane on BPL-activated carbon. This is an adsorption isotherms typical of data necessary for design. Here, the adsorption is expressed as a capacity by weight percent as a function of the temperature of

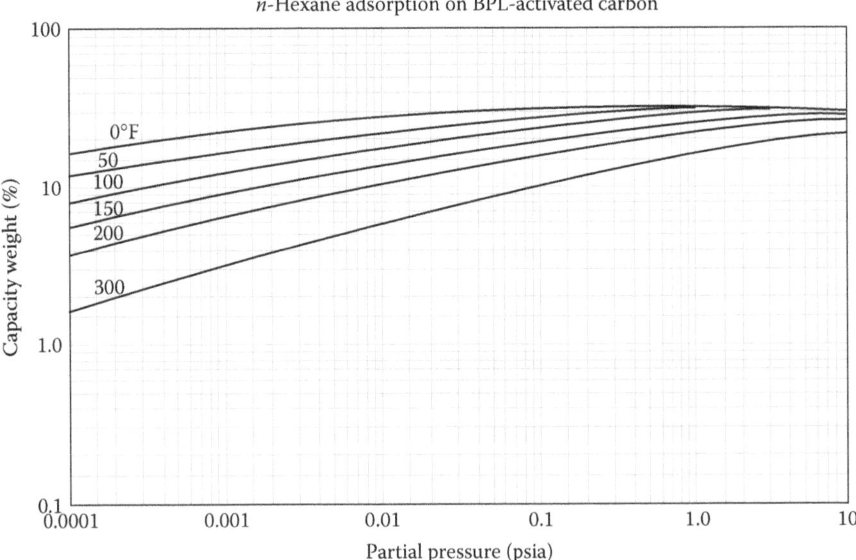

FIGURE 12.2 Equilibrium adsorption on activated carbon. (Reprinted by permission from Calgon Corporation, Pittsburgh, PA, 15205.)

adsorption, with the abscissa being the partial pressure of the adsorbate in the gas in psia. For example, if the partial pressure of hexane is 0.0147 psia and the temperature is 80°F, the capacity indicated is about 20%. This means that 100 lbm of carbon will adsorb $0.20 \times 100 = 20$ lbm of hexane at equilibrium.

As a result of the shape of the openings in capillaries and pores of the solid or of other complex phenomenon such as wetting the adsorbate, different equilibria result during desorption than was present in adsorption. The adsorption process exhibits hysteresis, in other words, desorption pressure is always lower than that obtained by adsorption. In some cases, it has been found that hysteresis disappears upon a thorough evaluation of the adsorbate. Thus, hysteresis must be due to impurities. Some experimenters have accepted the desorption curve as the true equilibria since it represents complete wetting after removal of impurities.

12.7 ADSORPTION ISOTHERMS

12.7.1 Freundlich's Equation

The Freundlich equation[8] is widely used for both liquid and gaseous adsorption. It is a simple equation but valid only for monomolecular layers. The equation may be written as

$$V = P^{1/n} \tag{12.5}$$

12.7.2 Langmuir's Equation

Due to the nature of Langmuir's derivation[9] of the adsorption isotherm, it is valid only for unimolecular layer adsorption. The isotherm can be written as

$$V = \frac{bPV_m}{1 + bP} \tag{12.6}$$

where:

- V is the volume adsorbed
- V_m is the volume adsorbed when surface is covered with a monomolecular layer
- b is a constant dependent upon molecular parameters and temperature
- P is the equilibrium pressure of adsorption

To determine b and V_m, rearrange Equation 12.6 as

$$\frac{P}{V} = \frac{1}{bV_m} + \frac{P}{V_m} \tag{12.7}$$

Measure P and V and plot P/V versus P. The slope is $1/V_m$, and the intercept is $1/(bV_m)$.

12.7.3 Brunauer, Emmett, Teller or BET Isotherm

The BET isotherm[10] can be used to describe all types of isotherms. Two forms of the isotherm are required, one neglecting capillary condensation and one accounting for capillary condensation.

12.7.3.1 Adsorption without Capillary Condensation

The BET theory assumes that the forces active in capillary condensation are similar to those active in ordinary condensation. It is also assumed that the heat of adsorption on each layer on the layers above the first is equal to the heat of condensation. The theory accounts for multilayered adsorption, and will reduce to the Langmuir mono-molecular layer equation in the limit of one layer. Equation 12.8 represents the BET isotherm in this case.

$$\frac{V}{V_m} = \frac{cx}{1 - x}\left[\frac{1 + nx^{n+1} - (n+1)x^n}{1 + (c-1)x - cx^{n+1}}\right] \tag{12.8}$$

where:

- n is the number of layers adsorbed
- E_L is the heat of condensation
- R is the gas constant
- T is the absolute temperature
- C_E is a constant related to condensation and evaporation
- c is a second constant related to the energy of condensation on the monomolecular layer and on the subsequent layers
- x is a function of pressure

In this case, x is related to pressure, temperature, and the energy of condensation.

$$x = \left(\frac{P}{C_E}\right) e^{(E_L/RT)} \tag{12.9}$$

For a monomolecular layer, $n = 1$, Equation 12.8 reduces to

$$\frac{V}{V_m} = \frac{cx}{1+cx} \tag{12.10}$$

Equation 12.6 is the Langmuir isotherm with $c = b$. When n approaches an infinite number of layers, Equation 12.8 becomes

$$\frac{V}{V_m} = \frac{cx}{(1+x)(1-x+cx)} \tag{12.11}$$

At this condition, the pressure is approaching the saturation pressure or vapor pressure of the gas, and $x \to 1.0$ or

$$x = \frac{(P/C_E)e^{E_L/RT}}{(P^o/C_E)e^{E_L/RT}} = \frac{P}{P^o} \tag{12.12}$$

where:
 P^o is the vapor pressure

Equation 12.11 can be rearranged under these conditions to give

$$\frac{x}{V(1-x)} = \frac{1}{V_m c} + \frac{(c-1)x}{V_m c} \tag{12.13}$$

Recall that $x = P/P^o$, and plot data in the form of $x/[V(1-x)]$ versus x. The intercept will then be $1/(V_m c)$, and the slope will be $(c-1)/(V_m c)$.

The method of fitting data from the plot suggested above is as follows:

1. Use Equation 12.11 up to $P/P^o = 0.35$, where it will begin to deviate from most data. Find V_m and c as suggested above.
2. Use Equation 12.8 with this value of V_m and c, and find n by trial and error.

12.7.3.2 Adsorption with Capillary Condensation

Some isotherms suggest that a complete or almost complete filling of the pores and capillaries of the adsorbent occurs at a pressure lower than the vapor pressure of the gas. This lowering of the vapor pressure indicates that as the pressure of the gas increases, an additional force appears to make the energy of binding in some upper layer to be greater than E_L, the heat of liquefaction of the gas. This extra energy is due to the last adsorbed layer, which is attracted to both sides of the capillary with an additional energy of binding.

Assume that $2n - 1$ layers build up in the capillary, then

$$\frac{V}{V_m} = \left[\frac{1+\left[1/(2nC_E)-n\right]x^{n-1}-\left(nC_E-n+1\right)x^n+\left(1/2nC_E\right)x^{n+1}}{1+(c-1)x+\left(1/2cC_E-c\right)x^n-\left(1/2cC_E\right)x^{n+1}}\right] \quad (12.14)$$

When the capillary forces are small, the binding energy is small, and $C_E \rightarrow 1.0$. Then Equation 12.14 reduces to Equation 12.11.

12.8 POLANYI POTENTIAL THEORY

The Polanyi potential theory states that the free-energy change in passing from the gaseous to the liquid state is a suitable criterion of the free-energy change for a gas passing to the adsorbed state.[11] The adsorption potential for a mole of material is given by

$$\varepsilon = RT\ln\left(f^{sat}/f_i\right) \quad (12.15)$$

This potential is a function of the volume of the adsorbed phase. The potential does change appreciably with temperature. For a given adsorbent and for a class of adsorbates, the following two parameters are plotted against each

$$\text{Ordinate} = 100\left[\frac{w_i}{(\rho_{L,i})}\right]$$

$$\text{Abscissa} = \left[\frac{T(\rho_{L,i})}{1.8\text{ MW B}}\right]\log\left(\frac{f^{sat}}{f_i}\right)$$

w_i is the weight of i adsorbed lb/lb
T is the temperature °R
$\rho_{L,i}$ is the liquid molar density at the normal boiling point
(f^{sat}/f_i) is the ratio of the saturation fugacity to the fugacity of component i

At low total pressure, this ratio can be replaced by the ratio of the vapor pressure to the partial pressure.

MW is the molecular weight of i
B is the affinity coefficient for use at temperatures above critical

Figure 12.3 from Grant and Manes[12] is a Polanyi potential theory equilibrium plot for adsorption of normal paraffins on BPL-activated carbon. The following example calculation is based on Figure 12.3.

12.8.1 Hexane Example of the Polanyi Potential Theory

For 1000 ppm hexane in air at 80°F and 1.0 atm, determine the amount of hexane adsorbed on BPL-activated carbon using the Polanyi potential theory. From the equilibrium plot of Figure 12.3 for BPL carbon

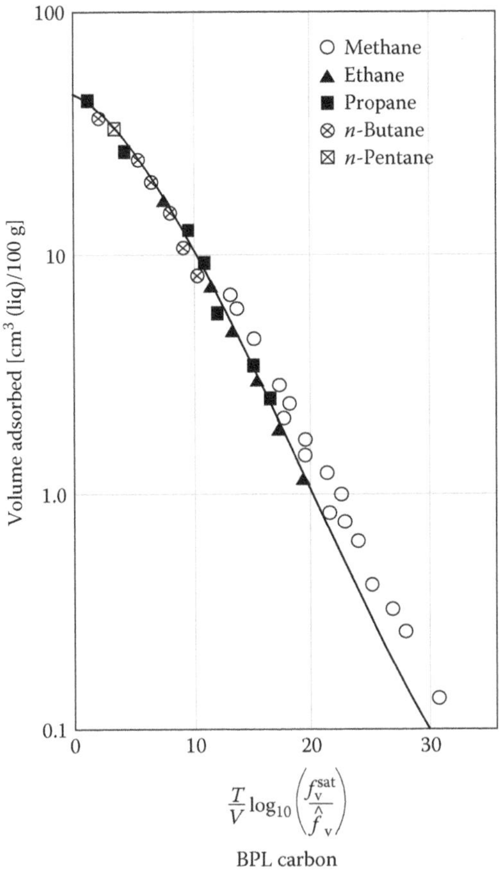

FIGURE 12.3 Polanyi potential theory equilibrium plot for adsorption of normal paraffins on BPL-activated carbon. (Reprinted with permission from Grant, R. J. and Manes, M., *Ind. Eng. Chem. Fundam.*, 3(3), 221–224. Copyright 1964 American Chemical Society.)

The absicca is

$$X = \frac{T}{V'}\log_{10}\left(\frac{f_V^{sat}}{\hat{f}_V}\right)$$

where:

$$\text{Specific gravity} = 0.66 \text{ and atomic weight} = 86$$

$$\rho_{hx}(\text{density of hexane}) = 0.66 \times 1.0 \text{ gm/cm}^3 = 0.66 \text{ gm/cm}^3$$

$$T = (\text{absolute temperature}) = 299.82 \text{ K}$$

$$V'(\text{specific volume}) = \frac{86 \text{ gm/gm-mole}}{0.66 \text{ gm/cm}^3} = 130.30 \frac{\text{cm}^3}{\text{gm-mole}}$$

At low pressures, the fugacities may be calculated as follows:

$$\hat{f}_V = \bar{p}_V \text{ (the partial pressure of the gas)} = \frac{1000}{1{,}000{,}000} \times 14.696 \frac{\text{psi}}{\text{atm}} = 0.0147 \text{ psi}$$

$$f_V^{sat} = p_V^{sat} \text{ (vapor pressure of the gas at the given temperature}$$

$$\text{for normal hexane at } 80^\circ\text{F}) = 3.0 \text{ psi}$$

$$X = \frac{T}{V'} \log_{10}\left(\frac{f_V^{sat}}{\hat{f}_V}\right) = \frac{299.82}{130.30} \log_{10}\left(\frac{3.0}{0.0147}\right)$$

$$X = 5.31$$

From Figure 12.3,

$$Y = 25.0 \frac{\text{cm}^3\text{ (liq.)}}{100 \text{ gm carbon}}$$

$$Y = \frac{0.66 \text{ gm/cm}^3 \times 25 \text{cm}^3}{100 \text{ gm carbon}} = \frac{16.5 \text{ gm hexane}}{100 \text{ gm carbon}}$$

Refer to Figure 12.2 of hexane on BPL carbon. At the partial pressure of 0.0147 and 80°F the capacity weight percent = 20 lbs hexane/100 lbs carbon. This is more than the above calculation, which is probably within the accuracy of the Polanyi potential theory. However, note that hexane is not included in the Polanyi plot of Figure 12.3, which is all for lower atomic weight organic molecules.

12.9 UNSTEADY-STATE, FIXED-BED ADSORBERS

In fixed-bed adsorbers, the fluid is passed continuously over the adsorbent, initially free of adsorbate. At first, the adsorbent contacts a strong solution entering the bed. Initially, the adsorbate is removed by the first portion of the bed and nearly all the solute is removed from the solution before it passes over the remaining part of the bed. Figure 12.4 illustrates the conditions for a downflow situation. In part (a) the effluent is nearly solute free. The uppermost part of the bed becomes saturated, and the bulk of the adsorption takes place over a relatively narrow portion of the bed in which concentration changes rapidly. This narrow adsorption zone moves down the bed as a concentration wave, at a rate much slower than the linear velocity of the fluid through the bed. As time progresses, the concentration of the solute in the effluent increases. When the effluent solute concentration reaches a predetermined value, set by emission standards, for example, the breakthrough point is reached. The solute concentration in the effluent now rises rapidly as the adsorption zone passes out the end of the bed, and the solute concentration in the effluent essentially reaches the initial concentration. The concentration volume of effluent curve in this portion is known as the *breakthrough curve*.

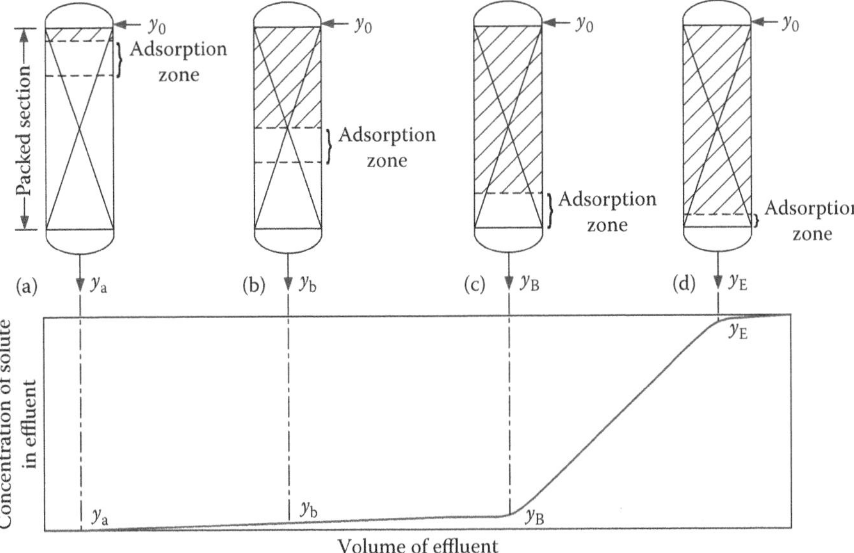

FIGURE 12.4 (a–d) Adsorption wave passing down through activated-carbon, bed-breakpoint curve. Changes in the location of the adsorption zone and the concentration of solute in the effluent from the time that the adsorber is first brought on-line to the time where regeneration is required are illustrated.

If a vapor is being adsorbed adiabatically from a gas mixture, the evolution of the heat of adsorption causes a temperature wave to flow through the bed similar to the adsorption wave. The rise in temperature of the effluent stream may be used to predict the breakpoint.

The shape and time of appearance of the breakthrough curve greatly influences the method of operating a fixed-bed absorber. The actual rate and mechanism of the adsorption process, the nature of the adsorption equilibrium, the fluid velocity, the concentration of the solute in the field, and the bed depth contribute to the shape of the curve produced for any system. The breakpoint is very sharply defined in some cases and poorly defined in others. Generally, the breakpoint time decreases with decreased bed height, increased particle size of adsorbent, increased rate of flow of fluid through the bed, and increased solute concentration in the feed. The design consists, in part, by determining the breakpoint curve.

12.10 FIXED-BED ADSORBER DESIGN CONSIDERATIONS

There are three main types of fixed-bed adsorbers: canisters, modular, and regenerative.

- *Canisters:* Canisters are easily installed and are returned for refilling with new adsorbent. Usual flow rates are about 100 cfm through 200 lb of adsorbent. Canisters are typically used in control of emissions from pesticide manufacture and to remove methyl and diethyl amine.

- *Modular:* Modular adsorbers remain in place. The adsorbent is replaced when the bed becomes saturated rather than regenerate the adsorbent on site. Service life is as much as three months service before replacement is required. Flow rates from 400 to 1100 cfm with beds from 2 to 4 ft in depth. With a 5-ft depth, flow rates of up to 10,000 cfm can be sustained
- *Regenerative:* With high flow rates and VOC concentrations up to 2000 to 10,000 ppm, it may be more economical to regenerate the adsorbent in place. Furthermore, exit concentrations in regenerative adsorbers as low as 50 to 100 ppm have been attained. Superficial velocities through the bed from 45 to 100 ft/min are common. As mentioned above, the relative humidity needs to be lowered to prevent interference with the adsorption process by the water molecules. Reduction to below 50% relative humidity is recommended, not only for regenerative beds but also for canisters and modular-type beds as well.

12.10.1 Safety Considerations

Especially with activated carbon, there is danger of fire due to the heat of adsorption. Ketones are especially bad. Cyclohexanone will oxidize, polymerize, and decompose with the result being high heat generation. Some activated carbons will liberate heat by reaction with oxygen. Thus in many cases adequate air flow is needed to keep the bed cool, or heat exchange coils may need to be installed in the bed.

Several safety measures listed below are often employed in adsorber design and operation to mitigate the concerns associated with high heats of adsorption that can lead to carbon bed fires.

1. Fixed bed adsorber cycles are optimally scheduled to minimize any bed idle time between the adsorption and regeneration cycles. A carbon bed fully loaded with recovered organics can more easily ignite if low air flow or no air flow is allowed
2. It has been observed that the outermost layer of carbon (e.g., the last 6 in. of a 36 in. deep carbon bed) that is the last layer exposed to the entering VOC laden air during the adsorption cycle should not be reactivated, especially if ketones are being adsorbed. This layer is still the *most active* layer within the full bed depth and should now become the first bed layer that is exposed to the VOC laden air when the carbon bed is reloaded with reactivated and/or virgin carbon. This provides an initial layer of carbon, that will be exposed to the VOC laden air, that is less active, adsorbs less ketones and, thus, generates less heat of adsorption. Virgin and reactivated carbon has the highest initial potential for adsorption, resulting in high heats of adsorption, and is thus more susceptible to bed fires. Any virgin carbon added to the adsorber should be the last layer loaded. Using this first *dead layer* of carbon approach can be useful for mitigating the effect of rapid temperature increases resulting from high heats of adsorption rates.
3. Continuous thermocouple strands can be routed through the adsorber in order to monitor any sudden unintended temperature rise in the adsorber

during the adsorption cycle. Thermocouple probes are often inserted at numerous fixed locations within the carbon bed; however, this has been ineffective since carbon is a good insulator. Essentially, a rapid temperature associated with a hot spot within an activated carbon bed has to occur within a couple of feet of a fixed thermocouple probe for a chemical operator to be able to mitigate the hot spot prior to it resulting in a carbon bed fire.

4. Many adsorbers include a nitrogen purge system that can provide an inert environment within the adsorber during idle cycles or no flow events. This inert nitrogen blanket will reduce the oxygen content in the adsorber below the level necessary to support combustion.
5. Adsorbers are often designed with a water flood spray system that can fill the adsorber quickly in order to rapidly quench a bed fire, if this occurs.

12.11 PRESSURE DROP THROUGH ADSORBERS

The Ergun[13] equation was designed to estimate pressure drop through granular solids and is, therefore, a reasonable method to determine pressure drop through adsorbers. In this equation, the pressure drop is determined as a function of the bed depth, the internal void volume, the particle diameter, the bed superficial velocity, the gas density, and a Reynolds number based on the gas properties and the particle diameter.

$$\left(\frac{\Delta P}{x}\right)\left(\frac{g_c d_p}{\rho_G V_G^2}\right)\left(\frac{\epsilon^3}{1-\epsilon}\right)=\frac{150(1-\epsilon)}{N_{Re}}+1.75 \tag{12.16}$$

where:

ΔP is the pressure drop in lbf/ft^2
x is the bed depth in feet
g_c is the gravitational constant = 32.174 ft-lbm/lbf-s^2
d_p is the particle diameter in ft^3
ρ_G is the gas density in lbm/ft^3
V_G is the superficial velocity in ft/s
ϵ is the internal void fraction

The Reynolds number is

$$N_{Re}=\frac{d_p V_G \rho_G}{\mu_G}$$

where:

μ_G is the viscosity in lbm/ft/s

12.11.1 Pressure Drop Example

An air stream at 77°F and 1.0 atm contains toluene at about 5000 ppm by volume. The stream is flowing at 20,000 acfm and must be treated by adsorption to remove the toluene down to the limit in which it may be emitted to the atmosphere. It has been determined that an adsorbent bed of activated carbon 12 by 6 ft and 3 ft deep of mesh size 4/6 will be suitable to reach the desired emission concentration. Determined the pressure drop through the bed using Equation 12.16.

$$\left(\frac{\Delta P}{x}\right)\left(\frac{g_c d_p}{\rho_g V_g^2}\right)\left(\frac{\epsilon^3}{1-\epsilon}\right)=\frac{150(1-\epsilon)}{N_{Re}}+1.75$$

The properties of 4/6 mesh activated carbon are as follows:

$$\text{Effective particle diameter } d_p = 0.0128 \text{ ft}$$

$$\text{Internal void fraction } \epsilon = 0.34$$

Here, properties of the gas are assumed to be the same as the properties of air at the same temperature and pressure.

$$\text{Density } \rho_g = 0.0740 \text{ lbm/ft}^3$$

$$\text{Viscosity } \mu_G = 0.0436 \text{ lbm/h/ft}$$

Additional data:

$$\text{Bed depth } x = 3 \text{ ft}$$

$$\text{Superficial velocity} = V_G = 20{,}000 \text{ ft}^3/\text{min} \times \frac{1}{10 \text{ ft} \times 20 \text{ ft}} \times \frac{60 \text{ min}}{\text{h}} = 6000 \text{ ft/h}$$

$$V_G = 20{,}000 \text{ ft}^3/\text{min} \times \frac{1}{10 \text{ ft} \times 20 \text{ ft}} \times \frac{1 \text{ min}}{60 \text{ s}} = 1.6667 \text{ ft/s}$$

$$N_{Re} = \frac{d_p V_G \rho_g}{\mu_G}$$

$$N_{Re} = \frac{(0.0128 \text{ ft})(6000 \text{ ft/h})(0.074 \text{ lbm/ft}^3)}{(0.0436 \text{ lbm/h/ft})} = 130.3486$$

$$\frac{\Delta P}{3 \text{ ft}} \frac{(32.174 \text{ ft/lbm/lbf/s}^2)(0.0128 \text{ ft})}{(0.0740 \text{ lbm/ft}^3)(1.6667 \text{ ft/s})^2}\left(\frac{0.34^3}{1-0.34}\right)=\frac{150(1-0.34)}{130.3486}+1.75$$

$$\Delta P = 63.10 \text{ lbf/ft}^2$$

12.12 ADSORBER EFFECTIVENESS, REGENERATION, AND REACTIVATION

Adsorption systems function as constant outlet concentration devices. Furthermore, the outlet concentration from an adsorber is a function of the amount of adsorbate left on the adsorbent after regeneration. This remaining quantity is called the *heel.*

The heel continues to build up until regeneration will no longer be effective in restoring the adsorptive surface. Until breakthrough, the outlet concentration is controlled by the amount of heel left on the adsorbent. The more vigorous the regeneration, the lower the heel that remains and the lower the outlet concentration will be before rising to breakthrough. Thus, the effectiveness of an adsorber is also controlled by the regeneration process.

The common methods of regenerating adsorptive beds are with steam or hot air. If the compound being removed is a valuable condensable organic material, regeneration by steam is usually followed by condensation and separation of the organic layer from the condensed steam layer (water) in a decanter. The organic layer can then be recovered for use again in the original process.

12.12.1 Steam Regeneration

The quantity, temperature, and pressure of the steam required for regeneration are dependent on the required outlet concentration and how much material is to be removed from the adsorption bed. The steam used must

1. Raise the bed to the regeneration temperature.
2. Provide the heat of desorption.
3. Act as a carrier gas for the desorbed product. (This is the major portion of the steam flow—60%–70%.)

For a solvent-recovery process, average steam requirement is in the range of 0.25–0.35 lb steam/lb carbon to achieve an adsorber outlet concentration of 70 ppm. For an adsorber operating at an outlet concentration in the 10–12 ppm range, steam requirements will be more in the range of 1.0 lb steam/lb carbon. Fair[15] recommends 3.0–5.0 lb steam/lb adsorbent. Peak steam demand at 2–2.5 times the average will occur in the first 10–15 min of the cycle. Figure 12.5 illustrates a fixed-bed VOC steam regeneration system with recovery of the organic material.

12.12.2 Hot Air or Gas Regeneration

Heated air or hot off gases from an incinerator are often used for regeneration. Figure 12.6 shows a schematic of a hot-gas regenerative system. If the off gases from the regeneration process contain enough organic compounds, they can be used as fuel to the incinerator. Additional fuel may have to be added to the incinerator to raise the temperature of the off gases from the incinerator to that required for regeneration. The incinerator might be a part of another waste gas cleanup process as well.

The use of steam or hot gases for regeneration may result in requiring that the adsorption bed be cooled before it is used again for adsorption. In addition, if steam is used for the regeneration, the bed will have to be dried as well as cooled. Therefore, time must be allowed for regeneration, cooling, and drying. Drying and cooling a bed with air may take no more than 15 min. Regeneration could be carried out by atmospheric air. However, it is not likely that the ambient temperature will provide the vigorous regeneration conditions needed.

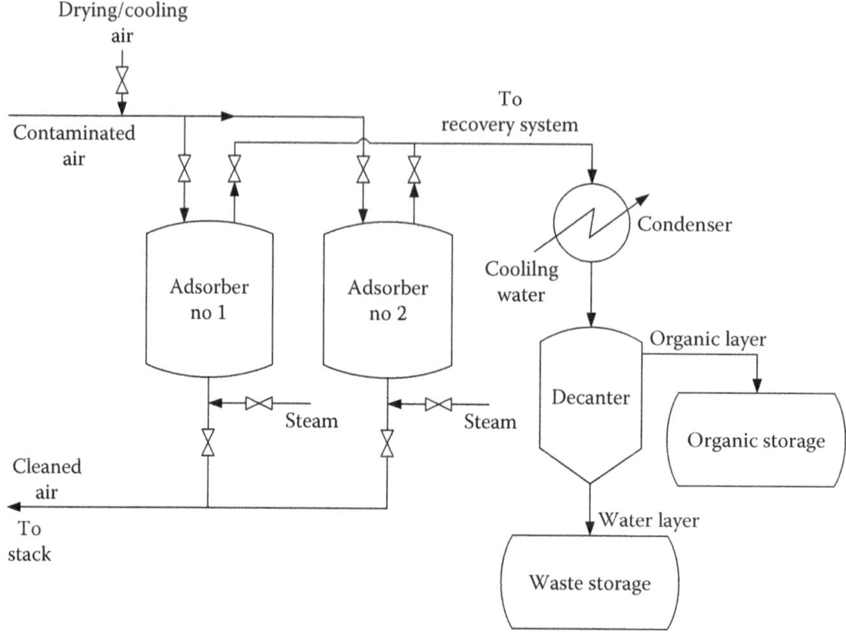

FIGURE 12.5 Fixed-bed activated carbon volatile organic recovery system.

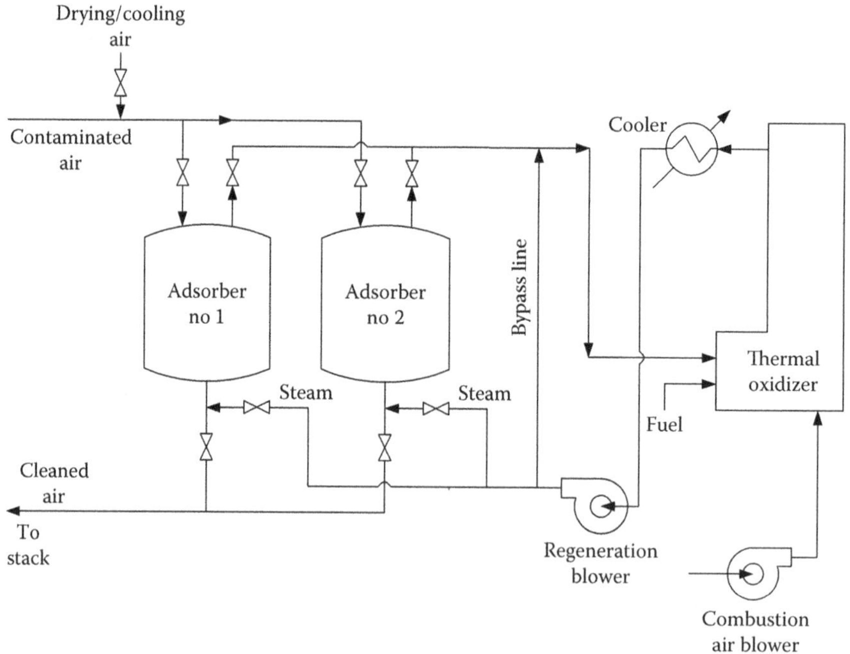

FIGURE 12.6 Fixed-bed activated carbon with hot air regeneration from a thermal oxidizer.

In order to provide more time for regeneration and cooling and to make use of the entire bed of adsorbent in the fixed-bed adsorber, a three-bed system might be used. Figure 12.7 shows a schematic of this kind of system. Operation of the system is illustrated by the breakthrough curves of Figure 12.8. Bed 1 is operated discharging to the atmosphere until breakthrough is encountered. Then the effluent from Bed 1 is fed to Bed 2. Bed 1 continues to operate until it reaches a predetermined effluent concentration near saturation. At this time, Bed 1 is put into regeneration and cooling. Bed 2 continues to operate to the breakthrough point. Then the effluent from Bed 2 is fed to Bed 3. Bed 2 continues to operate until it reaches the predetermined

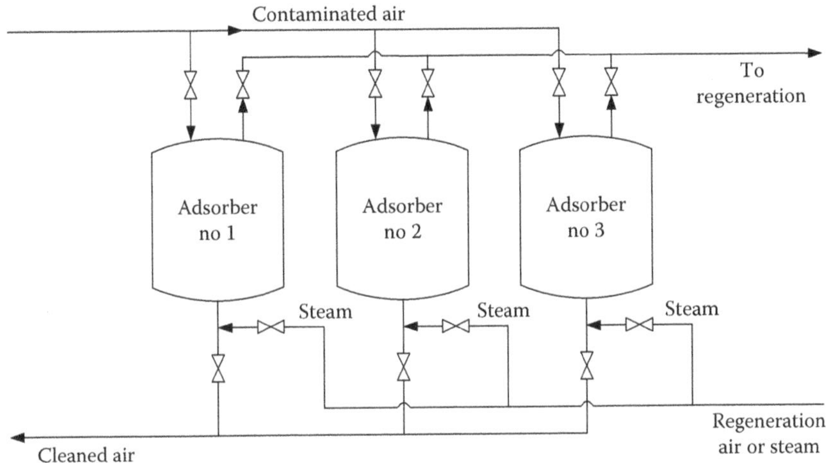

FIGURE 12.7 Three-fixed-bed activated carbon adsorption system.

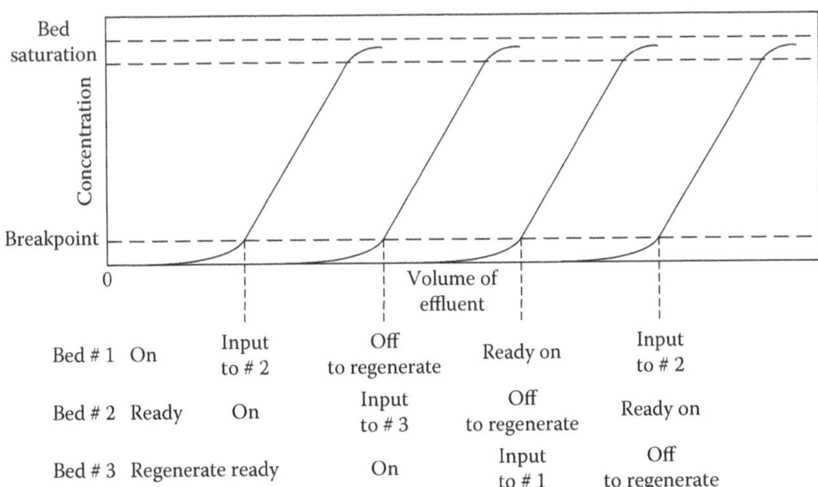

FIGURE 12.8 Three-fixed-bed activated carbon adsorption system cycle operation breakthrough curves.

effluent concentration near saturation. Now it is time for Bed 2 to be regenerated and cooled. Bed 3 continues to operate until it reaches the breakpoint. Bed 1 must be designed to be ready to receive Bed 3 effluent at this time. The process is then repeated until shutdown for replacement of the adsorbent is required.

12.12.3 Reactivation

In general, activated carbon adsorbents can be regenerated for a period of 6–12 months. The adsorbent gradually loses is effectiveness over this period of time and this degradation will be observed as higher breakthrough levels sooner in the adsorption cycle (see Section 12.13 for an explanation of breakthrough). At some point in the life of an activated carbon adsorbent, its effectiveness is determined to be unacceptable to achieve the recovery efficiencies desired. At that point, the activated carbon is generally either sent off for disposal to a chemically secure landfill or thermally reactivated. Thermal reactivation involves heating the activated carbon in a kiln to high temperatures often greater than 1500°F. This thermal reactivation allows the carbon to reused for another adsorption period of 6–12 months. Thermal reactivation at these high temperatures in kilns usually results in approximately a 10% loss of carbon (generation of carbon fines) and this loss is typically made up with virgin activated carbon.

12.13 BREAKTHROUGH MODEL

For a cyclic, fixed-bed process, breakthrough relationships determine the size of the adsorber. Hougen and Marshall[14] proposed a series of partial differential equation models to calculate the time-position-temperature-concentration conditions of dilute gases flowing through granular beds. The solutions to the partial differential equations are limited to isothermal conditions with a linear adsorption equilibrium relationship. The solutions are valid for single-component adsorption where mass transfer through the gas film is the controlling step. The fluid velocity is not a function of the direction of flow, and the velocity profile is uniform across the cross section of the column. Furthermore, there is no longitudinal diffusion or axial mixing in the gas stream where the pressure is assumed constant. Graphical solutions of the equations were presented. Later Fair[15] further developed this use of the model and graphical methods. More recently Ufrecht et al.[16] presented solutions of the equations through digital computer techniques using a standard numerical package. A computer model may be easily developed using Mathematica©, a mathematical programming language, which will readily integrate the Bessel function solution to the model equations. The solutions to the partial differential equations presented by Hougen and Marshall in dimensionless form are as follows:

$$\frac{y}{y_o} = 1 - e^{-b\tau} \int_0^{ax} e^{-\alpha} I_0\left(2\sqrt{b\tau\alpha}\right) d\alpha \tag{12.17}$$

$$\frac{w}{w_o} = e^{-ax} \int_0^{b\tau} e^{-\beta} I_0\left(2\sqrt{ax\beta}\right) d\beta \tag{12.18}$$

where:

$\alpha = ax$ is a dummy variable proportional to bed depth
$\beta = b\tau$ is a dummy variable proportional to adsorption time

The model variables and parameters are defined as follows:

x is the bed depth in feet
τ is the time of bed operation in hours
y is the adsorbate content of fluid stream at τ
w is the adsorbate content of solid at τ

Adsorbent characteristics

D_p is the effective particle diameter in feet
A_v is the external area of solid particles in ft^2/ft^3
ρ_B is the bulk density in lb/ft^3

Gas to be treated

G is the mass velocity in $lb/h/ft^2$
μ_G is the viscosity in lb/h/ft
ρ_G is the density lb/ft^3

Model parameters

$a = 1/H_d$ per feet
$b = G_c/_B H_d$ per feet
H_d is the height of a mass transfer unit in feet
$c = y_o/w_o$
y_o is the inlet adsorbate content of fluid stream lb/lb of fluid on an adsorbate-free basis
w_o is the adsorbate content of solid in equilibrium with y_o lb/lb of fluid on an adsorbate-free basis

Data for various adsorbents as reported by Fair is shown in Table 12.2.

12.13.1 Mass Transfer

Mass transfer is described by the height of a transfer unit method. The height of an overall mass transfer unit H_{OG} is given by

$$H_{OG} = \frac{\bar{G}}{K_o a_v M_G} \tag{12.19}$$

where:

$\bar{G}$ is the gas mass velocity based on total cross section
K_o is the overall mass transfer coefficient based on gas concentration
a_v is the external surface of packing material
M_G is the molecular weight of gas

Correlations are based on the Chilton-Colburn j_d factor, which is a function of the Schmidt number N_{Sc}, where

$$j_d = \left[\frac{K_o M_G}{\bar{G}}\right] N_{Sc}^{2/3} \tag{12.20}$$

$$N_{sc} = \frac{\mu_G}{\rho_G D_G} \tag{12.21}$$

μ_G is the gas viscosity
ρ_G is the gas density
D_G is the diffusion coefficient for gas

Redefining the height of an overall gas phase transfer unit in adsorption to be H_d,

$$H_d = \frac{N_{Sc}^{2/3}}{j_d a_v} \tag{12.22}$$

Fair[15] reports several correlations for the j_d mass transfer factor as a function of the Reynolds number N_{Re}. For large packings,

$$j_d = 1.30\left(N_{Re}\right)^{-0.45} \tag{12.23}$$

$$N_{Re} = \frac{d_p \rho_G V_G}{\mu_G} \tag{12.24}$$

where:
d_p is the effective particle diameter
V_G is the superficial gas velocity

However, this correlation produces mass-transfer rates, which may be two to three times greater than is realistic. Fair[15] then presents data in his paper, which he states is more appropriate for the kind of adsorbent found in the usual adsorbent bed used for removal of pollutants. The following equations are deduced from the data in Fair's paper and are a function of the mesh size of the adsorbent.

For 4/6 mesh adsorbent

$$j_d = 0.2189\left(N_{Re}\right)^{-0.2739}$$

For 6/8 mesh adsorbent

$$j_d = 0.1820\left(N_{Re}\right)^{-0.2762}$$

For 8/10 mesh adsorbent

$$j_d = 0.1424\left(N_{Re}\right)^{-0.2704}$$

For 20/28 mesh adsorbent

$$j_d = 0.1404\left(N_{Re}\right)^{-0.3144}$$

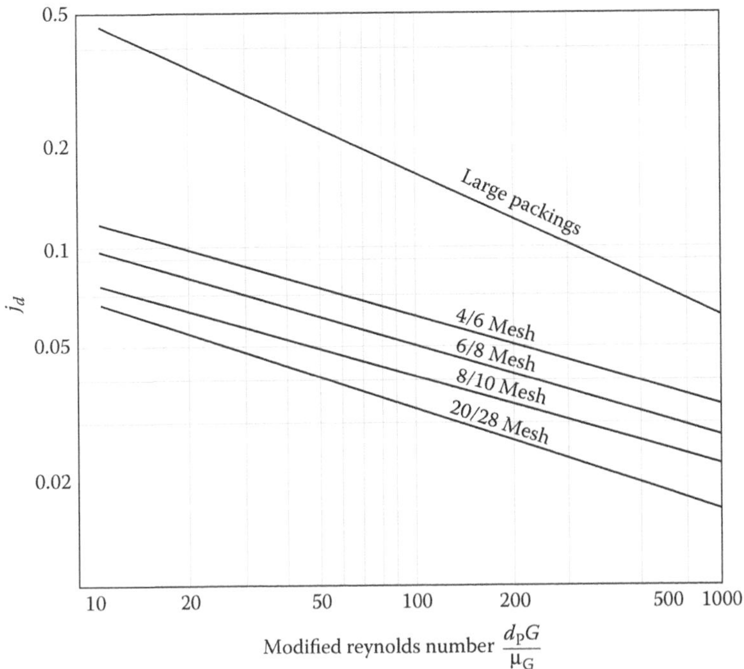

FIGURE 12.9 Mass transfer of gases through granular solids. *Note:* Figure abscissa should be $d_p G/\mu_G$.

The four equations above for the mass transfer factor j_d are plotted in Figure 12.9, which illustrates the large difference in the factor for large particles and the adsorbent size normally used in adsorption columns.

12.13.2 Breakthrough Curve Example

Adsorption of n*-hexane from natural gas*

Normal hexane is to be removed from natural gas (molecular weight 17.85) at 815 psia and 94°F by adsorption on silica gel. The entering gas contains 0.853 mol% hexane and has a superficial velocity of 11.4 ft/min. The bed is 43 in. deep and contains 3/8 mesh silica gel. The equilibrium concentration has been measured to be 0.17 lb hexane/lb of gel. Breakthrough data for this system have been reported by McCleod and Campbell.[17]

Hexane–Natural Gas Measured Breakthrough Data

Time (in min)	y/y_0	$b\tau$
17	0.02	15.87
21	0.04	19.60

(*Continued*)

Hexane–Natural Gas Measured Breakthrough Data

Time (in min)	y/y_0	$b\tau$
23.5	0.150	21.93
26.2	0.305	24.45
30	0.510	28.00
33	0.770	30.80
35.5	0.895	33.13
38.5	0.965	35.63
42	0.980	39.20

The breakthrough model can be tested with this data.

Physical properties:
Temperature = 94°F
Pressure = 1.0 atm
Hexane:
$D_G = 0.0057$ ft²/h (diffusivity in air)
Equilibrium data:
measured = 0.17 lbs hexane/lb silica gel
estimated = 0.25 lbs hexane/lb silica gel
Natural gas:
$\rho_G = 2.73$ lbm/ft³
$\mu_G = 0.029$ lbm/ft/h
Silica gel: 3/8 mesh
$\rho_B = 52$ lb/ft³
$d_p = 0.010$ ft
$a_v = 284$ ft²/ft³

Summary of Variable and Parameter Values

x is the bed depth in feet = 3.58 ft

τ is the time of bed operation in hours

y is the adsorbate content of fluid stream, lb/lb of fluid on an adsorbate-free basis at τ

w is the adsorbate content of solid, lb/lb of solid on an adsorbate-free basis at τ

y_o is the inlet adsorbate content of fluid stream, lb/lb = 0.415 of fluid on an adsorbate-free basis

w_o is the adsorbate content of solid in equilibrium with $y_o = 0.17, 0.25$ lb/lb of solid on an adsorbate-free basis

Adsorbent Characteristics

d_p is the effective particle diameter in feet = 0/010 ft

a_v is the external area of solid particles in ft²/ft³ = 284 ft²/ft³

ρ_B is the bulk density in lb/ft³ = 52 lb/ft³

Gas to Be Treated

V_G is the superficial velocity in ft/h = 684 ft/h
G is the mass velocity in lb/h/ft^2 = 1867 lbm/h/ft^2
μ_G is the viscosity in lb/h/ft = 0.029 lbm/ft/h
ρ_G is the density in lb/ft^3 = 2.73 lbm/ft^3
D_G is the diffusivity of vapor in gas in ft^2/h = 0.0057 ft^2/h

Mass-Transfer Correlation for Height of a Transfer Unit

$N_{Re} = \dfrac{d_p \rho_G V_G}{\mu_G}$, Reynolds number (unitless) = 644

j_d is the Chilton-Colburn analogy factor for mass transfer = 0.035

$N_{Sc} = \dfrac{\mu_G}{\rho_G D_G}$, Schmidt number (unitless) = 1.86

$$H_d = \frac{(N_{Sc})^{2/3}}{j_d a_v} = 0.1529\,\text{ft}$$

Parameters

H_d is the height of a mass transfer unit in feet = 0.1529 ft
$c = y_o/w_o = 0.24$
$a = 1/H_d$ per feet = 6.54/ft
$b = \dfrac{G_c}{\rho_B H_d}$ per hour = 56/h

The model was programmed using Mathematica. The resultant curve along with the measured data is plotted in Figure 12.10. Shown are curves for the measured equilibrium adsorption, the estimated equilibrium adsorption, and results for equilibrium adsorption amounts 10% and 20% greater than the measured value.

12.13.3 Second Breakthrough Curve Example: Hexane Problem

An off gas flowing at 10,000 acfm and containing 1.02% (mole%) of hexane is to be treated to reduce the concentration by 90% of the original concentration. The gas is flowing at a superficial velocity of 100 ft/min at 1.0 atm and 94°F. Activated carbon in an 8 × 10 mesh size is to be used in a fixed bed adsorption process to remove the hexane. For bed depths of 1, 2, 3 and 4 ft determine the breakthrough times and the amount of carbon required.

The *breakthrough model*, which follows from Equation 12.17, will be used to solve this problem. This model was used previously to compare breakthrough data to measured data for hexane being removed from natural gas. This case is an example of using the model to essentially design an absorber for removal of a pollutant gas from an air stream. In the statement above, four bed depths have been selected for the design. Time to recover 90% of the pollutant is determined. The process design

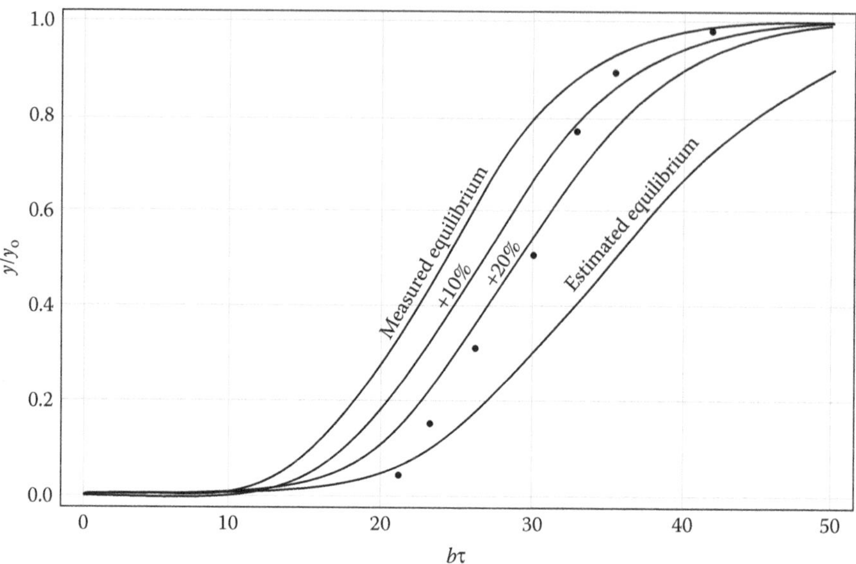

FIGURE 12.10 Adsorption of hexane from natural gas. (•, measured data).

could then be extended to determine how much time would be required to saturate the bed and then to clean it. A process such as discussed in Section 12.9 could then be designed. As in the example before Mathematica will be used to find the solution. The following information is prepared to use in the Mathematica solution.

$$\frac{y}{y_0} = 1 - e^{-b\tau} \int_0^{\alpha x} e^{-\alpha} I_0\left(2\sqrt{b\tau\alpha}\right) d\alpha$$

I_o is a Bessel function of the first kind and zeroth order

The parameters of this model to use in the Mathematica code are as follows:

α is a dummy variable = ax, the thickness modulus and is dimensionless and x is the bed depth

$b\tau$ is a dummy variable the time modulus and is dimensionless and t is the time. In the solution by Mathematica $b\tau$ is plotted versus y/y_o, which is calculated below to be

$$y/y_o = 0.099$$

Definition of variable and parameter values

x is the bed depth in feet = 1,2,3, and 4 ft

τ is the time of bed operation in hours—to be calculated

y is the adsorbate content of fluid stream, lb/lb of fluid on an adsorbate-free basis

$$\alpha t{\cdot}t = b\tau/b$$

90% Removal

Outlet stream mole fraction = 0.10 × 0.0102 = 0.00102

$$y = \frac{0.00102 \times 86}{0.9898 \times 29} = 3.028 \times 10^{-3} \text{ lbs hexane/lb air}$$

y_0 is the inlet adsorbate content of fluid stream, lb/lb = 0.415 of fluid on an adsorbate-free basis

$$y_0 = \frac{0.00102 \times 86}{0.9898 \times 29} = 3.056 \times 10^{-2} \text{ lbs hexane / lb air}$$

$$\frac{y}{y_0} = \frac{3.028 \times 10^{-3}}{3.056 \times 10^{-2}} = 0.099$$

w_0 is the adsorbate content of solid in equilibrium with y_0, 0.26 lb/lb of adsorbate free solid. From Figure 12.2 with P = 1.0 atm = 14.696 psia,

$$\overline{P} = 0.0102 \times 14.696 = 0.1499 \text{ psia}$$

Gas to Be Treated

Physical properties:
Temperature = 94°F
Pressure = 1.0 atm

Air
$\rho_G = 0.0717$ lbm/ft^3
$\mu_G = 0.046$ lbm/h/ft

Hexane
$\rho_G = 0.21314$ lbm/ft^3
D_G (diffusivity in air) = 0.2906 ft^2/h
Equilibrium data measured = 0.26 lbs hexane/lb of 8 × 10 carbon mesh

Superficial velocity
V_G = 100 ft/min × 60 min/h = 6000 ft/h

Absorbent characteristics
Activated carbon 8 × 10 mesh

$$\rho_B = 30 \text{ lb/ft}^3$$

$$d_p = 0.0064 \text{ ft} \qquad a_v = 645 \text{ ft}^2/\text{ft}^3$$

Mass-transfer correlation for height of a transfer unit
Average density of gas
Basis: 100 moles

	Moles	Atomic Weight	Mass (in lbs)	Mass (%)
Hexane	1.02	86.17	87.99	0.02974
Air	98.98	29	2820.42	0.97026
	100.00		2958.41	1.00000

$$\rho_{avg} = 0.97026 \times 0.0717 + 0.02974 \times 0.21324 = 0.07591\,/\,\text{lbm}\,/\,\text{ft}^3$$

Reynolds number

$$N_{Re} = \frac{D_p \rho_G V_G}{\mu_G} = \frac{0.0064\,\text{ft} \times 0.07591\,\text{lbm}/\text{ft}^3 \times 6000\,\text{ft}/\text{h}}{0.046\,\text{lbm}/\text{h}/\text{ft}} = 63.37$$

j_d is the Chilton-Colburn analogy factor for mass transfer from Figure 12.9 = 0.048

Schmidt number

$$N_{Sc} = \frac{\mu_G}{\rho_G D_G} = \frac{0.046\,\text{lbm}/\text{h}/\text{ft}}{0.07591\,\text{lbm}/\text{ft}^3 \times 0.2906\,\text{ft}^2/\text{h}} = 2.09$$

Parameters

H_d is the height of a mass transfer unit

$$H_d = \frac{(N_{Sc})^{2/3}}{j_d a_v} = \frac{2.09^{2/3}}{0.048 \times 645} = 5.28 \times 10^{-2}\,\text{ft}$$

$$c = y_o / w_o = 3.056 \times 10^{-2} / 0.26 = 0.1175$$

$$a = 1/H_d \text{ per feet} = 1/5.48 \times 10^{-2}\,\text{ft} = 18.94/\text{ft}$$

In the breathrough model calculation

X (in ft)	ax = α
1	18.94
2	37.88
3	56.82
4	75.76

Mass velocity

$$G = \rho_G V_G = 0.07591\ \text{lbm}/\text{ft}^3 \times 6000\ \text{ft}/\text{h} = 455.46\ \text{lb}/\text{h}/\text{ft}^2$$

$$b = \frac{Gc}{\rho_B H_d} = \frac{(455.46 \times 0.1175)\,\text{lbm}/\text{ft}^2/\text{h}}{30\ \text{lbm}/\text{ft}^3 \times 5.28 \times 10^{-2}\ \text{ft}} = 33.79\ \text{h}$$

For 10,000 acfm to be treated, calculate the area of the adsorption bed.

$$A_B = \text{area of bed in ft}^2$$

$$Q = A_b V_G$$

$$Q = 10{,}000 \text{ acfm}$$

$$A_b = \frac{Q}{V_G} = \frac{10{,}000 \text{ ft}^3/\text{min}}{100 \text{ ft/min}} = 100 \text{ ft}^2$$

$$\text{Activated carbon} = 30 \text{ lbs/ft}^3$$

Summary of results:

For $x = 1.0$ ft, $a = 18.94$/ft and $b = 33.79$/h remains constant. The resultant curves based on this data are plotted in Figure 12.11. Then $b\tau$ can be read from the curves.

X (ft)	ax	$b\tau$ at $y/y_o = 0.099$	T (h)	Bed Volume (ft³)	Carbon Required (in lbm)
1	18.94	11	0.326	100	3000
2	37.88	26	0.769	200	6000
3	56.82	42	1.243	300	9000
4	75.76	58	1.716	400	12,000

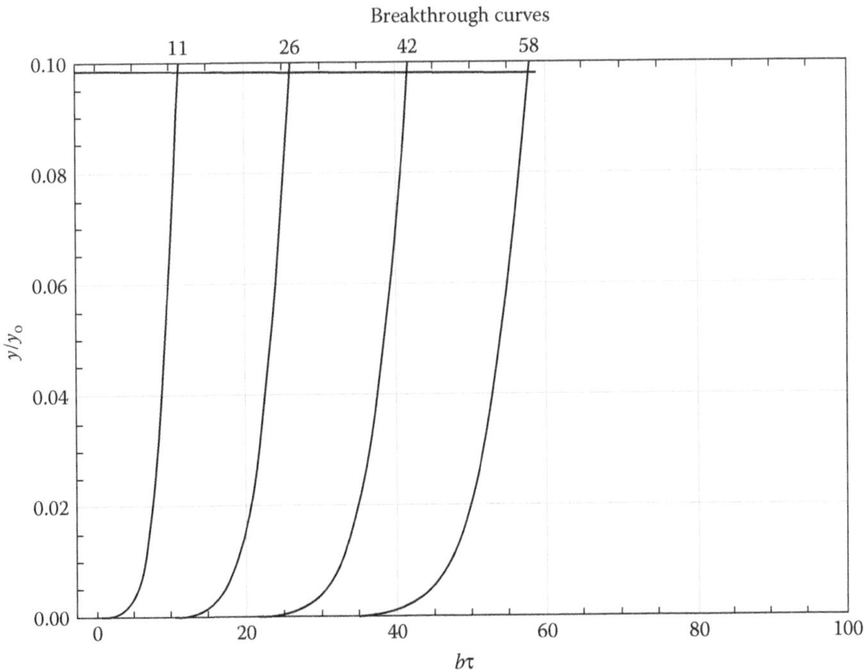

FIGURE 12.11 Adsorption of hexane from air to determine time to reach 90% removal of hexane.

12.14 REGENERATION MODELING

Fair[15] suggests that the design problem in hot-gas regeneration is the determination of how long it takes the bed temperature to approach the required regenerating gas temperature. This is essentially a problem in direct-contact heat transfer. The total heat required is the sum of the heat absorbed by the vessel and any appurtances, the heat absorbed by the bed, and the heat of desorption. The heat of desorption is equivalent to the heat of adsorption. The rate of heat addition must be determined. Vessel heat-up time should include estimated heat losses to the surroundings. Bed heat-up time includes the time for desorption. An equation similar to the breakpoint model could be used to estimate the temperature breakthrough.

After desorption is completed, the bed must be cooled back to an acceptable adsorption temperature. If steam is used, time must be allowed to dry the bed as well.

12.14.1 Steam Regeneration Example

This example is based on Section 12.13.3.

The bed depth of 4 ft will be used. This bed contains 12,000 lbm of activated carbon 8 × 10 mesh size.

Gas flow: 10,000 acfm × 0.07591 lbm/ft^3 = 759.1 lbm/min of gas

Mass percent hexane = 0.02974, 90% adsorbed in 1.716 h

Hexane flow = 759.1 × 0.02974 = 22.58 lbs hexane/min

Hexane adsorbed = 0.90 × 22.58 = 20.32 lbs/min

An average value of steam requirement for regeneration from LeVan and Schweiger[18] is suggested to be

$$\frac{0.30 \text{ lbs steam}}{\text{lb carbon}}$$

Therefore, from the 12,000 lbs of carbon

$$\frac{0.30 \text{ lbs steam}}{\text{lb carbon}} \times 12{,}000 \text{ lbm carbon} = 3600 \text{ lbs steam to regenerate}$$

Another method would be to consider the equilibrium concentration of hexane with carbon which is from the breakthrough curve example w_o = 0.26 lbs hexane/lb carbon.

$$\frac{0.30 \text{ lbs steam}}{\text{lb carbon}} \times \frac{1}{0.26 \text{ lbs hexane/lb carbon}} = 1.15 \text{lbs steam/lb hexane}$$

Then the steam needed for regeneration would be:

Hexane adsorbed:

$$20.32 \text{ lbs/min} \times 60 \text{ min/h} \times 1.716 \text{ h} \times 1.15 \text{ lbs steam/lb hexane}$$

$$= 2406 \text{ lbs of steam to regenerate}$$

LeVan and Schweiger[18] present a more detailed and refined method to model steam regeneration. It should also be noted that the heat of capacity of the bed and the vessel holding the bed of adsorbate need to be considered.

12.15 USING MASS EXCHANGE NETWORK CONCEPTS TO SIMULTANEOUSLY EVALUATE MULTIPLE MASS-SEPARATING AGENT (ABSORBENT AND ADSORBENT) OPTIONS

Systems-based methodologies also exist that allow the simultaneous evaluation of multiple adsorbents that could be used for the removal of contaminants from gaseous emission streams. Numerous adsorbents can be employed for the separation task. The task of simultaneous evaluation of these adsorbents uses the composition-interval-diagram and mass exchange network methodologies, previously discussed in Section 11.5.10. In fact, the mass exchange network methodology can be used to simultaneously screen both solid *adsorbents* and liquid *absorbents* for the separation task since both types of mass separating agents are direct-contact mass transfer operations that driven by a composition difference between and equilibrium line and an operating line. Synthesis of adsorption-based mass exchange networks are covered in detail in El-Halwagi's textbook.[19]

Of particular importance to the economics of the adsorption system design is the inclusion of the effectiveness of the regeneration process along with its impact on the operating cost of the system. El-Halwagi and Manousiouthakis have developed a methodology to simultaneous screen multiple potential adsorbents along with their regeneration networks and this approach can be employed during the simultaneous analysis of multiple adsorbent candidates.[20]

13 Thermal Oxidation for Volatile Organic Compounds Control

Volatile organic compounds (VOCs) generally are fuels that are easily combustible. Through combustion, which is synonymous with thermal oxidation and incineration, the organic compounds are oxidized to CO_2 and water, while trace elements such as sulfur and chlorine are oxidized to species such as SO_2 and HCl.

Three combustion processes that control vapor emissions by destroying collected vapors to prevent release to the environment are (a) thermal oxidation—flares, (b) thermal oxidation and incineration, and (c) catalytic oxidation. Each of these processes has unique advantages and disadvantages that require consideration for proper application. For example, flares are designed for infrequent, large volumes of concentrated hydrocarbon emissions, while thermal oxidizers are designed for high-efficiency treatment of continuous, mixed-hydrocarbon gas streams, and catalytic oxidizers are designed to minimize fuel costs for continuous, low concentration emissions of known composition. The design of the basic processes can be modified for specific applications, resulting in the overlap of the distinctions between processes. For example, ground flares are basically thermal oxidizers without heat recovery that frequently are used for intermittent flow of relatively low volumes of concentrated VOC streams.

13.1 COMBUSTION BASICS

As every Boy Scout, Girl Scout, and firefighter knows, combustion requires the three legs of the fire triangle illustrated in Figure 13.1. The oxidizer and fuel composition, that is, air-to-fuel ratio, is critical to combustion. If the fuel concentration in air is below the lower flammability limit, also known as the *lower explosive limit* (LEL), the mixture will be too lean to burn. If it is above the upper flammability limit, it will be too rich to burn. Fuels with a wide range of flammability limits burn more easily than those with a narrow range. With a narrow range, the flame is more unstable since the interior of the flame can easily be starved for air.

The heating value of the fuel—the amount of heat released by the combustion process—is determined by the heat of combustion and the concentration of the hydrocarbons in the gas stream. Values for the heat of combustion for common organic compounds are provided in Table 13.1. The heat of combustion is the same as the heat of reaction for the oxidation reaction, and therefore can be calculated from the heats of formation of the reactants and products. It is the net chemical energy that is released by the oxidation reaction when the reactants begin at 25°C and after the reaction products are cooled to 25°C. That the reactants are first heated

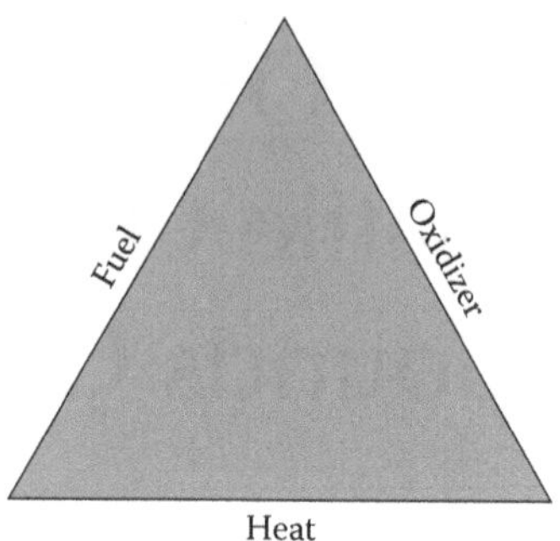

FIGURE 13.1 Fire triangle.

TABLE 13.1
Heat of Combustion for Various Compounds

Compound	Lower Heating Value (BTU/lb)
Acetaldehyde	10,854
Acetone	12,593
Acetylene	20,776
Ammonia	7,992
Benzene	17,446
Butane	19,697
Carbon monoxide	4,347
Chlorobenzene	11,772
Chloroform	1,836
Cyclohexane	18,818
Dichloroethane	4,990
Ethane	20,432
Ethanol	12,022
Ethylbenzene	17,779
Ethylene	20,295
Ethylene dichloride	5,221
Ethylene glycol	7,758
Formaldehyde	7,603
Heptane	19,443
Hexane	19,468
Hydrogen	51,623
Hydrogen sulfide	6,545
Methane	21,520
Methanol	9,168
Methyl ethyl ketone	13,671
Methylene chloride	2,264
Naphthalene	16,708
Octane	19,227
Pentane	19,517

(*Continued*)

TABLE 13.1 (*Continued*)
Heat of Combustion for Various Compounds

Compound	Lower Heating Value (BTU/lb)
Phenol	13,688
Propane	19,944
Propylene	19,691
Styrene	17,664
Toluene	17,681
Trichloroethane	3,682
Trichloroethylene	3,235
Vinyl chloride	8,136
Xylene	17,760

to the ignition temperature and the exhaust gases are hot does not affect the value for the heat of combustion, because the value includes the energy recovered by cooling the exhaust gases. Indeed, the *higher heating value* includes the energy recovered when water vapor is condensed to liquid at 25°C, while the lower heating value is based on water remaining in the gaseous state.

The flame temperature is determined by a heat balance including the energy produced by combustion, absorbed by the reactant gases, released to the exhaust gases, and lost to the surroundings by radiation. Therefore, factors such as the combustion air temperature, composition of the exhaust gases, and configuration of the combustion chamber affect the peak flame temperature.

Despite exposure to flame in the presence of oxygen, not all of a hydrocarbon pollutant will react. The destruction efficiency of VOC pollutants by combustion depends on the three Ts: temperature (typically 1200°F–2000°F), time (typically 0.2–2.0 s at high temperature), and turbulence. The required destruction efficiency often is expressed as 9s. Two 9s is 99% destruction efficiency, and five 9s is 99.999% destruction efficiency. Some VOCs burn easily and do not require extremely high destruction efficiency. Others, especially chlorinated hydrocarbons, do not burn as easily, and the required high destruction efficiency demands a good combination of high temperatures, adequate residence time at high temperature, and turbulence to promote mixing for good combustion of the entire gas stream. Table 13.2 lists the relative destructibility for some common VOCs.

13.2 FLARES

Flaring is a combustion process in which VOCs are piped to a remote location and burned in either an open or an enclosed flame. Flares can be used to control a wide variety of flammable VOC streams, and can handle large fluctuations in VOC concentration, flow rate, and heating value. The primary advantage of flares is that they have a very high turndown ratio and rapid turndown response. With this feature, they can be used for sudden and unexpected large and concentrated flow of hydrocarbons such as safety-valve discharges, as well as venting-process upsets, off-spec product, or waste streams.

TABLE 13.2
Relative Destructability of VOC Pollutants by Combustion

VOC	Relative Destructability
Alcohols	High
Aldehydes	↕
Aromatics	
Ketones	
Acetates	
Alkanes	
Chlorinated hydrocarbons	Low

Flares cannot be used for dilute VOC streams, less than about 200 BTU/scf, without supplemental fuel because the open flame cannot be sustained. Adding supplemental fuel, such as natural gas or propane, increases operating cost. Flammable gas sensors can be used to regulate supplemental fuel.

13.2.1 Elevated, Open Flare

The commonly known flare is the elevated, open type. Elevated, open flares prevent potentially dangerous conditions at ground level by elevating the open flame above working areas to reduce the effects of noise, heat, smoke, and objectionable odors. The elevated flame burns freely in open air. A schematic of a simplified flow of an elevated, open-flare system is shown in Figure 13.2. The typical system consists of a header to collect waste gases, some form of assist to promote mixing (frequently steam

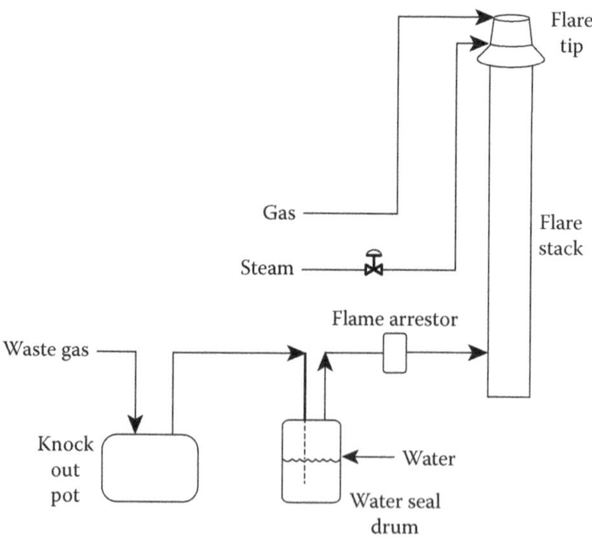

FIGURE 13.2 A schematic of a simplified flare.

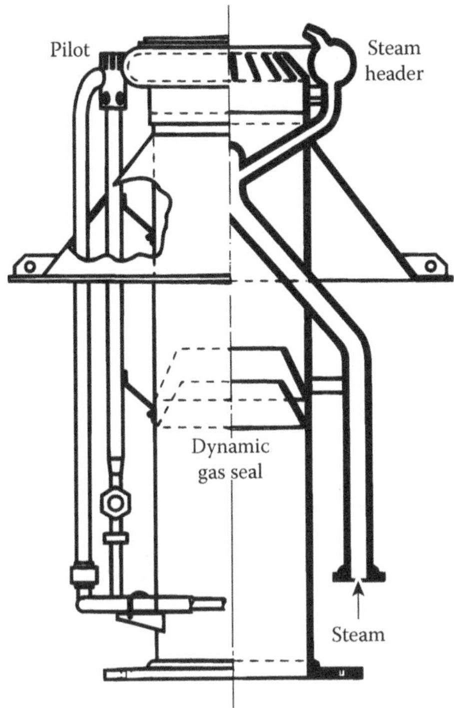

FIGURE 13.3 Steam-assisted smokeless flare tip. (Courtesy of Flare Industries, Austin, TX.)

is used), and an elevated burner tip with a pilot light. A typical burner tip is shown in Figure 13.3. Atmospheric combustion air is added by turbulence at the burner tip.

Although flares have a very high turndown velocity, exit velocity extremes determine the size of the flare tip. Maximum velocities of 60 ft/s and 400 ft/s are used for waste streams with heating values of 300 BTU/scf and 1000 BTU/scf, respectively, to prevent blowout of the flame. A correlation for maximum velocity with heating value is provided by Equation 13.1:

$$\log_{10}\left(V_{max}\right)=\frac{\left(B_v+1214\right)}{852} \tag{13.1}$$

where:

V_{max} is the maximum velocity, ft/s
B_v is the net heating value, BTU/scf

The design volumetric flow should give 80% of the maximum velocity.

13.2.2 Smokeless Flare Assist

Mixing and complete combustion can be improved at the flare tip by steam-assist, air-assist, or pressure-assist mechanisms. As shown in Figure 13.4, the supplemental assist can have a dramatic positive effect on preventing the production of black smoke.

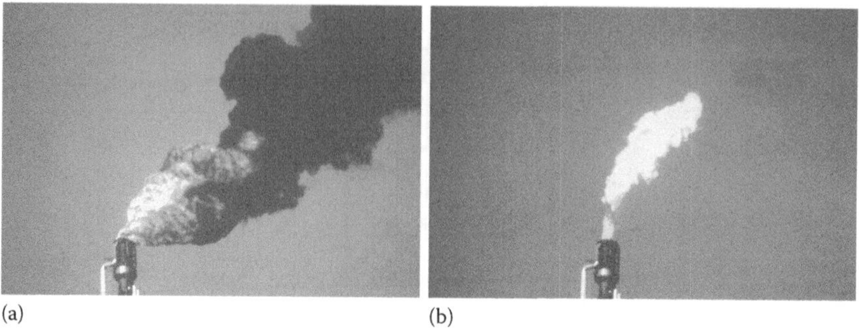

(a) (b)

FIGURE 13.4 Steam-assisted flare: (a) steam off and (b) steam on. (Courtesy of John Zink Company, LLC, Tulsa, OK.)

A large part of the effect can be attributed to turbulence that draws in combustion air. The water molecules in steam-assisted flare headers may contribute additional benefits. They may separate hydrocarbon molecules, which would prevent polymerization and formation of long-chained oxygenated compounds that burn at a reduced rate. And they may react directly with hot carbon particles through the water–gas reaction, forming CO, CO_2, and H_2 from soot.

Steam typically is added at a rate of 0.01–0.6 lb steam per lb of vented gas, depending on the carbon content of the flared gas. Typical refinery flares use about 0.25 lb steam per lb of vent gas, while many general VOC streams use about 0.4 lb steam per lb of vent gas. A useful correlation is 0.7 lb steam per lb of CO_2 in the flared gas.

Steam assist can produce a loud, high-frequency (above 355 Hz) jet noise in addition to the noise produced by combustion. Noise is reduced by using multiple small jets and by acoustical shrouding.

Air assist is accomplished by using a fan to blow air into an annulus around the flare gas stack center channel. The turbulent air is then mixed at the burner tip. Due to the fan power requirement, air assist is not economical for high gas volumes, but is useful where steam is not available.

Pressure assist relies on high pressure in the flare header and high pressure drop at the burner tip. This approach cannot be used with variable flow, greatly reducing the number of viable applications.

13.2.3 Flare Height

The required height of an elevated, open flare is determined primarily by limitation on thermal radiation exposure, although luminosity, noise, dispersion of combustion products, and dispersion of vented gases during flameout also are considerations. The maximum heat intensity for a very limited exposure period of 8 s is 1500–2000 BTU/h-ft^2. This may give one just enough time to seek shelter or quickly evacuate the area. Most flares are designed for extended exposure at a maximum heat intensity of 500 BTU/h-ft^2. The distance from the center of the flame to an exposed person is determined using Equation 13.2:

$$D^2 = \frac{\tau FR}{4\pi K} \tag{13.2}$$

where:

D is the distance from center of flame, in feet

τ is the fraction of radiated heat that is transmitted (assume 1.0, but could be less for smoky or foggy conditions)

F is the fraction of heat that is radiated, function of gas composition, burner diameter, and mixing (typical values are 0.1 for H_2 in a small burner to 0.3 for C_4H_{10} in a large burner)

R is the net heat release, BTU/h

K is the allowable radiation, BTU/h-ft^2

The distance from the center of the flame to an exposed person takes into account not only the height of the flare tip but also the length of the flame and the distortion of the flame in windy conditions. The length of the flame is determined by:

$$\log_{10} L = 0.457 \log_{10}(R) - 2.04 \tag{13.3}$$

where:

L is the flame length, in feet

Elevated flare stacks typically are supported in one of three ways: (1) self-supporting; (2) guy-wires; and (3) derrick. Self-supported stacks tend to be smaller, shorter stacks of about 30–100 ft, although stacks of 200 ft or more are possible, depending on soil conditions and the foundation design. Tall stacks can be supported more economically with the aid of guy-wires. Gas piping temperature fluctuations that cause expansion and contraction must be considered. A derrick structure is relatively expensive, but can be used to support the load of a very tall stack.

13.2.4 Ground Flare

It is possible to enclose a flare tip with a shroud and bring it down to ground level. In an enclosed ground flare, the burners are contained within an insulated shell. The shell reduces noise, luminosity, heat radiation, and provides wind protection. These devices also are known as once-through thermal oxidizers without heat recovery. This type of flare often is used for continuous-flow vent streams but can be used for intermittent or variable flow streams when used with turndown and startup/shutdown controls. A common application is vapor destruction at fuel loading terminals, where the vapor flow is intermittent, but predictable.

Enclosed ground flares provide more stable combustion conditions (temperature, residence time, and mixing) than open flares because combustion air addition and mixing is better controlled.

Maintenance is easier because the flare tip is more accessible. But a disadvantage is that ground flares cannot be used in an electrically classified area because it creates an ignition source at ground level.

Temperatures are generally controlled within the range of 1400°F–2000°F using air dampers. They may use single or multiple burner tips within a refractory-lined steel shell. Multiple burners allow the number of burners in use to be staged with the gas flow. Staging can be accomplished by using liquid seal diplegs at different depths or by using pressure switches and control valves.

A ground flare enclosure that contains multiple burner tips typically is sized for about 3–4 MBTU/h/ft^2 of open area within the refractory lining of the enclosure.[1] The height of the enclosure depends on the flame length, which is a function of a single burner size, rather than the total heat release. A typical height for 5 MBTU/h burner tips is about 32 ft.

13.2.5 Safety Features

Flashback protection must be provided to avoid fire or explosion in the flare header. Protection is provided by keeping oxygen out of the flare header using gas seals, water seals, and/or purge gas, and by using flame arrestors and actuated check valves.

Gas seals keep air from mixing with hydrocarbons in the vertical pipe of an elevated flare. Two types of gas seals, namely a dynamic seal and a density seal, are shown in Figure 13.5.

A density or molecular seal forces gas to travel both up and down to get through the seal, like a P-trap water seal, and high-density (high molecular weight) gas cannot rise through low-density gas in the top of the seal. A low purge flow of natural gas, less than 1 ft/s, ensures that the gas in the top of the seal is more buoyant than air, and can keep the oxygen concentration in the stack below 1% with winds up to 20 mph. Density seals are recommended in larger flares with tips greater than 36 in. diameter.[2]

A dynamic gas seal is designed to provide low resistance to upward flow and high resistance to air flowing downward. Natural gas can be used for purge flow at about 0.04 ft/s to keep the oxygen concentration in the flare stack below 6%. Nitrogen also can be used as purge gas, and eliminates the possibility of burn-back into the flare tip at low flow rates.

After high-temperature gas is flared, the stack is filled with hot gas that will shrink upon cooling, and that can tend to draw air into the stack. The purge flow compensates for the reduction in volume, and the required purge rate may be governed by the rate of cooling during this period.

Flame arrestors and liquid-seal drums also are used to prevent flashback into the flare header. Liquid-seal drums have the advantage of avoiding the potential for being plugged by any liquids that might collect and congeal in the system. In addition, they can be used as a back-pressure device to maintain positive pressure in the flare header. A disadvantage is the possibility of freezing if the liquid seal contains water. Steam coils can be used to heat the seal.

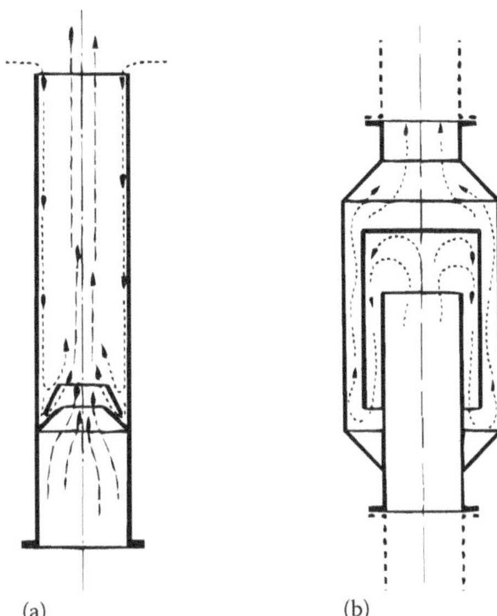

FIGURE 13.5 Types of gas seals: (a) dynamic seal and (b) density seal. (Courtesy of Flare Industries, Austin, TX.)

Hydrocarbon liquids must be kept out of flare stacks to prevent burning liquid droplets from being emitted from the stack. Knockout drums are used to separate and collect any liquid droplets larger than about 300–600 μm before gases are sent to the flare. They may be of either horizontal or vertical design. Generally, knockout drums are designed based on American Petroleum Institute Recommended Practices.[3]

13.3 INCINERATION

Incineration, sometimes referred to as *afterburning*, is used frequently to remove combustible air pollutants. It is a method to convert hazardous air pollutants (HAP) and VOC by oxidation to acceptable products for emission to the atmosphere. In the case of organic compounds containing chlorine, nitrogen, or sulfur, objectionable products of combustion might have to be removed by scrubbing (i.e., absorption) usually in a water stream before emission.

There are three major process classifications for incineration. Use of each of these processes is dictated by the type of source that is being treated.

Direct flame incineration: Used when the gas to be treated contains large quantities of highly combustible waste gases. If there are enough combustible materials, a waste heat boiler may be used to recover the energy. Frequently, in the case of refineries or petrochemical plants, the process takes place in a vertical, open-ended combustor called a *flare*.

Thermal incinerator: When the volume of gases is large and the level of combustible material is small, the material must be burned with an addition of fuel in a thermal incinerator. This is one of the best means to destroy organic odorous compounds and other offensive malodorous materials such as hydrogen sulfide, mercaptans, and cyanide gases. Organic aerosols that cause visible plumes may be emitted from places such as smoke houses, enamel baking ovens, and coffee roasters. These aerosols also can be controlled by thermal incineration.

Catalytic incineration: Another method to treat off gases low in combustible materials is through catalytic incineration. This process is similar to thermal incineration except that a catalyst is used to promote the reaction. Ignition temperatures are lowered requiring less fuel for combustion. Thus the cost of the catalyst is offset by the reduction in energy cost. The catalyst will also lower the time needed to complete the oxidation reaction, thus requiring smaller reaction chambers and reducing equipment cost.

All of the abovementioned processes are controlled by the three Ts of combustion—time, temperature, and turbulence (or mixing). The thermodynamics and the kinetics of the combustion reactions determine the temperature and residence time of the reactants in the combustion chamber. Proper turbulence results in improved combustion and is controlled by the manner in which the chamber is constructed, the flow rate of the gases, fuel (if required), and air and by the manner in which they are introduced into the chamber. The residence time determines the size of the chamber, and the temperature required determines the amount of auxiliary fuel that will need to be fired. At lower combustion temperatures, the reaction kinetics will control the process. However, as combustion temperatures go higher, the reaction will proceed faster. Eventually the reaction may become so fast that turbulent mixing will become the primary controlling process. However, higher temperatures will cause the formation of nitrogen oxides when excess air is available.

13.3.1 Direct Flame Incineration

A mixture of combustible gases and air within certain concentration limits is explosive. These limits are known as the *upper explosive limit* (UEL) and the LEL. The LEL is defined as the minimum concentration in air or oxygen above which the vapor burns upon contact with an ignition source and the flame spreads through the flammable gas mixture. Above the UEL, a flame will not propagate either. However, to transport gases commercially, the concentration in flammable mixtures is usually limited to 25% of the LEL to satisfy safety requirements imposed by insurance companies.

Direct flame incineration implies that a waste gas containing a combustible pollutant is to be burned directly in a combustor. In some cases, the pollutant-air mixture may not contain enough combustible material to burn, and natural gas is added to bring the mixture into the combustible range. In other cases, dilution air will be required to adjust the concentration within the flammability limits. Since the LEL will decrease with increasing temperature, gases that are below the LEL at their original temperature may be made combustible by heating them a few hundred degrees before entering the combustion chamber. For economic reasons, the combustible materials in an air stream must contribute a significant amount of the total energy for the air stream to be a candidate for direct flame incineration. Both the UEL and LEL limits can be found on the web in free encyclopedia Wikipedia, for example.

13.3.2 Thermal Incineration

Thermal incineration is a widely used air-pollution control technique that results in ultimate disposal of the pollutant compounds by converting them usually to H_2O and CO_2. The process is feasible when the concentration of combustible pollutants is so low that direct flame incineration would not be practical. Destruction efficiencies in the range of 99% are possible with streams of widely varying contaminants and concentrations. A typical thermal incineration process is shown below in Figure 13.6.

Typically, the waste gas stream is preheated by the discharge gases from the incinerator. The waste gases are then passed through the combustion zone of a burner supplied with supplemental fuel. The waste gases are brought above their autoignition temperatures. If the oxygen supply available to the stream is not sufficient, additional oxygen is added to the steam. The combustion products then pass through the heat exchanger, preheating the incoming waste gas. From this heat exchanger, the combustion products would go to an absorber or scrubber to remove any acid gases that might have formed from the combustion compounds such as sulfur and chlorine. A schematic of such a process follows.

Thermal incinerator design depends on residence time and combustion chamber temperature to set destruction efficiency. In addition, good mixing promoted by turbulence in the chamber, where oxygen, fuel, and combustible pollutants come into contact, is essential for proper operation. Complete mixing is especially important in odor control more so than in general hydrocarbon control. The Environmental Protection Agency suggests the following guidelines in Table 13.3 for permit evaluation.

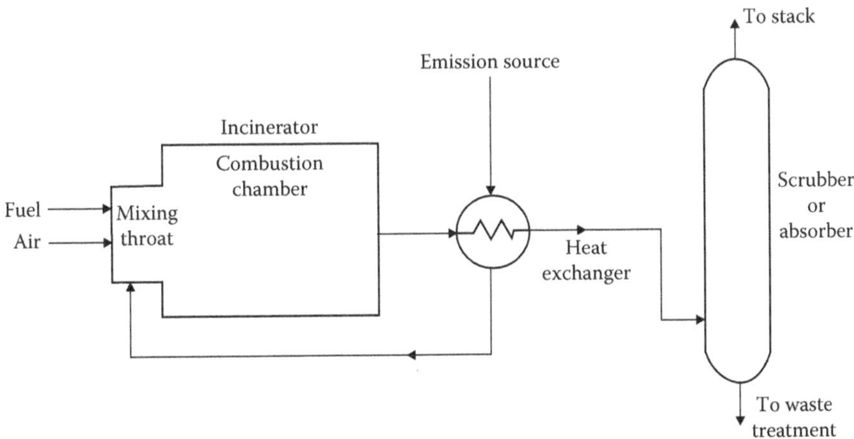

FIGURE 13.6 Typical thermal incineration process.

TABLE 13.3
Thermal Incinerator System Design Variables

	Nonhalogenated System		Halogenated System	
Required Destruction Efficiency DE (%)	**Combustion Temperature T_c (°F)**	**Residence Time t_r (s)**	**Combustion Temperature T_c (°F)**	**Residence Time t_r (s)**
98	1600	0.75	2000	1.0
99	1800	0.75	2000	1.0

TABLE 13.4
Average Temperature Range for Normal Operation

Material to Be Incinerated	Temperature (°F)
Hydrocarbons	950–1400
Carbon monoxide	1250–1450
Odor control	900–1300

These guidelines are general values designed to be effective with a variety of compounds and are conservatively high. Normal residence times are in the range of 0.2–0.8 s, with 0.5 s being as reasonable guideline. Gas velocities in the range of 15–30 ft/s in the throat regions of the incinerator will suffice to promote the desired degree of turbulence to mix the combustion products and pollutant gasses. Combustion chamber velocities in the range of 10–14 ft/s are sufficient in the combustion chamber. Temperature ranges to match these residence times for less severe operation are given in Table 13.4.

Typical operational problems that occur in thermal incinerators include low burning firing, rates, poor fuel atomization, poor air/fuel ratios, inadequate air supply, and quenching of the burner flame. The result of these problems is to lower the destruction efficiency for the pollutant material. A smoky effluent may result. An example calculation follows.

MACCOOKER Huge Meat Smoker

The smokehouse for all of the MACCOOKER (Figure 13.7) must install an afterburner to incinerate the contaminants discharged to eliminate the visible emissions and odors. The maximum rate of discharge is 1200 scfm (60°F and 1.0 atm) at 1.0 atm and 180°F. No heating value need to be assigned to the contaminants due to their low concentration. The afterburner is to be operated at 1200°F and will use natural gas with the following concentration as fuel:

The air enters at 70°F. The fuel enters at 60°F and is supplied with 20% excess air. Heat loss to the surroundings amounts to 10% of the energy released by combustion. Since the pollutant concentration is small in the inlet and exhaust gases, assume that these gases have the properties of air.

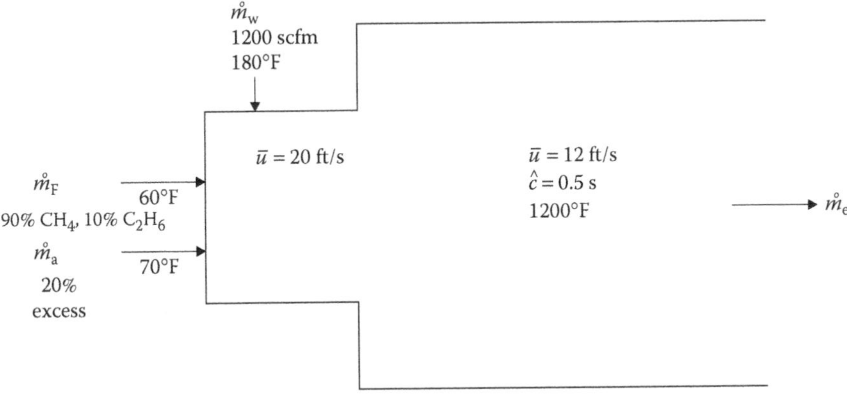

FIGURE 13.7 MACCOOKER huge meat smoker.

Component	Volume Percent
CH_4	90
C_2H_6	10

$$\text{Lower heating value} = 980\ \text{BTU/ft}^3$$

The calculation will be based on the mass and energy balance.

- Mass balance

$$\dot{m}_e = \dot{m}_w + \dot{m}_a + \dot{m}_f$$

- Energy balance

$$\dot{m}_e\left(h_{1200} - h_{600}\right)_e = \dot{m}_w\left(h_{180} - h_{60}\right)_w + \dot{m}_a\left(h_{70} - h_{60}\right)_e + Q + \dot{m}_f\,\Delta h_e$$

h_f is the enthalpy of air at temperature t
Q is the heat loss in BTU (= $-0.10\ \dot{m}_f \Delta h_e$)
Δh_c is the heat of combustion in BTU/lb or heating value of fuel

- Air requirements
  Combustion

$$\frac{CH_4}{\text{mw} = 16} + 2O_2 \rightarrow 2H_2O + CO_2$$

$$\frac{C_2H_6}{\text{mw} = 30} + \frac{7}{2}O_2 \rightarrow 3H_2O + 2CO_2$$

Molecular weight of fuel

$$0.9 \times 16 = 14.4$$

$$0.1 \times 30 = 3.0$$

$$\text{Sum} = 17.4$$

$$\frac{\text{Moles air}}{\text{Moles fuel}} = \frac{n_a}{n_f} \; 1.2\left(0.90 \times \frac{2.0}{0.21} + 0.10 \times \frac{3.5}{0.21}\right) = 12.2852$$

$$\frac{m_a}{m_f} = \frac{n_a \times 29}{n_f \times 17.4} = 20.4762 \frac{\text{lbm air}}{\text{lbm fuel}}$$

- Mass flow rate of contaminated gas

$$1200 \text{ scfm at } 60°\text{F}, \quad \rho_{air} = 0.0763 \text{ lbm/ft}^3$$

$$m_w = 1200 \text{ ft}^3\text{/min} \times 0.0763 \text{ lbm/ft}^3 = 91.56 \text{ lbm/min}$$

- Mass flow rate of exhaust gas

$$\dot{m}_e = 91.56 + \dot{m}_a + \frac{\dot{m}_a}{20.4762}$$

- Heating value of fuel at 60°F
  Fuel density by ideal gas

$$\rho_f = \frac{m}{V} = (\text{mw})\frac{P}{RT}$$

$$\rho_f = 17.4 \frac{\text{lbm}}{\text{mole}} \times \frac{14.696 \text{ psia}}{10.73 \frac{\text{ft}^3 \text{ psia}}{\text{mole °R}} \times 520 \text{ °R}} = 0.0458 \frac{\text{lbm}}{\text{ft}^3}$$

Heating value

$$\Delta h_c = 980 \frac{\text{BTU}}{\text{ft}^3} \frac{1}{0.04581 \text{ lbm/ft}^3} = 21{,}383.60 \frac{\text{BTU}}{\text{lbm}}$$

- Energy balance
  Enthalpies

$$h_{60} = 124.3 \text{ BTU/lbm} \quad h_{70} = 126.7 \text{ BTU/lbm}$$

$$h_{1200} = 411.8 \text{ BTU/lbm} \quad h_{1800} = 153.1 \text{ BTU/lbm}$$

$$\left(91.56 + \dot{m}_a + \frac{\dot{m}_a}{20.4762}\right)(411.8 - 124.3) = 91.56(153.4 - 124.3)$$

$$+ \dot{m}_a(126.7 - 124.3) + 0.90\left(\frac{\dot{m}_a}{20.4762}\right)21{,}383.60$$

$$\dot{m}_a = 37.0\ \text{lbm/min}$$

$$q_a = 37.0\ \text{lbm/min}/0.07631\ \text{ft}^3/\text{lbm} = 484.9\ \text{ft}^3/\text{min}$$

$$\dot{m}_f = \frac{37.0\ \text{lbm/min}}{20.4762\ \text{lbm air/lbm fuel}} = 1.81\ \text{lbm/min}$$

$$q_f = 1.81/0.0458 = 39.45\ \text{ft}^3/\text{min}$$

$$\dot{m}_e = 91.56\ \text{lbm/min} + 37.0\ \text{lbm/min} + 1.81\ \text{lbm/min} = 130.37\ \text{lbm/min}$$

$$\text{Density of air at } 1200°\text{F} = \rho = 0.0238\ \text{lbm/ft}^3$$

$$q_e = 130.37\ \text{lbm/min}/0.0238\ \text{lbm/ft}^3 = 5477.7\ \text{ft}^3/\text{min} = 91.30\ \text{ft}^3/\text{s}$$

- Throat diameter

$$\bar{u} = 20\ \text{ft/s}$$

$$q = A\bar{u}$$

$$A = \frac{91.3\ \text{ft}^3/\text{s}}{20\ \text{ft/s}} = 4.56\ \text{ft}^2$$

$$D = \sqrt{\frac{4A}{\Pi}} = 2.41\ \text{ft}$$

- Combustion chamber diameter

$$\bar{u} = 12\ \text{ft/s}$$

$$A = \frac{91.3\ \text{ft}^3/\text{s}}{12\ \text{ft/s}} = 7.61\ \text{ft}^2$$

$$D = \sqrt{\frac{4A}{\Pi}} = 3.11\ \text{ft}$$

- Combustion chamber length

$$\tau = \text{residence time} = 0.5\ \text{s}$$

$$L = \tau\bar{u}$$

$$L = 0.5\ \text{s} \times 12\ \text{ft/s} = 6\ \text{ft}$$

13.3.3 Catalytic Incineration

Catalytic incineration has the advantage of taking place at a lower temperature, thus reducing the call on outside energy to affect the oxidation. The catalyst bed is usually a metal mesh-mat, ceramic honeycomb, or other ceramic matrix structure designed to maximize catalyst area. Catalysts typically used for the process include platinum and palladium. Other formulations such as metal oxides such as chrome/alumina, cobalt oxide, and copper oxide/manganese oxide are used for emission streams containing chlorinated compounds. Recent advances in catalysts have developed catalysts that are more resistant to poisoning by sulfur compounds. There are platinum-based catalysts that can now be used for sulfur-containing HAP and VOC. However, these catalysts are still sensitive to chlorine. Atoms such as lead, arsenic, and phosphorus do not lend themselves well to catalytic oxidation.

A catalyst accelerates the rate of chemical reaction. This results in time requirements for catalytic action in the order of a few hundredths of a second. Thus, residence time has very little meaning as a means for design of catalytic incinerators. The parameter space velocity defined as the volumetric flow rate of the emission stream plus the supplemental fuel stream. The combustion air divided by the volume of the catalyst bed is the variable used to describe catalytic incineration. Typical destruction efficiencies are shown in Table 13.5.

Catalysts also lower the ignition temperature in the oxidation process. The temperature for reaction may be lowered by as much as 500°F. Table 13.4 indicates some typical catalytic incineration temperature ranges. This results in lower external energy requirements and lessens the cost of the process. Waste gases must be heated up to a temperature between 600 °F and 1000 °F before the catalyst will become effective. This is below the autoignition temperature of the gases for thermal incineration. However, the oxidation reaction will take place on the surface of the catalyst.

TABLE 13.5
Catalytic Incinerator System Design Variables

Required Destruction Efficiency DE (%)	Temperature at Catalyst Bed Inlet T_0 (°F)	Temperature at Catalyst Bed Outlet T_0 (°F)	Space Velocity SV (h^{-1})	
			Base Metal	Precious Metal
95	600	1000–1200	10,000–15,000	30,000–40,000
98–99	600	1000–1200		

Note: Minimum temperature of combined gas stream (emission stream + supplementary fuel combustion products) entering the catalyst bed is designated as 600°F ensure an adequate initial reaction rate.

Minimum temperature of the flue gas leaving the catalyst bed is designated as 1000°F ID ensure an adequate overall reaction rate ID achieve the required destruction efficiency. Note that this is a conservative value; it is in general a function of the HAP concentration (or heal content) and a temperature lower than 1000°F may be sufficient ID achieve the required destruction level. Maximum temperature of flue gas leaving lhe catalyst bed is limited ID 1200°F to prevent catalyst dea by overheating. However, base metal catalysis may degrade somewhat taster at these temperatures than precious metal catalysis.

- The space velocities given are designed to provide general guidance not definitive values. A given application may have space velocities that vary from these values. These values are quoted for monolithic catalysts. Pellet-type catalysts will typically have lower space velocities.
- In general, design of catalytic incinerator systems in this efficiency range is done relative to catalytic incinerator-specific process conditions.

The catalyst may glow, but there will be no observable flame structure. Efficiency of the order of 95%–98% is readily obtainable.

Design of a catalytic incinerator depends largely upon information from the catalyst manufacturer. Table 13.5 provides information on temperature ranges required for the indicated pollutants. Performance can be judged based on cost information from the manufacturer. The performance depends upon maintaining the temperature indicated by the manufacturer and the pressure drop across the bed. As the catalytic incinerator is used, bits of the catalyst will become entrained in the steam leaving the apparatus. The result will be a decrease in pressure drop across the catalyst bed. Therefore, both outlet temperature and pressure from the apparatus should be monitored. Furthermore, as operation proceeds catalysts will slowly deactivate. The catalyst will have to be replaced usually every two or three years.

13.3.4 Energy Recuperation in Incineration

Gases leaving an incinerator may vary in temperature from 700°F to 1000°F in a catalytic process to 2000°F in a direct flame process. Furthermore, these gases are relatively clean having just passed through the incinerator. Therefore, the gases are an excellent high-grade energy source. One of the most obvious ways to use the gases is to preheat the incoming contaminated gases. Figure 13.8 shows a schematic of such a recuperative process.

The efficiency of a process can be measured by the following Equation 13.4.

$$\eta_{\text{eff}} = \frac{T_B - T_A}{T_C - T_A} \tag{13.4}$$

The temperatures, T, are taken at positions indicated in Figure 13.8. The reduction in fuel costs must be weighed against an appreciable capital investment for the extra heat exchange equipment. The gases leaving the recuperative heat exchanger may still have enough energy (i.e., high enough temperature) to be used for additional heating assignments in other parts of the plant.

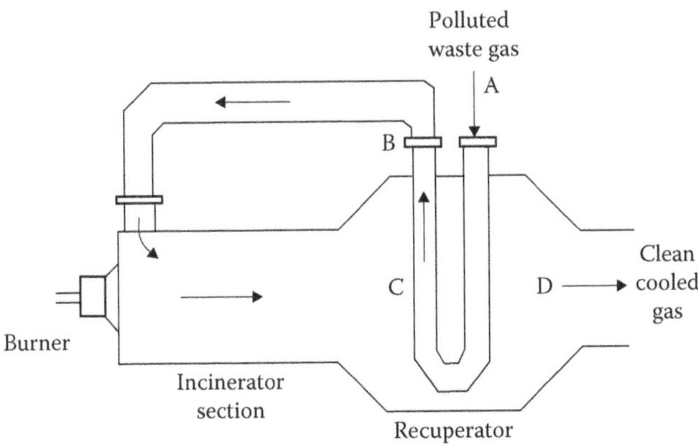

FIGURE 13.8 Energy recuperation in incineration process.

14 Control of Volatile Organic Compounds and Hazardous Air Pollutants by Condensation

14.1 INTRODUCTION

Condensation of a vapor from an air stream can take place as a film of the condensed material on the wall of the condenser tube or as a series of drops that form at various points on the surface. Film-type condensation is the more common mechanism encountered in a condenser. The film uniformly coats the surface, and the thickness of the film increases with the extent of the surface. In dropwise condensation, the surface is not uniformly covered. The individual drops form and grow on the surface and tend to coalesce with neighboring drops. Adhesion of the drops is then overcome by gravitational forces, and the coalesced drops run off the surface. Impurities in the vapor stream promote dropwise condensation, which results in higher heat transfer coefficients. Unfortunately, there is not much information available on dropwise condensation. Therefore, design methods are limited to the film-type case.

Condensers are best applied for removal of volatile organic compounds (VOC) and hazardous air pollutants (HAC) from emission streams when the concentration is greater than 5000 ppmv. Removal efficiencies range from 50% to 90%. The upper end of efficiencies is practically achievable for concentrations in the range of 10,000 ppmv or greater. With high concentrations of pollutant, condensers are frequently employed as preliminary air-pollution control devices prior to other devices such as incinerators, absorbers, and adsorbers. Flows up to 2000 scfm can be handled in condensers.

In condensation, one or more volatile components of a vapor mixture are separated from the remaining vapors through saturation followed by a phase change. The phase change from gas to liquid can be achieved by increasing the system pressure at a given temperature, or by lowering the temperature at a constant pressure. The lower the normal boiling point, the more volatile the compound, the more difficult to condense, and the lower the temperature required for condensation. Refrigeration must often be employed to obtain the low temperatures required for acceptable removal efficiencies.

A brief summary of condensation-based separations containing a general description, advantages, and disadvantages for the removal of volatile organic compounds from gaseous emission streams is provided below.[2–4]

14.1.1 Description

- VOC gaseous emissions are cooled below the stream dew point to condense the stream VOC.
- Cooling occurs in indirect-contact heat transfer equipment (i.e., shell and tube heat exchangers and finned heat exchangers).
- Cooling mediums are usually cooling water, chilled water, and refrigerants.
- The cooling medium is recycled, re-cooled, and reused for additional VOC condensation.
- Pressurization of the gaseous VOC stream is used to enhance condensation effectiveness.
- De-pressurization is used for energy recovery.
- *VOC-free* gas exits the condenser.

14.1.2 Advantages

- Can achieve moderate recovery efficiencies (50%–90%). Including stream pressurization can achieve higher recovery efficiencies (>90%) than conventional condensation processes.
- It is a simple process that does not require contacting the VOC gas stream with other streams (i.e., oils and activated carbons), thus minimizing contamination concerns.
- Efficiency improves as VOC concentration in inlet gas increases.
- Good for low volatility (high boiling point) VOCs.
- Pressurization improves effectiveness for high volatility (low boiling point) VOCs.

14.1.3 Disadvantages

- It may result in the generation of a wastewater stream.
- It may require inert gas blanketing if inlet gas exceed the upper explosive limit to eliminate explosion hazards.
- The liquid produced via condensation may require treatment for water removal or may require additional separation (typically distillation) if multiple VOCs are recovered.
- Cryogenic temperatures may be necessary and special equipment designs for these temperatures will be required.
- Condensation is typically used for low to moderate inlet gas flow rates (<20,000 cfm).
- Extensive cooling is required for low concentration VOC gaseous streams.
- Ice formation in heat transfer equipment may occur.

14.2 VOLATILE ORGANIC COMPOUNDS CONDENSERS

The two most common types of condensers used are surface and contact condensers. In surface condensers, the coolant does not contact the gas stream. Most surface condensers are the shell and tube type as shown in Figure 14.1. Shell and tube condensers

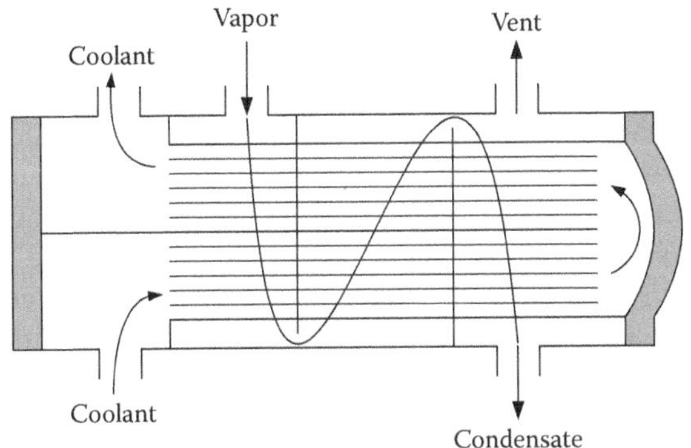

FIGURE 14.1 Schematic of shell- and tube-type surface condenser.

circulate the coolant through tubes. The VOCs condense on the outside of the tubes within the shell. Plate- and frame-type heat exchangers are also used as condensers in refrigerated systems. In these condensers, the coolant and the vapor flow separately over thin plates. In either design, the condensed vapor forms a film on the cooled surface and drains away to a collection tank for storage, reuse, or disposal.

In contrast to surface condensers, where the coolant does not contact with either the vapors or the condensate, contact condensers cool the volatile vapor stream by spraying either a liquid at ambient temperature or a chilled liquid directly into the gas stream. Spent coolant containing the VOCs from contact condensers usually cannot be reused directly and can be a waste disposal problem. Furthermore, spent coolant is then contaminated with the VOC, and therefore, must undergo further treatment before disposal.

14.2.1 Contact Condensers

In contact condensers, a coolant, frequently water, is sprayed into the gas stream. Condensation proceeds as a heat exchange process, where the air stream containing the condensable materials is first cooled to its condensation temperature, then loses its heat of condensation. The coolant first gives up its sensible heat, then its heat of vaporization. The balancing of the heat exchange between the two streams will determine the amount of coolant needed.

Design of contact condensers is based on the gas–liquid stage concept. However, spray systems operate with a high degree of back mixing of the phases. This practically limits spray chamber performance to a single equilibrium stage. For a direct contact device, this means that the temperatures of the exiting gas and liquid would be the same. Backmixing results because the chief resistance to flow is only the liquid drops. There is no degree of stabilization of the flow such as one that would happen in a packed tower. Anything less than perfect liquid distribution will induce large eddies and bypass streams. Thus, special care must be taken to obtain a uniform spray pattern.

14.2.2 Surface Condensers

In the shell and tube heat exchangers, the coolant typically flows through the tubes and the vapors condenser on the outside of the tubes. In these units, the pollutant gas stream must be cooled to the saturation temperature on the material being removed. The problem of design is complicated by the fact that most pollutant gas streams are essentially air with a small amount of VOC or HAP included. Therefore, condensation takes place from a gas in which the major component is noncondensable. In the case of a simple air stream, where the other component is condensable, condensation occurs at the dew point when the partial pressure of the condensable equals its vapor pressure at the temperature of the system. Since the coolant from surface condensers does not contact the vapor stream, it is not contaminated and can be recycled in a closed loop. Surface condensers also allow for direct recovery of VOCs from the volatile gas stream. This chapter addresses the design of surface condenser systems only.

Figure 14.2 shows some typical vapor pressure curves. The more volatile the component, that is, the lower the normal boiling point, the larger the amount that will remain uncondensed at a given temperature, hence the lower the temperature that is required to reach saturation. Condensation for this type of system typically occurs nonisothermally. The assumption of constant temperature conditions in the design of surface condensers does not introduce large errors into the calculations.

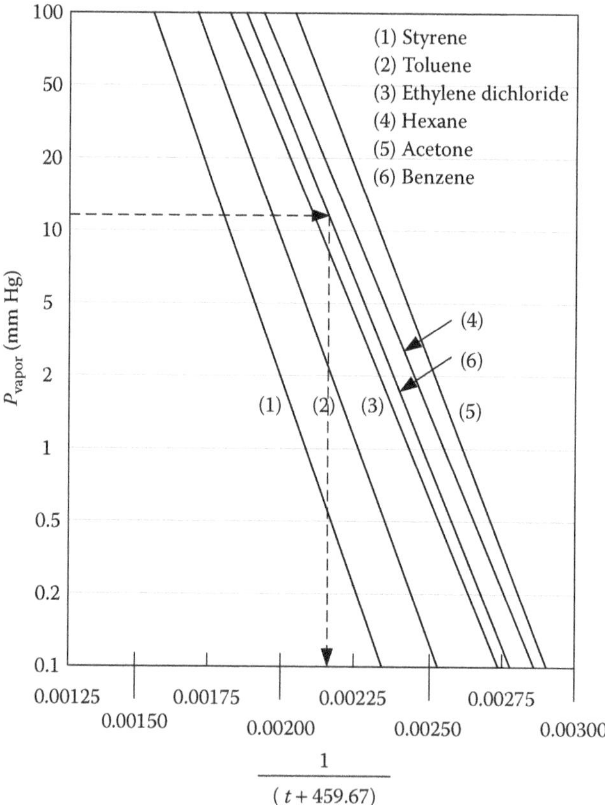

FIGURE 14.2 Typical vapor pressure curves.

14.2.2.1 Example—Design Condensation Temperature to Achieve Desired Volatile Organic Compounds Recovery

Consider an air stream flowing at 771 scfm containing 13.00 mol% benzene in which 90% removal of the benzene is required. The air flow entering the condenser is at 1.000 atm. What temperature is necessary to achieve this percent removal? The condenser is depicted in Figure 14.3, which is Figure 14.1 labeled with the conditions of operation specific to this example. The partial pressure of the benzene at the outlet of the condenser can be calculated as follows:

Basis: 1.000 moles of air stream including the benzene
Assumption: Condenser operates at 1.000 atmosphere or 760 mm of Hg
Moles of benzene entering = 0.1300
Moles of benzene leaving = (1 − 0.90) × 0.1300 = 0.0130
Moles of air + benzene leaving = 1.000 − (0.1300 − 0.0130) = 0.8830
Partial pressure of benzene leaving = (0.0130/0.8830) × 760 = 11.19 mm of Hg
Refer to Figure 14.2, the vapor pressure curve

$$\text{For benzene the value of the abscissa } \frac{1}{(t_{CON}+459.67)}=\frac{0.00214}{R^{\circ}}$$

Solving for t_{CON} = 6.80°F
Therefore, the condenser temperature must be below 6.80°F.

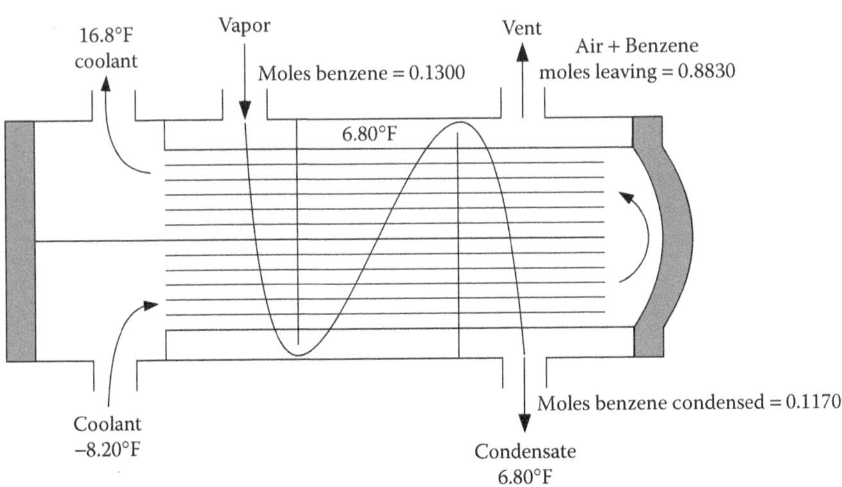

FIGURE 14.3 Shell- and tube-type surface condenser schematic, an example calculation.

14.3 COOLANT AND HEAT EXCHANGER TYPE

The next step is to select the coolant based on the condensation temperature required. Table 14.1 summarizes some possible coolants. In the case of water, chilled water, and brine solutions, all remain in the liquid phase as they condense the VOC. Refrigerants usually condense the VOC by absorbing the heat as they change phase from liquid to vapor. The usual refrigeration cycle is used for these refrigerants. The vapor is compressed and condensed at a higher pressure and a higher temperature by a fluid at a temperature lower than the condensation temperature. Frequently, water can be used to condense the high pressure refrigerant. Most thermodynamics textbooks contain a description of a refrigeration cycle.

The problem now is to determine the size and design of the particular type of heat exchanger that is required to carry out the heat transfer needed. Figure 14.1 illustrates a horizontal shell and tube heat exchanger with the coolant inside the tubes and the condensing vapor outside the tubes. Vertical shell and tube heat exchanger arrangements are shown in Figure 14.4. The advantages and disadvantages of each type are listed in Table 14.2.

The design of this type of equipment requires the knowledge of suitable heat transfer coefficients. These coefficients are highly dependent on the condensing material, the coolant used, and the particular arrangement of the heat exchanger. They range from 10 to 300 BTU/(h-ft^2-°F). Finally, the design procedure would include determining the amount of coolant needed.

14.3.1 Example—Heat Exchanger Area and Coolant Flow Rate

For the heat exchanger discussed previously, where the flow is 771.0 scfm, the number of moles would be 2.0 moles/min or 120 moles/h. Data are as follows:

$$\text{Heat of condensation of benzene} = 13{,}230 \text{ BTU/lb-mole at 1 atm}, 176\,^\circ\text{F}$$

$$\text{Specific heats, at } 77\,^\circ\text{F} = 25\,^\circ\text{C}$$

$$\text{Air} = C_{\text{PA}} = 6.96 \text{ BTU/lb-mole-}^\circ\text{F}$$

$$\text{Benzene} = C_{\text{PA}} = 19.65 \text{ BTU/lb-mole-}^\circ\text{F}$$

$$\text{Heat transfer medium } C_{\text{PM}} = 0.65 \text{ BTU/lb-}^\circ\text{F}$$

TABLE 14.1
Coolant Selection

Required Condensation Temperature (°F)	Coolant
80 to 100	Water
45 to 60	Chilled water
−30 to 45	Brine solutions
−90 to −30	Refrigerants

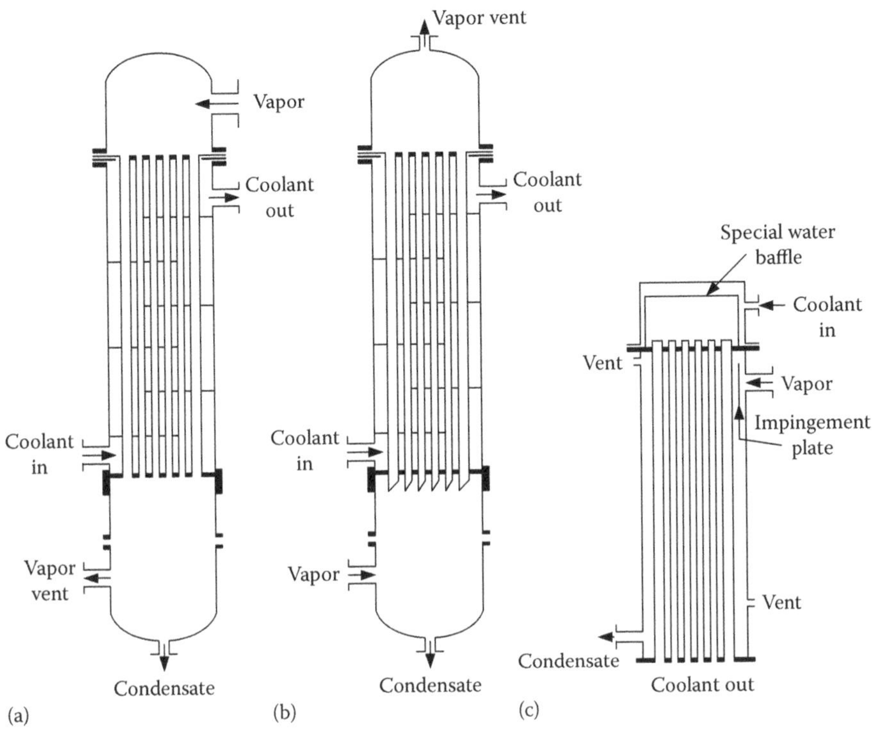

FIGURE 14.4 Vertical shell and tube heat exchangers arrangements: (a) condensation inside tubes, downflow vapor; (b) condensation inside tubes, upflow vapor; and (c) condensation outside tubes, downflow vapor.

On the basis of 1 h of operation,

$$\text{Benzene in} = 0.13 \quad 120 = 15.60 \text{ moles}$$

$$\text{Air in and out} = 0.87 \quad 120 = 104.40 \text{ moles}$$

$$\text{Benzene condensed} = 0.90 \quad 15.6 = 14.04 \text{ moles}$$

$$\text{Benzene out} = 15.60 - 14.04 = 1.56 \text{ moles}$$

$$\text{Air + benzene out} = 104.4 + 1.56 = 105.96 \text{ moles}$$

Refer to Figure 14.3 for temperatures.

Condenser heat load, assuming no heat loss from the heat exchanger to the atmosphere,

$$W_S = 0.0; \text{ therefore, } \Delta H = Q$$

$$\Delta H = \Delta H_{\text{uncon voc}} + \Delta H_{\text{air}} + \Delta H_{\text{cond voc}}$$

$$\Delta H_{\text{uncon voc}} = 1.56 \times 19.65 \times (80 - 6.8) = 2244 \text{ BTU/h}$$

$$\Delta H_{\text{air}} = 104.4 \times 6.96 \times (80 - 6.8) = 53{,}189 \text{ BTU/h}$$

TABLE 14.2
Advantages and Disadvantages of Shell and Tube Heat Exchanger Types Used in Condensation

Shell and Tube Condenser Types	Advantages	Disadvantages
Horizontal exchanger		
Condensate outside tubes	May be operated partially flooded	Free draining
Condensate inside tubes	Liquid builds up causing slugging	
Vertical exchanger		
Condensate inside tubes, vertical downflow	Positive venting of noncondensables	Wet tubes retain light-soluble components
Low pressure may require large tubes		
Condensate inside tubes, vertical upflow	Used for refluxing Usually partially condensing Liquid and vapor remain in intimate contact	
Condensate outside tubes, vertical downflow	High coolant side heat transfer coefficient Ease of cleaning	Requires careful distribution of coolant

To estimate the heat of condensation at $T_2 = (6.8 + 460) = 466.8°R$, use the Watson equation.[1]

$$\Delta H_{\text{voc at } T_2} = \Delta H_{\text{voc at } T_1}\left(\frac{1 - T_2/T_C}{1 - T_1/T_C}\right)^{0.38}$$

where:

$T_C = 1012°R$, the critical temperature of benzene
$T_1 = (176 + 460) = 636°R$

$$\Delta H_{\text{voc at } T_2} = 13{,}236\ \text{BTU/lb-mole}$$

$$\Delta H_{\text{voc at } T_2} = 13.236\left(\frac{1 - 466.8/1012}{1 - 636/1012}\right)^{0.38} = 15{,}243\ \text{BTU/lb-mole}$$

$$\Delta H_{\text{cond voc}} = \Delta H_{\text{voc to cond temp}} + \Delta H_{\text{cond}}$$

$$\Delta H_{\text{cond voc}} = 14.04 \times 19.65 \times (80 - 6.8) + 14.04 \times 15{,}243$$

$$\Delta H_{\text{cond voc}} = 234{,}207\ \text{BTU/h}$$

$$\Delta H = 2244 + 53{,}180 + 234{,}207 = 289{,}640\ \text{BTU/h}$$

Calculate the heat transfer area,

$$Q = U \times A \times \Delta T_{\text{log mean}}$$

where:

U is the heat transfer coefficient

U = 40 BTU/h-ft^2-°F

$\Delta T_{\text{log mean}}$ is the log-mean temperature difference

For a derivation and discussion of the log-mean temperature difference (LMTD), see Perry and Green[2] or a text on heat transfer.

$$\Delta T_{\text{log mean}} = \frac{(80-16.8)-[6.8-(-8.2)]}{\ln\{(80-16.8)/[6.8-(-8.2)]\}} = 33.51°\text{F}$$

$$A = \frac{289{,}640}{0.65[6.8-(-8.2)]} = 216\,\text{ft}^2$$

Calculate the coolant flow.

$$Q = W_{\text{cool}} C_{\text{PM}} (T_{\text{in}} - T_{\text{out}})$$

$$W_{\text{cool}} = \frac{289{,}640}{0.65[6.8-(-8.2)]} = 29.707\ \text{lbs/h}$$

14.4 MIXTURES OF ORGANIC VAPORS

The condensation of mixtures of organic vapors occurs over a range of temperatures, thereby complicating the design of heat exchangers for condensing these mixtures. The process for a binary mixture is illustrated in Figure 14.5, where it is presumed that the pressure remains constant. A vapor at point A is cooled until it reaches its dew point at point B. Further reduction of temperature will cause the mixture to form two phases. At the temperature at point C, the vapor composition is given by point D, and the liquid composition by point E. A constant temperature flash calculation could determine not only the compositions at this temperature but also the quantity of vapor and liquid. Continued coolant to the bubble point temperature F will produce 100% liquid with the same composition as the initial vapor. Therefore, as an organic mixture cools from its dew point to its bubble point, the condensing liquid is changing composition. This results in the heat of condensation varying throughout the cooling process. This variation in the heat of condensation should be accounted for in the determination of the area for heat exchange and will result in a greater area than would be calculated from the LMTD method. In some cases, it can make a major difference in the area and, if not accounted for, can result in poor performance of the heat exchanger. For a more detailed description of the dew point, bubble point, and

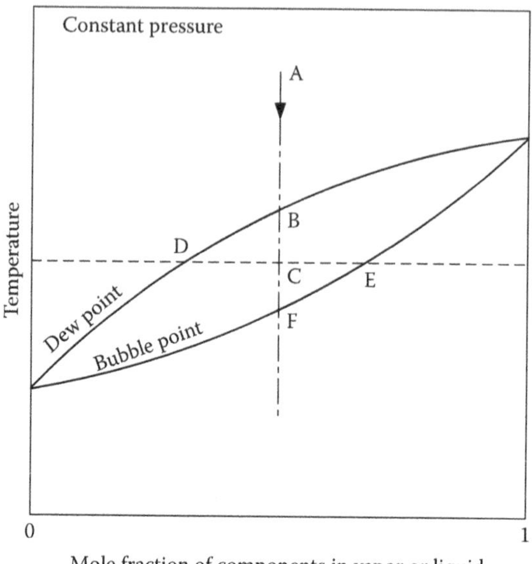

FIGURE 14.5 Equilibrium dew point–bubble point curve for a binary mixture.

flash calculation methodology, refer to a thermodynamics textbook like Smith et al.[1] The original model for sizing a condenser for mixtures of vapors was presented by Colburn and Hougen.[3] This method was elaborated upon by Silver[4] and Bell and Ghaly.[5]

14.4.1 Example—Condensation of a Binary Mixture

As an example to illustrate the methodology, consider the condensation of an isopropyl alcohol (IPA)—water mixture in a vertical, countercurrent, upflow condenser at 1 atm. The heat exchanger is to be sized to totally condense the mixture. The total flow rate is 608 lb/h at 214.4°F (101.33°C) and atmospheric pressure at a mole fraction of IPA = 0.128. Cooling water is available at 80°F with a 10°F temperature rise allowed. In the vertical, countercurrent, upflow heat exchanger, the vapor is condensing inside the tubes. Figure 14.6 is the bubble point–dew point curve for the IPA–water system. It shows this system to be an azeotrope. The mixture composition we are considering is to the left or the lower IPA composition side. From Figure 14.7 for the IPA mole fraction of 0.128, the dew point is 95.8°C (204.4°F), and the bubble point is 82.6°C (180.7°F). Overall heat transfer coefficient U_0 = 100 BTU/h-ft²-°F, and the gas film heat transfer coefficient h_g = 7 BTU/h-ft²-°F.

First calculate the heat exchange area from the LMTD method with the overall heat transfer coefficient. The total heat transferred can be approximated from the latent heats of vaporization and the molar composition. On a mole basis, both latent heats of vaporization for IPA and water are about equal to 17,400 BTU/lb-mole. The 608 lb/h flow rate is 25.9758 lb-moles/h. Therefore, the heat transferred is

$$Q_T = 17{,}400 \times 25.9758 = 451{,}979\,\text{BTU/h}$$

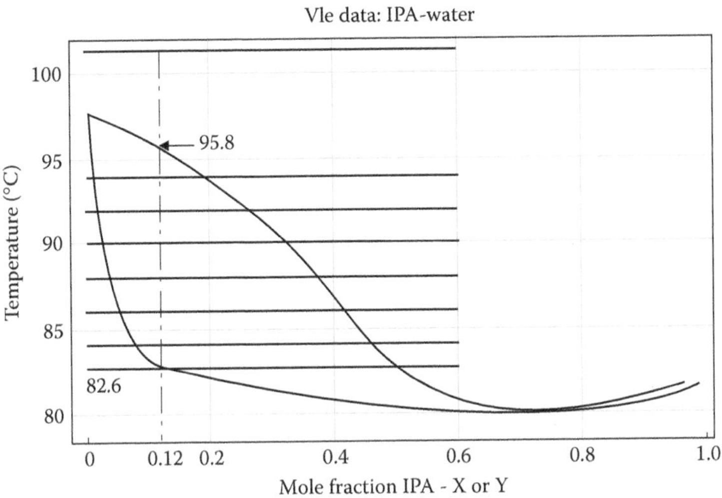

FIGURE 14.6 Bubble point–dew point curve for the isopropyl alcohol (IPA)–water system at 1 atm.

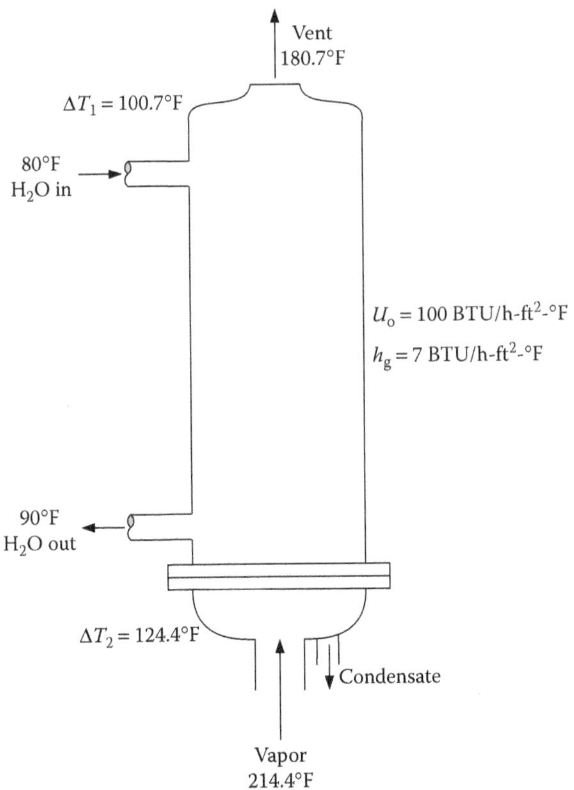

FIGURE 14.7 Vertical upflow total condenser, an example calculation.

Figure 14.7 is a schematic of the vertical heat exchanger. From Figure 14.7,

$$\text{LMTD} = \frac{124.4 - 100.7}{\ln(124.4/100.7)} = 112.13\ °\text{F}$$

$$\text{Area} = \frac{451{,}979\ \text{BTU/h}}{100\ (\text{BTU/h} - \text{ft}^2\ °\text{F}) \times 112.13\ °\text{F}} = 40.25\ \text{ft}^2$$

The more correct method is from Silver,[4] and Bell and Ghaly.[5] Based on these references, the following model is derived in Appendix 14A of this chapter.

$$\text{Area} = \int_0^{Q_T} \left[\frac{\left[1 + \left(zU_o/h_g\right)\right]}{U_o\left(T_G - T_L\right)} \right] dQ_T \qquad (14.1)$$

The temperature difference at a mole fraction of IPA = 0.128 between the entering temperature of 214.4°F (101.33°C) and the final temperature of 180.7°F (82.6°C) at the top of the condenser, is divided into seven segments as shown in Figure 14.6. To evaluate the area, the argument of the integral is calculated over each segment and plotted as a function of the heat transferred, Q_T. Figure 14.8 shows gas, T_G, and liquid coolant, T_L, temperature profiles as a function of the heat transferred Q_T as determined by the algorithm shown in Appendix 14B of this chapter. These temperature profiles seem to suggest that the LMTD should suffice because they look almost linear. However, examination of the temperature difference as a function of Q_T in Figure 14.9 shows a large variance from a linear curve. Therefore, the LMTD approach would not give reasonable results. Further calculations are made according to the algorithm in Appendix 14B. Figure 14.10 presents the numerical results of the

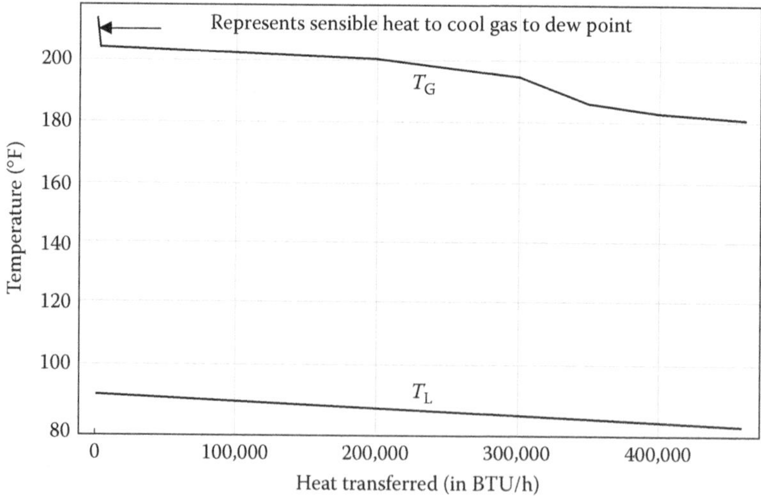

FIGURE 14.8 Temperature profiles through heat exchanger. T_G is the gas and T_L is the liquid coolant.

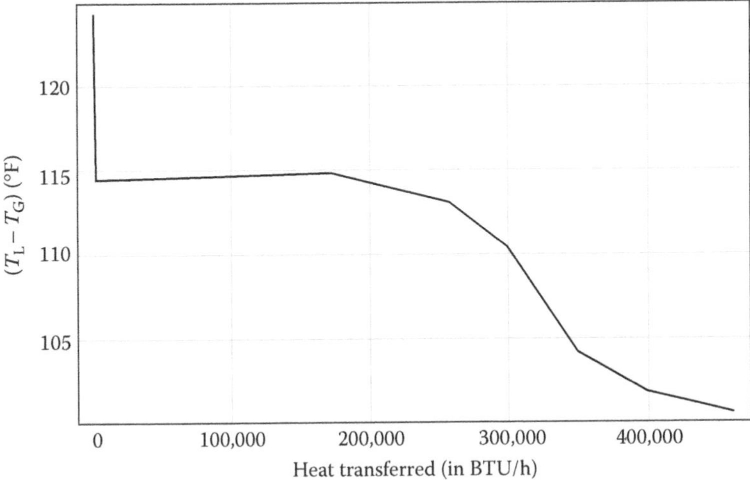

FIGURE 14.9 Gas–liquid temperature difference through the heat exchanger.

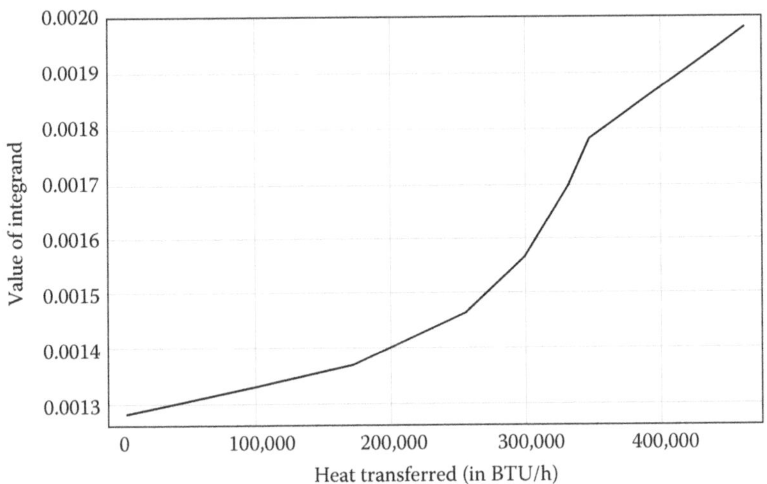

FIGURE 14.10 Area integral to evaluate Equation 14.1.

calculation produced by the algorithm for the heat exchanger of this example. Here, the heat transferred, Q_T, is plotted versus the integrand of the area equation. A *counting of squares* produces an area = 178 ft^2, considerably different from the LMTD result of 40.31 ft^2.

14.5 AIR AS A NONCONDENSABLE

A significant problem with polluted air streams is the large amount of noncondensable species included in the air. In the case where water is in the stream, the problem is even more complex. The total pressure is now the sum of the partial pressures of the air, the miscible organics, and the water. The equilibrium of the miscibles is unaffected by the presence of the other components except that the equilibrium exists only at the sum of the partial pressures and not the operating pressure of the condenser. Since the water–air and miscibles–air mixtures attain their separate equilibrium relationships, the system may have two dew points: one for water and the other for the miscible organics. For multicomponent mixtures, the calculation could become quite complex.

14.6 SYSTEMS-BASED APPROACH FOR DESIGNING CONDENSATION SYSTEMS FOR VOLATILE ORGANIC COMPOUNDS RECOVERY FROM GASEOUS EMISSION STREAMS

Systems-based methodologies have been developed for designing cost-effective heat integrated systems for condensation-based VOC recovery systems. Initially, El-Halwagi, Srinivas, and Dunn identified condensers as heat-induced separators.[6] A heat-induced separator is any indirect-contact unit operation that employs an energy separating agent for the separation of certain species via phase change. Examples of heat-induced separators are surface condensers, evaporators, crystallizers, and dryers. Examples of energy separating agents for VOC separation are refrigerants and coolants. A heat-induced network is a system of one or more heat-induced separators along with heat exchangers and other support equipment. An illustration of a condensation-based heat-induced network for the recovery of VOCs from a gaseous waste stream is shown in Figure 14.11.

For the network shown, it is desired to determine the flow rate and type of refrigerant needed to condense a prespecified quantity of VOC from the gaseous waste stream(s). The operating temperature and cost ($/kJ) are known for each refrigerant considered for the separation task. Details for calculating this value for a refrigerant have been provided by Dunn and El-Halwagi.[7]

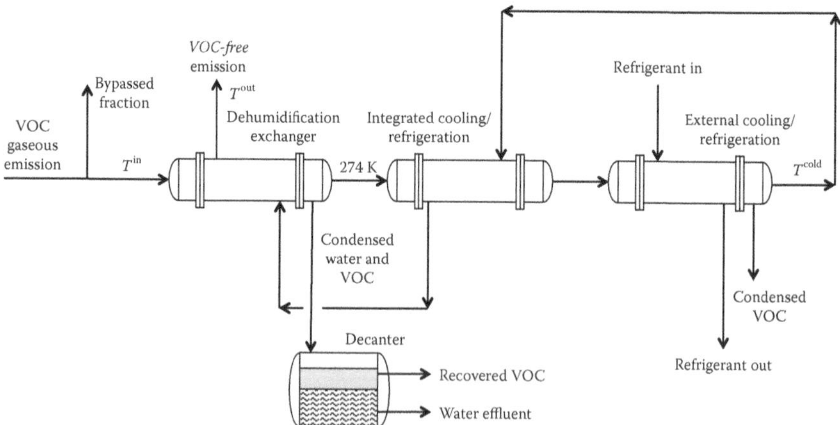

FIGURE 14.11 A schematic representation of a condensation-based heat-induced network for the recovery of VOC gaseous emissions.

Initially, the design problem is converted from a separation task to a heat transfer duty. In general, it is necessary for the designer to predetermine the temperature, denoted T^{cold} in Figure 14.11, that the waste stream must be cooled to in order to satisfy the desired removal rate. This temperature can be determined based on the stream total pressure and the stream vapor pressure at the temperature that the stream is cooled to and is calculated as previously shown in Section 14.2.2.1. After the stream temperature is determined, a minimum temperature driving force is preset by the designer for each exchanger. Next, a heat-induced pinch diagram developed by El-Halwagi, Srinivas, and Dunn is established by plotting enthalpy versus temperature for the VOC gaseous waste (shown as the T-scale) and for the recycled cold portion of this gas (shown as the T-scale).[6] An example pinch diagram for VOC condensation is shown on Figure 14.12.

The recycled cold gas stream is moved vertically until it touches the gaseous VOC stream and the point at which the streams contact is called the *pinch point* and represents a thermodynamic bottleneck for the design. Since these streams cannot cross, moving the recycled cold gas stream until the pinch point is established allows the designer to determine the cooling duty required by an external refrigerant as indicated on the y-axis. It also allows the designer to determine the outlet gas temperature from the network, which has been denoted T^{out}.

The next step in the design process is to select the external refrigerant to be used. This is accomplished by selecting the refrigerant operating below T^*, which possesses the minimum operating cost ($/kJ). Finally, tradeoffs between fixed and

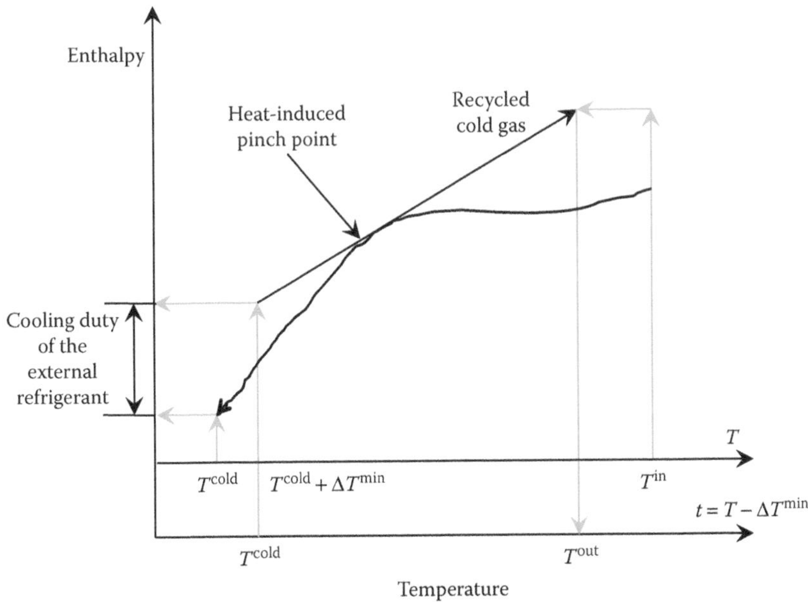

FIGURE 14.12 A heat-induced pinch diagram for a condensation process for VOC gaseous emissions.

operating costs are made by calculating the overall network cost (operating plus capital cost) for several values of ΔT^{min}. It has been established that some value of ΔT^{min} represents the most cost-effective design. Dunn and El-Halwagi have provided case studies applying this technique to the design of condensation systems for VOC recovery from gaseous emission streams containing a single VOC and multicomponent VOCs.[7,8]

In addition, Dunn et al. extended this design task by defining energy-induced separating networks. An energy-induced separating network is a system of one or more heat-induced separators along with pressurization and depressurization devices, such as compressors and turbines.[9] Pressure is increased to augment VOC recovery efficiency and potentially eliminate cryogenic temperatures and depressurization is incorporated for energy recovery. An illustration of a condensation-based energy-induced separating network for the recovery of VOCs from a gaseous waste stream is shown in Figure 14.13.

The solution technique is similar to that previously described except that a double iteration process is necessary. For a preselected value of P_w, the pressure of the gas stream exiting the compressor, network costs (operating plus capital) are determined at several values of ΔT^{min}. This procedure is repeated for several values of P_w to determine the value of P_w and ΔT^{min}, which results in the minimum network cost. A flowsheet summarizing this systems-based design procedure is shown in Figure 14.14.

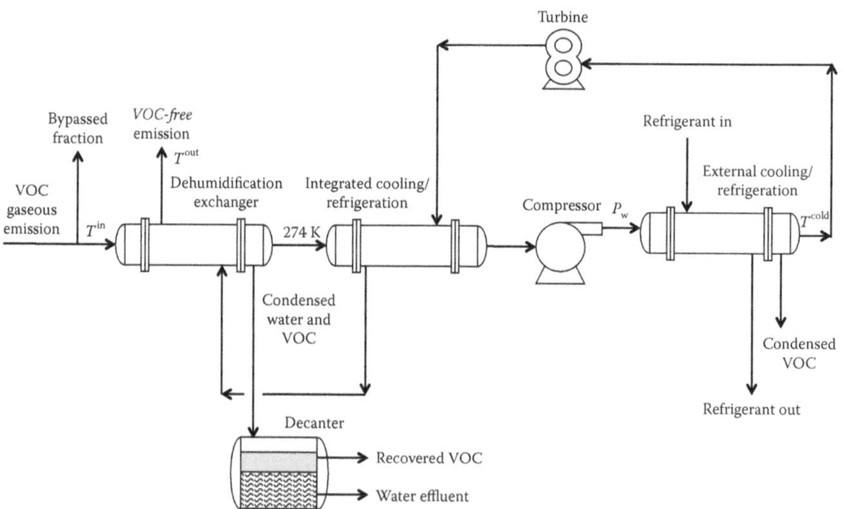

FIGURE 14.13 A schematic representation of a condensation-based energy-induced separating network with pressurization and depressurization for the recovery of VOC gaseous emissions.

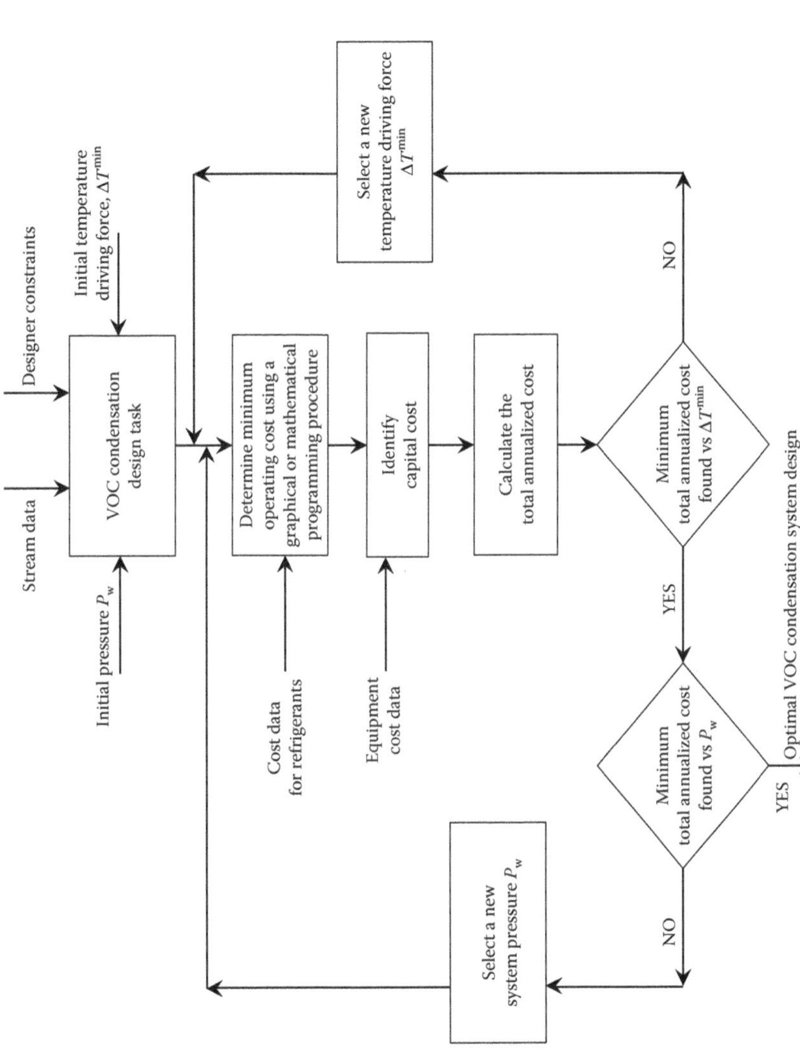

FIGURE 14.14 Design strategy for identifying an optimal energy-induced separating network condensation process for VOC gaseous emissions.

APPENDIX 14A: DERIVATION OF THE AREA MODEL FOR A MIXTURE CONDENSING FROM A GAS

Refer to Figure 14A.1 for a schematic of heat transfer through a condensing film.

dQ_{SV} is the sensible heat of vapor transferred through the gas film
dQ_T is the latent heat and sensible heats of both vapor and liquid
T_G is the gas temperature in main gas stream (condensing vapor)
T_L is the liquid temperature in main liquid stream (coolant)
T_i is the temperature at interface between gas film and condensing film
h_g is the heat transfer coefficient of gas film
U_o is the overall heat transfer coefficient through the condensing film, the tube wall, and the coolant film

This simplified model uses the two heat transfer coefficients defined above. A more complex model would take into account the conduction through the tube wall and the coolant film with separate heat transfer coefficients.

Define the heat transfer in terms of the two heat transfer coefficients.

$$dQ_{SV} = h_g(T_G - T_i)dA$$

$$dQ_T = U_o(T_i - T_L)dA$$

where:
A is the heat transfer area

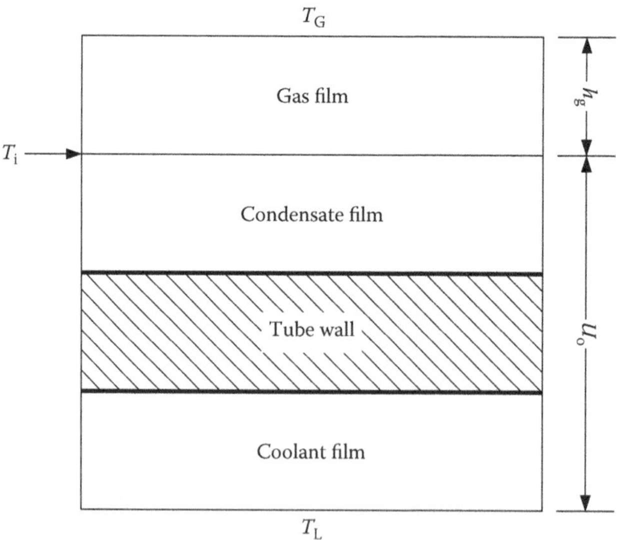

FIGURE 14A.1 Heat transfer through a condensing film.

Solve the above two equations for $T_i dA$.

$$T_i dA = -\frac{dQ_{SV}}{h_g} + T_G dA = \frac{dQ_T}{U_o} + T_L dA$$

Define z,

$$z \equiv \frac{dQ_{SV}}{dQ_T}$$

Note that $z = 0.0$ for a pure component condensing isothermally. Then

$$\frac{dQ_T}{U_o} + \frac{zU_o}{h_g U_o} dQ_T = \left(T_G - T_L\right) dA$$

Solve for dA,

$$dA = \frac{\left[1 + \left(zU_o/h_g\right)\right]}{U_o\left(T_G - T_L\right)} dQ_T$$

Then integrate for the final model,

$$\text{Area} = \int_0^{Q_T} \left\{ \frac{\left[1 + \left(zU_o/h_g\right)\right]}{U_o\left(T_G - T_L\right)} \right\} dQ_T \tag{14A.1}$$

APPENDIX 14B: ALGORITHM FOR THE AREA MODEL FOR A MIXTURE CONDENSING FROM A GAS

- This algorithm applies specifically to binary mixtures without air but could be extended to multicomponent mixtures. Consult Bell and Ghaly[5] for extensions to include noncondensables like air
- Data required:
  - Binary vapor liquid equilibria
  - Latent heat of the components of the mixture
  - Vapor and liquid specific heat
  - Coolant specific heat
  - Heat transfer coefficients U_o and h_g
  - Amount of vapor to be condensed (molar flow rate required for molar latent heats and specific heats)
- Assumptions:
  - Pressure near atmospheric
  - Molal average latent heats and specific heats are good approximations
  - Approximate $z = dQ_{SV}/dQ_T$ by $\Delta Q_{SV}/\Delta Q_T$

14B.1 Calculation Procedure

14B.1.1 For Section Zero

1. At the initial concentration, cool superheated vapor to the bubble point (removal of sensible heat)
2. Calculate the sensible heat removed
3. Since this is simply cooling a vapor with no condensation, assume that $z = 1.0$
4. Divide the temperature between the bubble point and the dew point into five or more equal segments

14B.1.2 For Section One: First Temperature Segment

1. Cool vapor mixture to first temperature by removal of sensible heat Q_{SV}
2. Flash mixture isothermally at this temperature and the system pressure to determine the amount of liquid formed and vapor remaining. Since the temperature is known, the vapor and liquid compositions are known. Therefore, this calculation is only a material balance to determine the split between the vapor and liquid
3. Calculate the latent heat removal required to condense the liquid amount determined in the flash calculation = Q_{SL}
4. Calculate the total heat removed, in this section $\Delta Q_T = Q_{SV} + Q_{SL}$
5. Calculate $Q_{SV}/\Delta Q_T$, z, and the integrand for the section

14B.1.3 For Section Two: Second Temperature Segment

1. Cool vapor mixture to the second temperature by removal of sensible heat Q_{SV}
2. Repeat steps 2 to 5 above, and add the integrand and the ΔQ_T for this section to the Q_T calculated so far for the previous sections

14B1.4 For Each Succeeding Section

Repeat steps indicated above until the bubble point is reached. The integrands and the ΔQ_T have now been summed, so that a plot can be made to determine the area. Also, the total heat removed during the condensation process, Q_T, is now known. The coolant flow rate, m, can be determined if the allowable temperature rise, ΔT, is set.

$$m = \frac{Q_T}{C_p \Delta T}$$

Knowing m and ΔQ_T for each section, the temperature of the coolant may now be determined for each section.

15 Control of Volatile Organic Compounds and Hazardous Air Pollutants by Biofiltration

15.1 INTRODUCTION

A biofilter consists of a bed of soil or compost beneath, which is a network of perforated pipe. Contaminated air flows through the pipe and out of the many holes in the sides of the pipe, thereby being distributed throughout the bed. A biofilter works by providing an environment in which microorganisms thrive. The organic substrate provides the salts and trace elements for the bacteria, and the volatile organic compounds (VOC) provides the food source. This action is an adaptation of biogdegradation in which the air cleanses itself naturally. The microorganisms are the same that degrade organic wastes in nature and in wastewater treatment plants. These microorganisms in a moist environment oxidize organic compounds to CO_2 and water. The soil or compost beds provide a network of fine pores with large surface areas. In soils, the pores are smaller and less permeable than in compost. Therefore, soil requires larger areas for biofiltration.

Biofiltration is a relatively new technology in the United States used for effectively controlling VOC emissions, organic and inorganic air tonics, and odor from gaseous streams. This technology has been used in Europe for many years and is considered to be a best available control technology for treating contaminated gaseous streams. Biofilters function efficiently and economically for removing low concentrations (less than 1000 to 1500 ppm as methane) of VOCs, air tonics, and odor. Biofiltration offers many potential advantages over existing control technologies, such as low installation and operation costs, low maintenance requirements, long life for the biofilter, and environmentally safe operation. Also, many of the existing control technologies cannot be economically applied to dilute gas stream treatment.

Biofilters are effective systems for removing pollutants from gaseous streams. The percent removals possible vary in the literature. For odorous compounds, 98%–99% removal has been reported.[1] Reported removal efficiencies for VOCs vary, but are generally in the range of 65%–99%.[1–3] The removal rates depend on the characteristics of the biofilter, such as the media, temperature, pH, moisture content, and gas residence time, as well as on the properties of the compounds being removed by the biofilter. Biofilters can also remove particulates and liquids from gas streams.

However, care must be taken because particulates or greasy liquids can function to plug the biofilter.

Biofilters have been applied for many uses. Their historical use has been for odor control. Industries including chemical manufacturing, pharmaceutical manufacturing, food processing, wastewater facilities, and compost operations have successfully used this technology for odor control. Biofiltration has also been used to reduce VOC emissions in aerosol propellant operations.[4] Fuel and solvent operations have also successfully used biofiltration to control emissions.

15.2 THEORY OF BIOFILTER OPERATION

A biofilter is a bed of media, which supports the growth of microorganisms. Biotransformations act along with adsorption, absorption, and diffusion to remove contaminants from the gaseous stream. The gaseous stream is pumped through perforated pipes located at the bottom of the biofilter. The gas passes upward through the biofilter media. The contaminants in the gas are either adsorbed onto the solid particles of the media or absorbed into the water layer that exists on the media particles. The rate of adsorption is related to the contaminant type in the gas stream. The media of the filter functions both to supply inorganic nutrients and as a supplement to the gas stream being treated for organic nutrients.

As the contaminated air flows upward through the bed, VOCs sorb onto the organic surface of the soil or compost. The sorbed gases are oxidized by the microorganisms to CO_2. The volatile inorganics are also sorbed and oxidized to form calcium salts. The biofilters are actually a mixture of activated carbon, alumina, silica, and lime combined with a microbial population that enzymatically catalyzes the oxidation of the sorbed gases. The sorption capacity is relatively low, but the oxidation regenerates the sorption capacity.

Gases are inherently more biodegradable than solids and liquids because they are more molecularly dispersed. Removal and oxidation rates depend upon the biodegradability and reactivity of the gases. Half-lives of contaminants range from minutes to months. Table 15.1 lists compounds in order of their degradability. In the case of hydrocarbons, aliphatics degrade faster than aromatics. Even though pollutants are being put into the ground, loading rates are low, gases degrade rapidly, oxygen is in excess, and the soil does not become contaminated.

The microorganisms exist in the slime layer, or biolayer on the surface of the media particles. Diffusion occurs through the water layer to the microorganisms. Once the contaminants are adsorbed or absorbed, the microorganisms begin to function. The contaminants provide a food source for the microorganisms. Through biotransformation of the food source, end products are formed, including carbon dioxide, water, nitrogen, mineral salts, and energy to produce more microorganisms.

Oxidation of adsorbed compounds allows the biofilter to self-regenerate. Adsorption sites are continually becoming available as oxidation by microorganisms occurs. Overloading of the biofilter results when adsorption occurs faster than oxidation. The result of overloading is to allow the contaminants to pass through the biofilter.

TABLE 15.1
Gases Classified according to Their Degradability

Rapidly Degradable VOCs	Rapidly Reactive VOCs	Slowly Degradable VOCs	Very Slowly Degradable VOCs
Alcohols	H_2S	Hydrocarbons	Halogenated hydrocarbons
Aldehydes	NO_X (not N_2O)	Phenols	
Ketones		Methylene chloride	Polyaromatic hydrocarbons
Ethers	SO_2		
Esters	HCl		CS_2
Organic acids	NH_3		
Amines	PH_3		
Thiols	SiH_4		
Other molecules with O, N, or S functional groups	HF		

15.3 DESIGN PARAMETERS AND CONDITIONS

A biofilter can have many design configurations, but they all function similarly, as described by the process theory discussed in Section 15.2. A biofilter can be open or enclosed, it can be built directly into the ground or in a reactor vessel, and it can be single or multiple beds. Figure 15.1 presents a typical open biofilter configuration. The basic components of a biofilter system include the filter media packed bed, an air distribution system under the filter bed, a humidifier to saturate the incoming gas stream, and a blower to move the gas stream through the system. Optional components include a heat exchange chamber to cool or heat the gas stream to optimal temperature for the filter bed and a water sprinkler system to apply moisture directly to the filter media surface.

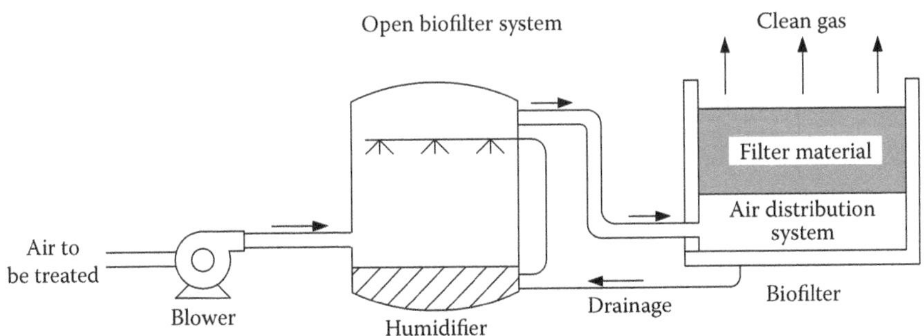

FIGURE 15.1 Typical open biofilter configuration.

15.3.1 Depth and Media of Biofilter Bed

The depth of biofilter media range from 0.5 to 2.5 m, with 1 m being the typical depth of a biofilter. Many different media types have been used in biofilters. Some examples include soil, compost, sand, shredded bark, peat, heather, volcanic ash, and a mixture of these components. Figure 15.2 shows a typical biofilter bed. Often polystyrene spheres or peat granules may be added to increase the structural support of the system and to increase the adsorptive capacity of the media. The two most commonly discussed media in the literature are soil and compost. Typical parameters required for the media, regardless of which media type is chosen, include a neutral pH range, pore volumes of greater than 80%, and a total organic content of 55% or greater. The properties of the media are important to the successful operation of the biofilter.

Soil is a stable choice for media in that it does not degrade. However, it contains fewer and less complex microorganisms than compost media. Compost has more and more complex microorganisms than sand. It also has higher air and water permeability. The buffering capacity of compost is also very good. However, it does not have the stability of sand. With time compost decomposes, and the average particle sizes of the filter media decrease. The choice of media material depends on availability, desired characteristics of the biofilter, and the compounds that are to be removed.

The useful life of the media is typically up to five years. After this time period, replacement is usually necessary. Fluffing, or turning, of the media material in the biofilter may be required at shorter intervals to prevent excessive compaction and settling. The turning and occasional replacement of the biofilter material are the two major components of maintenance required by this treatment system.

15.3.2 Microorganisms

Three types of microorganisms are generally present in a biofilter. These include fungi, bacteria, and actinomycetes. Actinomycetes are organisms that resemble

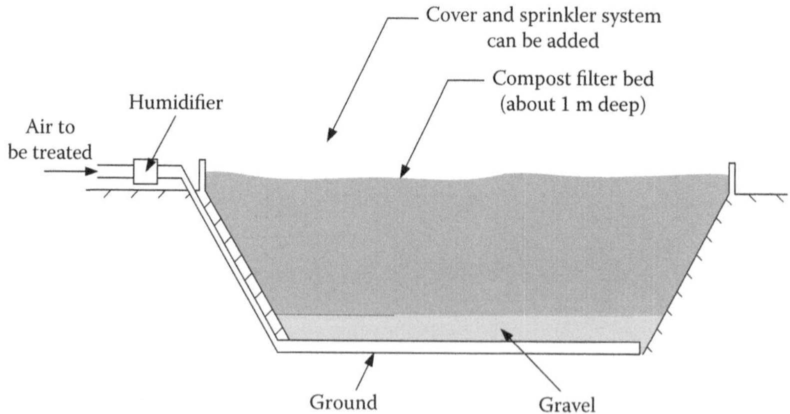

FIGURE 15.2 Typical biofilter bed.

both bacteria and fungi. The growth and activity of the microorganisms is dependent on the environment of the biofilter. The environment relies on such parameters as ample oxygen supply, absence of toxic materials, ample inorganic nutrients for the microorganisms, optimum moisture conditions, appropriate temperatures, and neutral pH range. All of these parameters must be controlled in the biofilter to ensure successful removal efficiencies. The microorganisms are fairly hearty, so staying in an acceptable range of these parameters allows microorganism survival.

The start-up of a biofilter process requires some acclimation time for the microorganisms to grow specific to the compounds in the gaseous stream. For easily degradable substances, this acclimation period is typically around 10 days.[2] For more complex compounds or for mixtures, this acclimation process may require additional time. The acclimation process also allows the microorganisms to develop tolerance or acceptance for compounds they may find to be toxic in nature.

Often, biofilters are not used continuously in the treatment process. They may be employed intermittently or seasonally, depending on the treatment process. The biomass has been shown to be able to be viable for shut downs of approximately two weeks. Also, if nutrient and oxygen supplies are continued, the biomass may be maintained for up to two months.

15.3.3 Oxygen Supply

A major factor that can limit the bio-oxidation rate is the oxygen available to the microorganisms. Typically, a minimum of 100 parts of oxygen per part of gas must be supplied. This is usually easily supplied in a biofilter since the gaseous streams being pumped through the biofilter usually contain excess oxygen, enough to keep the biofilter well supplied. Anaerobic zones need to be avoided to ensure that the compounds are biotransformed and to prevent any anaerobic zone odors (primarily hydrogen sulfide) from forming.

15.3.4 Inorganic Nutrient Supply

Besides oxygen, microorganisms require certain nutrients to sustain their activity and growth. These are typically nitrogen, phosphorous, and some trace metals. Trace metals are almost always well supplied in the media material. Nitrogen and phosphorous may need to be added, depending on the media characteristics. For aerobic microorganisms, the O/N/P ratio is estimated at 100/5/1.

15.3.5 Moisture Content

The literature agrees that moisture content is the most critical operational parameter for the successful operation of a biofilter. The gaseous streams tend to dry out the biofilter media. Too little water will result in decreased activity of the microorganisms, and perhaps transfer of the adsorbed contaminants out of the filter and into the atmosphere. Too much water can also cause problems, such as anaerobic zones, with the potential of producing odors, and increases in the head loss of the system. Optimal water contents vary in the literature, but generally the range of 20%–60% by weight is accepted.

Moisture can be added to the system in two ways: humidification of the gas stream and direct application of water to the biofilter surface. Humidification of the gas stream occurs as a pretreatment to the biofilter. The humidifier process location is shown in Figure 15.1. Typically, the degree of saturation suggested is at least 95%, with saturation percentages of 99% and 100% quoted as the optimum.[2] Surface sprays to the biofilter surface can also be used. Care must be taken that the water droplets are small. Typically, water droplet diameters of less than 1 mm are suggested, in order to prevent compaction of the biofilter. The maximum water loading rate suggested[3] is 0.5 gal/ft^2-h.

15.3.6 Temperature

The microorganisms' activity and growth is optimal in a temperature range of 10°C–40°C. Higher temperatures will destroy the biomass, while lower temperatures will result in lower activities of the microorganisms. In winter, heating of the off gas streams may be required before passing the streams through the biofilter. This will ensure an acceptable rate of degradation by the microorganisms. High temperature off gases may need to be cooled before the biofilter to ensure the survival of the microorganisms.

15.3.7 pH of the Biofilter

The pH in the biofilter should remain near neutral, in the range of 7–8. When inorganic gases are treated, inorganic acids may be produced. For example, treated H_2S will produce H_2SO_4. Other inorganic acids that can be formed include HCl and HNO_3. These acids can cause lowered pH in the media over time. Carbon dioxide production by the microorganisms can also lower the pH over time. The media typically has some inherent buffering capacity to neutralize small changes in the pH. However, if the buffering capacity is not sufficient, lime may need to be added to the biofilter for pH adjustment.

15.3.8 Loading and Removal Rates

The loading rate of the biofilter can be used to determine the size of the system required. Loading rates can be expressed in three ways: (1) flow rates of gases through the bed, (2) gas residence times, and (3) removal rates. Flow rates of gas into the bed range from 0.3 to 9.5 m^3/min-m^2. The typical range[5] is 0.3–1.6 m^3/min-m^2. Off gas rates are typically[2] around 1,000–150,000 m^3/h. Higher loading rates can result in overloading the bed and pass through of contaminants.

Gas residence times, the time the gas actually spends in contact with the biofilter material, is the time available for adsorption and absorption to occur. Suggested gas residence times are a minimum of 30 sec for compost media[5] and a minimum of one min for soil media.[6] Slightly longer residence times are suggested for inorganic gases.

Removal rates depend on the compound and the media type in the biofilter. They are typically reported in units of g/kg of dry media/day. Williams and Miller[5] provide a list of removal rates for various compounds and various media types. Generally, the lower molecular weight, less-complex compounds are more easily degraded and more quickly removed in a biofilter.

15.3.9 Pressure Drop

The pressure drop through the filter bed depends on the media type, porosity, moisture content, and compaction of the media. The porosity of the media can change with time as the filter media becomes more compacted. Fluffing or replacing the media over time can help to prevent compaction and higher pressure drops. Higher pressure drops result in more energy required to overcome the back pressure of the filter bed. Typical pressure drops range from 1 to 3 in. of water.[7] Typical power consumption[2] for a biofilter is in the range of 1.8–2.5 KWh/1000 m^3. The pressure drop is related to the surface load of contaminants and the media type.

15.3.10 Pretreatment of Gas Streams

As discussed in Section 15.3.5, humidification of the gas stream may be required prior to the biofilter to provide adequate moisture for the microorganisms. Other pretreatment necessary may include removing particulates. Though the biofilter is capable of removing particulates, the solid matter can cause clogging of the biofilter and gas distribution system. The gas stream also may need to be heated or cooled to meet the temperature requirements of the microorganisms.

15.4 BIOFILTER COMPARED TO OTHER AVAILABLE CONTROL TECHNOLOGY

Other control technology for the control of VOCs and air toxics include incineration, carbon adsorption, condensation, and wet scrubbing. The advantage that biofilters have over all of these technologies is their ability to treat dilute gas streams in a cost-effective manner. Shortcomings associated with the other available technologies include high fuel use, high maintenance requirements, high capital costs, and the pollution of wash water or air streams in the removal process. Other technologies often take the pollution from one form and place it in another, for example, removing contamination from an air stream and placing it in the wash water. Biofilters allow the biotransformation of the pollution to less- or nontoxic forms and reduced volumes.

Incineration works as a control technology for highly concentrated waste streams. For dilute streams, the process is too energy intensive to be practical. It is also more expensive to install and operate than a biofilter system. Also, there has been much public opposition to incineration treatment methods.

Carbon adsorption is a very effective technology. However, it is very expensive to use, which is especially prohibitive to small operations. If the carbon is regenerated on site, the costs will be less than if it is not regenerated on site.

Condensation is an effective technology for treating concentrated and pure off gases. As with incinerators, the treatment of dilute streams is too energy intensive to accomplish cost effectively. Wet scrubbing technology is also more expensive than biofilter systems.

15.5 SUCCESSFUL CASE STUDIES

Throughout the literature, there are successful case studies quoted for a variety of applications of the biofilter. For example, gases from an animal rendering plant process were treated in a soil biofilter for odor removal. Removal rates of 99.9% were obtained.[7] Another application used a sludge compost biofilter to treat a gas stream containing volatile amine compounds. Removals exceeding 95% were obtained.[1] A prototype biofilter with soil media was used to treat light aliphatic compounds and trichloroethylene from aerosol propellant releases. Reduction rates of 90% were obtained.[4] The various case studies reinforce that fact that this technology can be successfully applied to various gas stream treatments.

15.6 FURTHER CONSIDERATIONS

Biofiltration offers a cost-competitive and competent technology alternative for the treatment of gas streams containing VOCs, air toxics, and odors. Its success has been proven in Europe and is starting to be applied more in the United States. As the U.S. Clean Air Act begins to focus more on smaller generators of VOCs and air toxics, the biofilter will most likely see more and broader applications. Adler[8] discusses current usage of biofiltration in the United States and Europe. He presents guidelines for scale-up and design of biofiltration processes. He also presents economic data for two cases comparing biofiltration to other means of control. In the case of methanol, a readily biodegradable molecule, biofiltration appears to be exceptionally good when compared to other technologies. The second case involved toluene. Here, biofiltration is only in the middle range of costs compared to other technologies.

Design of a biofilter may require a pilot study, especially if the gas stream contains a mixture of compounds. Often, co-metabolism will occur, which will affect the degradation rates and perhaps some of the design parameters. Successful biofiltration requires a design to ensure a proper environment for the microorganisms. This includes being able to control and monitor parameters such as moisture content, pH, temperature, and nutrient supply.

16 Membrane Separation

16.1 OVERVIEW

Membrane separation has developed into an important technology for separating volatile organic compounds (VOCs) and other gaseous air pollutants from gas streams during the last 15 years. The first commercial application was installed in 1990, and more than 50 systems have been installed in the chemical process industry worldwide.[1]

The technology utilizes a polymeric membrane that is more permeable to condensable organic vapors, such as C^{3+} hydrocarbons and aromatics, than it is to noncondensable gases such as methane, ethane, nitrogen, and hydrogen. The air stream to be treated is separated into a permeate that contains concentrated VOCs and a treated residue stream that is depleted of VOC.

Because the technology concentrates the VOC gas stream, it can be used with a condenser to recover the VOC. It is best suited for relatively low-flow streams containing moderate VOC concentration.

The typical overall VOC recovery process consists of two steps: (1) compression and condensation and (2) membrane separation.[2] A mixture of vapor and air is compressed to about 45–200 psig. The compressed mixture is cooled and condensed vapor is recovered. Uncondensed organics are separated from the gas stream and concentrated in the permeate by the membrane. The treated gas is vented from the system and the permeate is drawn back to the compressor inlet. A schematic representation of the system is provided as Figure 16.1.

A brief summary of membrane separations containing a general description, advantages, and disadvantages for the removal of volatile organic compounds from gaseous emission streams is provided below.[2–4]

16.1.1 Description

- VOC gaseous emissions are concentrated using organic selective (VOC permeable) membranes.
- Air and VOCs permeate through the membrane at rates determined by their relative permeabilities and the pressure difference across the membrane.
- Membranes are typically 10–100 times more permeable to VOCs than air, depending on the specific VOCs under evaluation.
- Based on the system design, the exit membrane stream VOC concentration can be increased 5–50 times the inlet membrane stream concentration.
- Concentrated gas streams are then compressed (typically to pressures between 45 and 200 psig) followed by the use of conventional condensation technology.

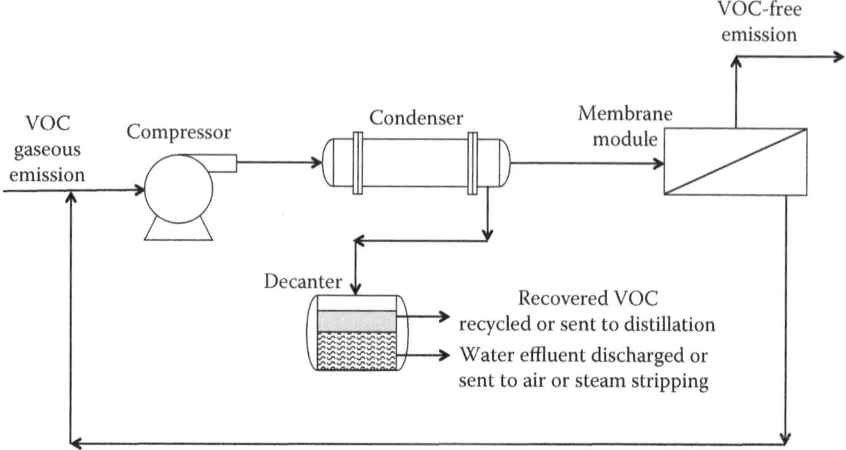

FIGURE 16.1 A schematic representation of a membrane-hybrid process for VOC gaseous emissions.

16.1.2 Advantages

- Can achieve high recovery efficiencies when combined with condensation (>90%) without operating at cryogenic temperatures.
- Can allow efficient recovery of high volatility (low boiling point) VOCs.
- Can efficiently handle low concentration gaseous emission streams.

16.1.3 Disadvantages

- Membranes may require annual replacement due to fouling.
- Typically used on high concentration (>10,000 ppm) inlet gas streams.
- Generally not cost-effective for high flow rate gas streams.

16.2 POLYMERIC MEMBRANES

Polymeric membrane consists of a layer of nonwoven fabric that serves as the substrate, a solvent-resistant microporous support layer for mechanical strength, and a thin film selective layer that performs the separation. It is manufactured as flat sheet and is wrapped into a spiral-wound module. Feed gas enters the module and flows between the membrane sheets. Spacers on the feed and permeate side of the sheets create flow channels.[5]

16.3 PERFORMANCE

Membranes are best suited for treating VOC streams that contain more than 1000 ppmv of organic vapor, where recovered product has value. Typical VOC recovery using membrane separation ranges from 90% to 99%, and can reduce the

TABLE 16.1
VOCs That Can Be Captured with Membrane Technology

Acetaldehyde
Acetone
Acetonitrile
Benzene
Butane
Carbon tetrachloride
CFC-11
CFC-12
CFC-113
Chlorine
Chloroform
Ethylene dichloride
Ethylene oxide
HCFC-123
Hexane
Methanol
Methyl bromide
Methyl chloride
Methyl chloroform
Methyl isobutyl ketone
Methylene chloride
Perfluorocarbons
Propylene oxide
Styrene
Toluene
Trichloroethylene
Vinyl chloride
Xylenes

Source: Simmons, V. et al., *Chem. Eng.*, 101(9), 92, 1994.

VOC content of the vented gas to 100 ppm or less.[5,6] A list of VOCs that can be recaptured using membrane separation is provided in Table 16.1.

16.4 APPLICATIONS

Applications for membrane separation include vent streams from the production and processing of vinyl chloride, ethylene, propylene, methylene chloride, and ammonia. In polyolefin plants, purification of ethylene and propylene feedstock in a splitter column is a common first step. When nitrogen, hydrogen, and methane are present in the feed, they build up in the column overhead stream and must be vented. Vent streams from reactor recycle and reactor purge also must be treated.

The vent streams may be fed to a membrane separator where valuable feedstock is recovered as the permeate. In several membrane applications, where the primary objective is product recovery, the residue stream, which still contains a small amount of VOC that did not permeate through the membrane, is sent to a flare.

Vent gases from ammonia plant reactors typically contain hydrogen, nitrogen, methane, and argon. Glassy polymer membranes, such as polysulfone, are much more permeable to hydrogen than to the other components. Approximately, 87% of the hydrogen can be recovered from the vent gas and recycled.[7]

16.5 MEMBRANE SYSTEMS DESIGN

Two design methodologies have been provided by Crabtree, El-Halwagi, and Dunn to identify the optimal design and operating conditions for membrane hybrid processes to recover VOCs from gaseous emission streams.[8] The first approach developed uses a predefined network structure combined with a mathematical program to determine system parameters (pressures, flow rates, etc.) that represent the minimum operating and capital cost for this network. The second approach is more mathematically rigorous and utilizes the state-space methodology that envisions all potential system configurations. This approach does not require a predefined network structure and results in the identification of the network configuration, and operating parameters, which possesses a minimum total annualized cost. Application of this methodology for the recovery of cresol from a tricresyl phosphate production process has been included in this publication.

17 NO_x Control

There are a number of oxides of nitrogen, including nitrous oxide (N_2O), nitric oxide (NO), nitrogen dioxide (NO_2), nitrogen trioxide (N_2O_3), and nitrogen pentoxide (N_2O_5), that are referred to collectively as NO_x. The two oxides of nitrogen that are of primary concern to air pollution are NO and NO_2. NO is a colorless gas that is a precursor to NO_2 and is an active compound in photochemical reactions that produce smog. NO_2 is a reddish brown gas that gives color to smog and can contribute to opacity in flue gas plumes from stacks.

NO_2 is a criteria pollutant with a National Ambient Air Quality Standard of 100 $\mu g/m^3$, or 0.053 ppm, annual average. It is also a precursor to nitric acid, HNO_3, in the atmosphere and is a major contributor to acid rain, although less important than SO_2, which is discussed in Chapter 18. Nitric acid contributes only one proton per molecule, while sulfuric acid has two protons per molecule, and mass emissions of sulfur compounds are larger than oxides of nitrogen. Finally, NO_x and volatile organic compounds react photochemically in a complex series of reactions to produce smog, which includes ozone, NO_2, peroxyacetyl nitrate, peroxybenzoyl nitrate, and other trace oxidizing agents.

By far the largest source of NO_x is combustion, although there are other industrial sources such as nitric acid manufacturing. Figure 17.1 shows the relative contribution from NO_x emission sources. The large amount of NO_x generated at coal-fired electric power plants is evident, and the very large contribution from motor vehicles and other forms of transportation, including ships, airplanes, and trains, is pronounced.

Figure 17.2 shows that total NO_x emissions in the United States have been fairly steady at about 23 tons/year, despite industrial growth and a growing number of vehicles on the road. Preventing an increase in total NO_x emissions can be attributed to the increased use of NO_x controls, especially in automobiles and in industrial fuel consumption.

17.1 NO_X FROM COMBUSTION

NO_x is generated during combustion from three mechanisms: (1) thermal NO_x, (2) prompt NO_x, and (3) fuel NO_x. Understanding these mechanisms enables one to utilize control methods for NO_x emissions.

17.1.1 Thermal NO_X

The thermal NO_x mechanism was first proposed by Zeldovich[1] and involves radicals to produce the overall reaction of combining oxygen and nitrogen:

$$O_2 \leftrightarrow 2O \tag{17.1}$$

$$O + N_2 \leftrightarrow NO + N \tag{17.2}$$

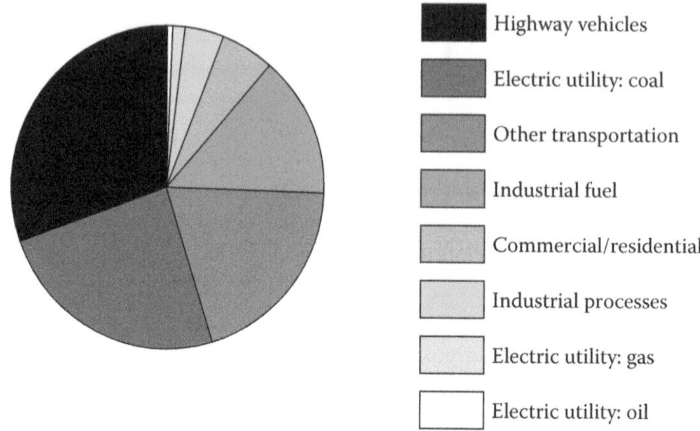

FIGURE 17.1 NO_x emission sources.

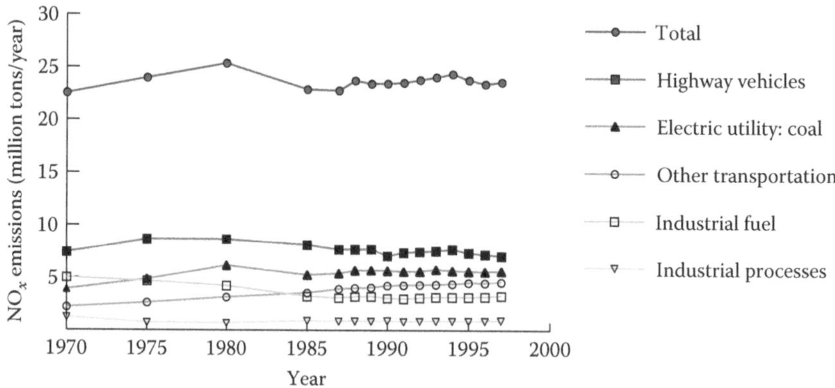

FIGURE 17.2 NO_x emission trends in the United States.

$$N + O_2 \leftrightarrow NO + O \tag{17.3}$$

$$N_2 + O_2 \leftrightarrow 2NO \tag{17.4}$$

The overall reaction that produces NO_2 is

$$NO + \frac{1}{2}O_2 \leftrightarrow NO_2 \tag{17.5}$$

Both thermodynamics and kinetics are important to the formation of thermal NO_x, so both concentration and temperature influence the amount of NO_x produced. The thermodynamic equilibrium concentrations for reactions in Equations 17.4 and 17.5 are as follows:

$$K_{P1} = \frac{\left[P_{NO}\right]^2}{\left[P_{N_2}\right]\left[P_{O_2}\right]} \tag{17.6}$$

$$K_{P2} = \frac{\left[P_{NO_2}\right]}{\left[P_{NO}\right]\left[P_{O_2}\right]^{1/2}} \tag{17.7}$$

Equilibrium constants at different temperatures are listed in Table 17.1. Equilibrium concentrations calculated from Equations 17.6 and 17.7 are presented in Table 17.2 for oxygen and nitrogen concentrations in air and in a typical combustion source flue gas.

Based on thermodynamic equilibrium alone, calculated NO_x concentrations in flame zones at 3000°F–3600°F would be about 6000–10,000 ppm, and the NO to NO_2 ratio would be 500:1 to 1000:1. At typical flue gas exit temperatures of 300°F–600°F, the NO_x concentration would be very low at less than 1 ppm, and the NO to NO_2 ratio would be very low at 1:10 to 1:10,000. Yet, typical NO_x emissions from uncontrolled natural-gas fired boilers are 100–200 ppm. Typical NO_x emissions from uncontrolled fuel-oil-fired and coal-fired boilers, which have higher flame temperatures, are 200–400 ppm and 300–1200 ppm, respectively. And the typical NO to NO_2 ratio in boiler emissions is 10:1 to 20:1.

The reason for the observed NO_x concentrations being so different from equilibrium expectations is the effect of kinetics. The reaction rate is a strong function of

TABLE 17.1
Equilibrium Constants for NO and NO_2 Formation

	$N_2 + O_2 \leftrightarrow 2NO$	$NO + ½O_2 \leftrightarrow NO_2$
°F	K_{P1}	K_{P2}
80	1.0×10^{-30}	1.4×10^{6}
1340	7.5×10^{-9}	1.2×10^{-1}
2240	1.1×10^{-5}	1.1×10^{-2}
3500	3.5×10^{-3}	2.6×10^{-3}

TABLE 17.2
Equilibrium Concentrations

	Air 21% O_2, 79% N_2		Flue Gas 3.3% O_2, 76% N_2	
°F	NO (ppm)	NO_2 (ppm)	NO (ppm)	NO_2 (ppm)
80	3.4×10^{-10}	2.1×10^{-4}	1.1×10^{-10}	3.3×10^{-5}
980	2.3	0.7	0.8	0.1
2060	800	5.6	250	0.9
2912	6100	12	2000	1.8

temperature. Gases reside in the flame zone of a burner for a very short time, less than 0.5 s. The time required to produce 500 ppm NO at 3600°F is only about 0.1 s, but at 3200°F the required time is 1.0 s. Once the gases leave the flame zone, reaction rates are reduced by orders of magnitude, so NO formation stops quickly. Also, the reversible reactions shown by Equations 17.4 and 17.5 slow to nearly a halt, thereby *freezing* the NO_x concentration and the ratio of NO to NO_2.

17.1.2 Prompt NO_x

NO_x concentrations near the flame zone for hydrocarbon fuels demonstrate less temperature dependence than would be expected from the thermodynamic and kinetics considerations of the Zeldovich mechanism discussed above for thermal NO_x. Near the flame zone, radicals such as O and OH enhance the rate of NO_x formation. Hence, some NO_x will form despite aggressive controls on flame temperature and oxygen concentration.

17.1.3 Fuel NO_x

Some fuels contain nitrogen, for example, ammonia or organically bound nitrogen in hydrocarbon compounds. For coal-fired burners, fuel NO_x typically falls in the range of 50%–70% of the total NO_x emissions.[2] Nitrogen in the fuel reacts with oxygen regardless of the flame temperature or excess oxygen concentration in the combustion air. Carbon–nitrogen bonds are broken more easily than diatomic nitrogen bonds, so fuel NO_x formation rates can be higher than thermal NO_x. Combustion control techniques that aim at reducing thermal NO_x formation by reducing flame temperature may not be effective for fuels that have high nitrogen content.

17.2 CONTROL TECHNIQUES

Two primary categories of control techniques for NO_x emissions are (1) combustion controls and (2) flue gas treatment. Very often more than one control technique is used in combination to achieve desired NO_x emission levels at optimal cost. When evaluating control technology, it is desirable to quantify the capability for percent reduction of NO_x. This can be difficult, however, because the baseline operations may or may not be established at good combustor operation, and because the performance of individual technologies is not additive. In addition, there are a number of techniques that can be used in a wide variety of combinations.

17.2.1 Combustion Control Techniques

A variety of combustion control techniques are used to reduce NO_x emissions by taking advantage of the thermodynamic and kinetic processes described above. Some reduce the peak flame temperature; some reduce the oxygen concentration in the primary flame zone; and one, reburn, uses the thermodynamic and kinetic balance to promote reconverting NO_x back to nitrogen and oxygen.

17.2.1.1 Low Excess Air Firing

Combustors tend to be easier to operate when there is plenty of oxygen to support combustion, and operators like to adjust the air-to-fuel ratio to produce a stable and hot flame. By simply cutting back the amount of excess air, the lower oxygen concentration in the flame zone reduces NO_x production. In some cases, where too much excess air has become normal practice, thermal efficiency is improved. However, low excess air in the resulting flame may be longer and less stable, and carbon monoxide emissions may increase. Tuning the combustion air requires minimal capital investment, possibly some instrumentation and fan or damper controls, but it does require increased operator attention and maintenance to keep the system in optimal condition. Depending on the prior operating conditions, combustion air tuning can produce NO_x reduction of 0%–25%.[3] Tuning ranges from simple adjustments to advanced modeling that incorporates neural networks. Applying advanced optimization systems at four coal-fired power plants resulted in NO_x emission reductions from 15% to 55%.[4]

17.2.1.2 Overfire Air

The primary flame zone can be operated fuel rich to reduce oxygen concentration, then additional air can be added downstream. This overfire air provides oxygen to complete combustion of unburned fuel and oxidizes carbon monoxide to carbon dioxide, creating a second combustion zone. Because there is so little fuel in this overfire zone, the peak flame temperature is low. Thus, NO_x formation is inhibited in both the primary and overfire combustion zones.

17.2.1.3 Flue Gas Recirculation

In this technique, some of the flue gas, which is depleted in oxygen, is recirculated to the combustion air. This has two effects: (1) the oxygen concentration in the primary flame zone is decreased and (2) additional nitrogen absorbs heat, that is, acts as a heat sink, and reduces the peak flame temperature. NO_x reduction as a function of the amount of recirculated flue gas is plotted in Figure 17.3.[5]

17.2.1.4 Reduce Air Preheat

Combustion air often is preheated in a recuperator with the heat from the flue gas. This conserves energy by recovering the heat in the flue gas. However, it also raises the peak flame temperature because the combustion air absorbs less heat from the combustor prior to reacting with the fuel. Reducing air preheat lowers the flame temperature to reduce the formation of thermal NO_x.

17.2.1.5 Reduce Firing Rate

Peak flame temperature is determined by the complete heat balance in the combustion chamber, including radiant heat losses to the walls of the chamber. Reducing both air and fuel proportionately would result in the same flame temperature if only fuel, air, and combustion products were considered. However, reducing fuel and air in a fixed size chamber results in a proportionately larger heat loss to the chamber walls and peak flame temperature is reduced.

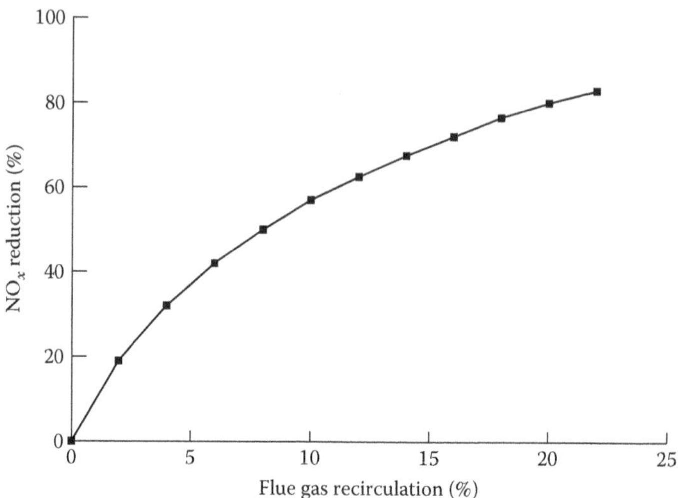

FIGURE 17.3 Effectiveness of flue gas recirculation. (Reproduced with permission of the American Institute of Chemical Engineers, Copyright 1994 AIChE. All rights reserved.)

17.2.1.6 Water/Steam Injection

Injecting water or steam into the combustion chamber provides a heat sink that reduces peak flame temperature. However, a greater effect is believed to result from the increased concentration of reducing agents within the flame zone as steam dissociates into hydrogen and oxygen. Compared to standard natural draft, in natural gas-fired burners, up to 50% NO_x reduction can be achieved by injecting steam at a rate up to 20%–30% of the fuel weight.[6]

17.2.1.7 Burners out of Service

In a large, multiburner furnace, selected burners can be taken out of service by cutting their fuel. The fuel is redistributed to the remaining active burners, and the total fuel rate is not changed. Meanwhile, combustion air is unchanged to all burners. This becomes an inexpensive way to stage the combustion air. The primary flames operate fuel rich and depleted in oxygen, reducing NO_x formation. Outside the hot flame zone, but still inside the combustion chamber, where combustion goes to completion, the additional combustion air burns the remaining fuel at a reduced flame temperature. Test results on a coal-fired boiler demonstrated NO_x emission reduction of 15%–30%.[7]

17.2.1.8 Reburn

A second combustion zone after the primary flame zone can be established by adding additional hydrocarbon fuel outside of the primary flame zone. NO_x is reduced by reaction with hydrocarbon radicals in this zone.[8] Overfire air is added after reburn to complete the combustion process at a low temperature flame. Results from five coal-fired boilers from 33 to 158 MW net capacities show NO_x reduction from baseline levels of 58%–77%.[9]

17.2.1.9 Low NO_x Burners

Low NO_x burners are designed to stage either the air or the fuel within the burner tip. The principle is similar to overfire air (staged air) or reburn (staged fuel) in a furnace. With staged-air burners, the primary flame is burned fuel-rich and the low oxygen concentration minimizes NO_x formation. Additional air is introduced outside of the primary flame where the temperature is lower, thereby keeping the thermo-dynamic equilibrium NO_x concentration low, but hot enough to complete combustion. The concept of a staged-air burner is illustrated in Figure 17.4.

Staged-fuel burners introduce fuel in two locations. A portion of the fuel is mixed with all of the combustion air in the first zone, forming a hot primary flame with abundant excess air. NO_x formation is high in this zone. Then additional fuel is introduced outside of the primary flame zone, forming a low oxygen zone that is still hot enough for kinetics to bring the NO_x concentration to equilibrium in a short period of time. In this zone, NO_x formed in the primary flame zone reverts back to nitrogen and oxygen. A staged fuel burner is illustrated in Figure 17.5. Low NO_x burners can reduce NO_x emissions by 40%–65% from emissions produced by conventional burners.

Because low NO_x burners stage either the air or the fuel, the flame zone is lengthened. The typical flame length of low NO_x burners is about 50%–100% longer than that of standard burners. This can cause a problem in some retrofit applications if the

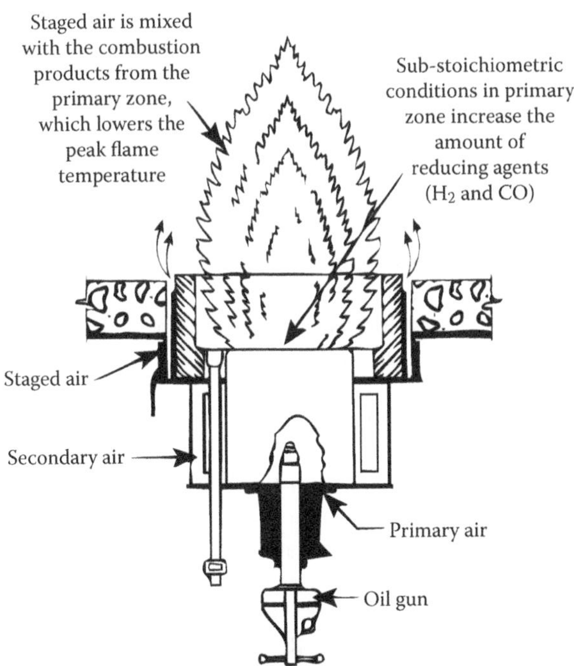

FIGURE 17.4 Staged air low NO_x burner. (Courtesy of John Zink Company, LLC, Tulsa, OK.)

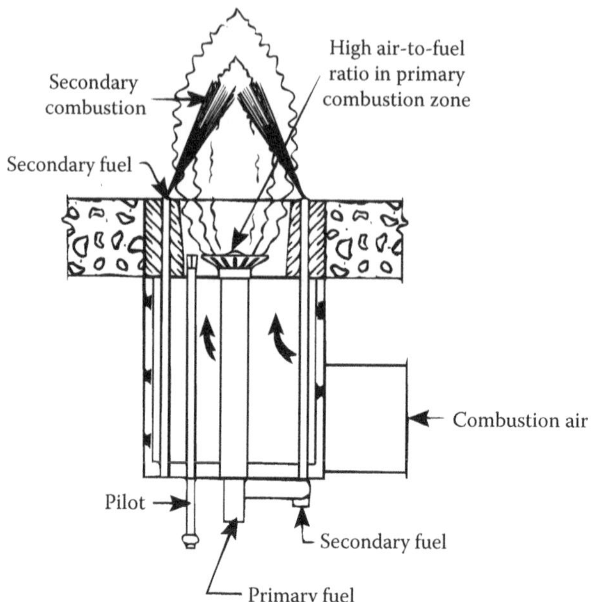

FIGURE 17.5 Staged fuel low NO_x burner. (Courtesy of John Zink Company, LLC, Tulsa, OK.)

longer flame impinges on the walls of the combustion chamber. Flame impingement can cause the chamber walls to erode and fail. While burner replacement may be an easy retrofit technique for NO_x control for many furnaces, it cannot be used in all situations. It is recommended that the flame length should be kept to a third of the firebox height for long vertical cylindrical heaters, and to no more than two-thirds of the firebox height for low-roof cabin heaters.[10]

Low NO_x burners also have been developed for coal-fired applications, where NO_x concentrations are significantly higher than those produced in liquid and gaseous fuel applications. Using low NO_x burners alone, an NO_x emission level of 90–140 ppm at 3% O_2 can be achieved.[11]

17.2.1.10 Ultra Low NO_x Burners

Ultra low NO_x burners have been developed that incorporate mechanisms beyond simply staging air or fuel as designed in low NO_x burners. They may incorporate flue gas recirculation within the furnace that is induced by gas flow and mixing patterns, and use additional levels of air and/or fuel staging. High fuel gas pressure or high liquid fuel atomization pressure is used to induce recirculation. Also, ultra low NO_x burners may use inserts to promote mixing to improve combustion despite low oxygen concentrations in the flame. Remember the *three Ts* of good combustion discussed in Chapter 13—time, temperature, and turbulence. Good combustion at low oxygen concentration is essential to balance low NO_x formation while avoiding soot and excess CO emissions.

Ultra low NO_x burners for gas-fired industrial boilers and furnaces have demonstrated the capability of achieving 10–15 ppm NO_x on a dry basis corrected to 3% oxygen.[12]

17.2.2 Flue Gas Treatment Techniques

17.2.2.1 Selective Noncatalytic Reduction

Selective noncatalytic reduction (SNCR) uses ammonia (NH_3) or urea (H_2NCONH_2) to reduce NO_x to nitrogen and water. The overall reactions using ammonia as the reagent are as follows:

$$2NH_3 + 2NO + {}^1\!/_2O_2 \leftrightarrow 2N_2 + 3H_2O \quad 1600-1900°F \tag{17.8}$$

$$2NH_3 + 2NO + O_2 + H_2 \leftrightarrow 2N_2 + 4H_2O \quad 1300-1900°F \tag{17.9}$$

The intermediate steps involve amine (NH_i) and cyanuric nitrogen (HNCO) radicals. When urea is used, it first dissociates to the primary reactants of ammonia and isocyanic acid.[13]

No catalyst is required for this process; just good mixing of the reactants at the right temperature and some residence time. The key to this process is operating within the narrow temperature window. Sufficient temperature is required to promote the reaction. The presence of hydrogen in the flue gas, if there is a source of it such as dissociation of steam, increases the operable temperature range at the cooler end. At higher temperatures, ammonia oxidizes to form more NO, thereby wasting ammonia reagent and creating the pollutant that was intended to be removed. Above 1900°F, this reaction dominates.

$$4NH_3 + 5O_2 \leftrightarrow 4NO + 6H_2O \tag{17.10}$$

The effect of temperature on SNCR performance is illustrated in Figure 17.6. The critical dependence of temperature requires excellent knowledge of the temperature profile within the furnace for placement of reagent injection nozzles. Computational fluid dynamic models often are used to gain this required knowledge.

A significant complication for a practical system is designing the system for the variable temperature profiles with turndown of a boiler. Operating a furnace at half load obviously will impact the temperature profile. Nozzles may be installed at multiple locations, then reagent is injected at only the locations appropriate for the load conditions. A second complicating factor is the availability of residence time at the proper temperature. The available residence time also may change with load conditions due to the flue gas flow and the boiler configuration.

In a typical application, SNCR produces about 30%–50% NO_x reduction. Some facilities that require higher levels of NO_x reduction take advantage of the low capital cost of the SNCR system, then follow the SNCR section with a selective catalytic reduction (SCR) system (discussed in Section 17.2.2.2). Under these conditions of the lower NO_x removal requirement, the SCR bed will be smaller than the SNCR system alone because the SNCR system has removed a significant portion of the NO_x. Capital costs then would most likely be lower due to the smaller size of the SCR system required.

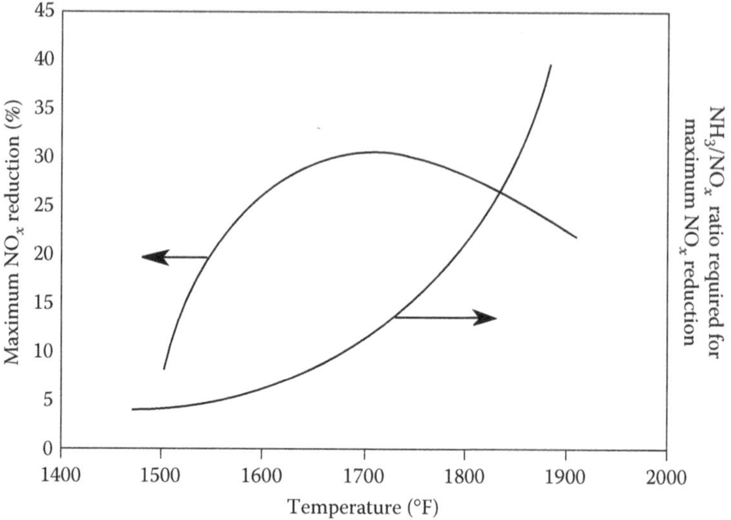

FIGURE 17.6 SNCR temperature window.

17.2.2.2 Selective Catalytic Reduction

A catalyst bed can be used with ammonia as a reducing agent to promote the reduction reaction and to lower the effective temperature. An SCR system consists primarily of an ammonia injection grid and a reactor that contains the catalyst bed. A simplified sketch of the system is shown in Figure 17.7.

The following reactions result in reducing NO_x in an SCR system. Reaction 17.11 is dominant. Since the NO_2 concentration in the flue gas from combustion systems usually is low, then reactions in Equations 17.13 through 17.15 are not particularly significant to the overall NO_x reduction or to the reagent requirement.

$$4NO + 4NH_3 + O_2 \rightarrow 4N_2 + 6H_2O \tag{17.11}$$

$$6NO + 4NH_3 \rightarrow 5N_2 + 6H_2O \tag{17.12}$$

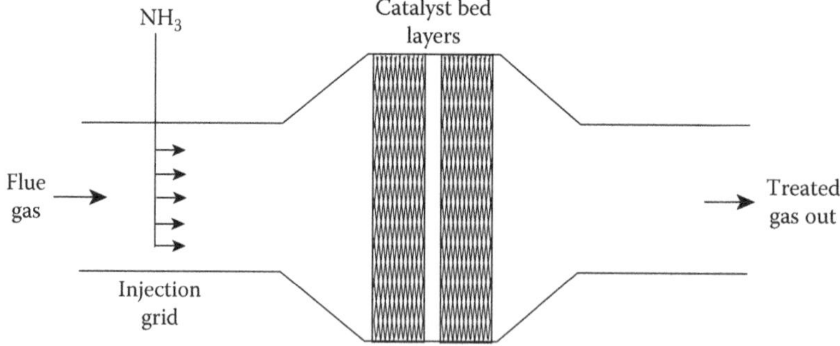

FIGURE 17.7 A schematic of the selective catalytic reduction process.

$$2NO_2 + 4NH_3 + O_2 \rightarrow 3N_2 + 6H_2O \quad (17.13)$$

$$6NO_2 + 8NH_3 \rightarrow 7N_2 + 12H_2O \quad (17.14)$$

$$NO + NO_2 + 2NH_3 \rightarrow 2N_2 + 3H_2O \quad (17.15)$$

SCR operating considerations include ammonia storage and handling. Ammonia can be in either the anhydrous form or the aqueous form. A small amount of ammonia, about 5–20 ppm, will pass, or *slip*, through the catalyst, which creates an emission of a small amount of a hazardous air pollutant in exchange for reducing NO_x.

A variety of catalyst types are used for SCR. Precious metals are used in the low temperature ranges of 350°F–550°F. Vanadium pentoxide supported on titanium dioxide is a common catalyst for the temperature range of 500°F–800°F. Zeolites, which are various aluminosilicates, are used as high temperature catalysts in the range of 850°F–1100°F.

Besides temperature, another catalyst issue is oxidation of SO_2 to SO_3 in flue gases from fuels that contain sulfur. As discussed in Chapter 18, SO_3 results in sulfuric acid mist emissions, which can create opacity that is expensive to control. Tungsten trioxide and molybdenum trioxide are catalysts that minimize sulfur oxidation.

Dust loading and catalyst fouling is a problem when there is a significant amount of particulate in the gas stream. SCR systems currently are being retrofit into some coal-fired power plants as a result of stringent NO_x control regulations. The catalysts are configured into structured grids to minimize dust accumulation. Some catalyst beds are fit with steam-operated sootblowers to remove dust.

SCR systems are capable of 70%–90% NO_x reduction. In the common power-generation application of gas-fired turbines, less than 5 ppm NO_x at 15% O_2 can be achieved.

17.2.2.3 Low Temperature Oxidation with Absorption

A recently commercialized, proprietary technology for NO_x removal is low temperature oxidation of NO_x species to highly soluble N_2O_5, followed by absorbing the N_2O_5 in a wet absorption tower. Ozone is used as the oxidizing agent for the reactions:

$$NO + O_3 \rightarrow NO_2 + O_2 \quad (17.16)$$

$$NO_2 + O_3 \rightarrow N_2O_5 + O_2 \quad (17.17)$$

$$N_2O_5 + H_2O \rightarrow HNO_3 \quad (17.18)$$

A schematic of the simplified process is shown in Figure 17.8. Oxidation with ozone takes place at a low temperature of about 300°F, in the temperature range after the combustion air heater and/or economizer in a typical boiler. At high temperatures above 500°F, ozone decomposes rapidly.

Ozone can be generated from either ambient air or pure oxygen. For typical small boilers, it is often economical to generate oxygen on-site rather than use ambient air.

The absorption step is accomplished in a spray tower, as described in Chapter 11. Caustic is used to neutralize the nitric acid that is formed.

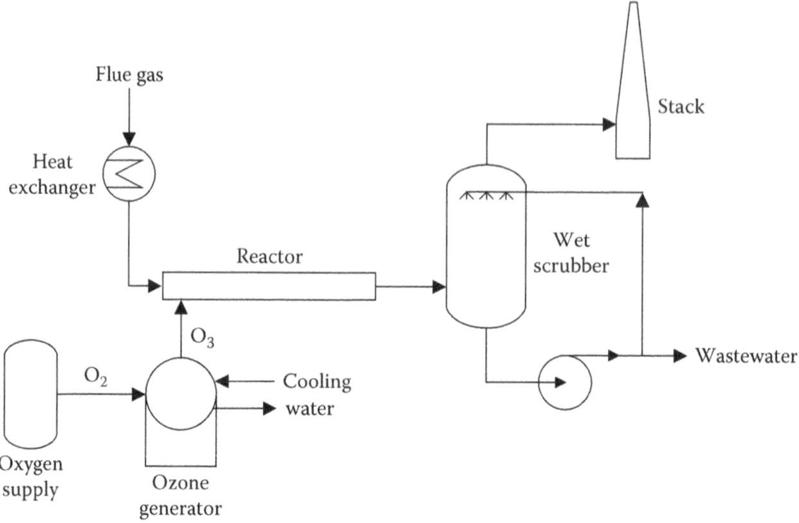

FIGURE 17.8 A schematic of the low temperature oxidation process. (Courtesy of BOC Gases, Murray Hill, NJ.)

Kinetics is a key to this process. Oxidation of NO_2 is faster than oxidation of NO; therefore, NO_2 concentration is not increased. Also, competing oxidation reactions with CO and SO_2 in combustion exhaust gases are relatively slow and do not compete for ozone consumption.

$$CO + O_3 \rightarrow CO_2 + O_2 \tag{17.19}$$

$$SO_2 + O_3 \rightarrow SO_3 + O_2 \tag{17.20}$$

An advantage of this process over other downstream treatment processes is very high NO_x removal efficiency, with 99% removal being reported for an industrial boiler.[14] Some facilities may already have a wet caustic absorber for SO_2 control, so only the ozone generation and injection systems need to be added. A possible disadvantage for some facilities would be additional nitrate in the wastewater discharge, since NO_x is removed in the form of nitric acid.

17.2.2.4 Catalytic Absorption

A proprietary catalytic absorption technology called $SCONO_x$™ also has been commercialized recently. The system utilizes a single catalyst for removal of both NO_x and CO. First, NO, CO, and hydrocarbons are oxidized to NO_2 and CO_2. Then NO_2 is absorbed in a coating of potassium carbonate on the catalyst.

$$2NO_2 + K_2CO_3 \rightarrow CO_2 + KNO_2 + KNO_3 \tag{17.21}$$

After the carbonate coating has been depleted, it is regenerated by passing a dilute stream of hydrogen and carbon dioxide over the catalyst in the absence of oxygen.

$$KNO_2 + KNO_3 + 4H_2 + CO_2 \rightarrow K_2CO_3 + 4H_2O + N_2 \tag{17.22}$$

Because the regeneration cycle must take place in an oxygen-free environment, a section of catalyst undergoing regeneration must be isolated from the exhaust gases. This is accomplished by dividing the catalyst bed into approximately 5–15 or more sections using louvers upstream and downstream of each section to isolate the section for regeneration. Eighty percent of the bed remains on-stream, while 20% is being regenerated. One section is always in the regeneration cycle, which lasts about 3–5 min, so each oxidation/absorption cycle lasts for about 9–15 min.

The process can operate effectively at temperatures ranging from 300°F to 700°F. Significant advantages of this process include the simultaneous removal of CO, hydrocarbons, and NO_x to very low levels, and the lack of ammonia storage and emissions. Less than 2 ppm NO_x has been demonstrated when used in conjunction with other NO_x control technologies, for example, water injection, that limit the $SCONO_x$ inlet NO_x concentration to 25 ppm.[15]

17.2.2.5 Corona-Induced Plasma

Nonthermal plasma consists of ionized gas that can be generated by corona-discharge reactors or electron beams. Plasmas produce chemically active radicals that can oxidize NO to NO_2 and N_2O_5. As discussed above, these NO_x species are water soluble and can be removed by absorption. One form of plasma generation is the conventional electrostatic precipitator, which is discussed in Chapter 24. Another specialized generator is the pulse-corona-discharge reactor.[16]

Corona discharge technology is in the advanced stages of development. However, it has not yet achieved commercial application.

18 Control of SO_x*

This chapter focuses primarily on sulfur dioxide emissions because they are the largest source of sulfur emissions and the primary contributor to acid rain. Other forms of sulfur emissions, including H_2S, SO_3, and sulfuric acid mist, will be discussed. In addition, because some of the SO_2 control technologies are directly applicable to HCl control, there will be brief discussion of HCl emissions.

By far the largest source of sulfur emissions is from burning coal to generate electricity. An approximate distribution of sources that generate SO_2 is presented in Figure 18.1. The declining trend in SO_2 emissions, which is largely as a result of efforts to control SO_2 as a source of acid rain, is plotted in Figure 18.2. Note in particular the sharp decline that occurred in 1995, which was a direct result of implementing the Phase I acid rain controls to comply with Title IV of the Clean Air Act amendments.

Sulfur occurs naturally in fuels. In coal, it is bound as iron pyrite, FeS_2, mineral sulfates, elemental sulfur, and in organic compounds and mercaptans. High sulfur coals typically contain 2%–5% sulfur. Low sulfur coals have less than 1% sulfur. Besides burning coal, sources of sulfur emissions include petroleum refining, oil and gas production, sulfur and sulfuric acid manufacturing, ore smelting, waste incineration, and petroleum coke calcining.

Sulfur emissions as a source of air pollution can be avoided by using processes that remove sulfur from coal before it is burned. Washing coal removes some of the mineral sulfur, sometimes as much as 30%–50% of sulfur in coal. But washing can be a relatively costly process, and research continues for froth flotation, magnetically enhanced washing, sonic enhancement, chemical oxidation, and selective agglomeration.

18.1 H_2S CONTROL

H_2S is a common pollutant in oil and natural-gas processing facilities. Many facilities have sulfur recovery units that convert H_2S to elemental sulfur using the Claus process. A simplified flow diagram for this process is illustrated in Figure 18.3. In this two-step process, a fraction of the H_2S is burned to SO_2. The combustion process must be carefully controlled to obtain the correct molar ratio of 2 moles of H_2S for every mole SO_2. The second step is to pass the mixture of H_2S and SO_2 over catalyst, where H_2S is oxidized and SO_2 is reduced to produce elemental sulfur by the reaction:

$$2H_2S + SO_2 \rightarrow 2H_2O + 3S \tag{18.1}$$

* Various sections of this chapter are based on Brown, C., *Chem. Eng. Progr.*, 94(10), 63, 1998.

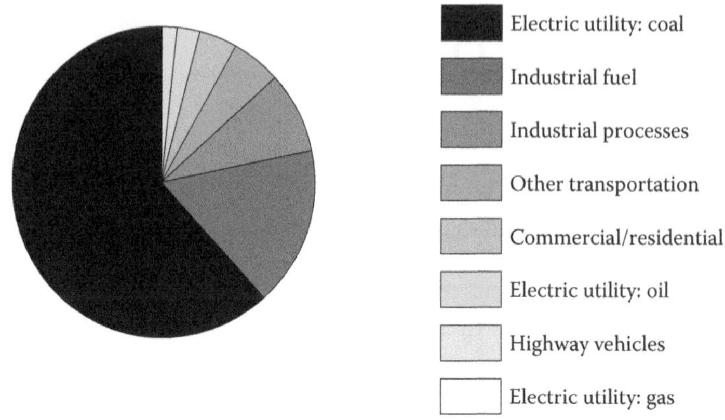

FIGURE 18.1 SO_2 emission sources.

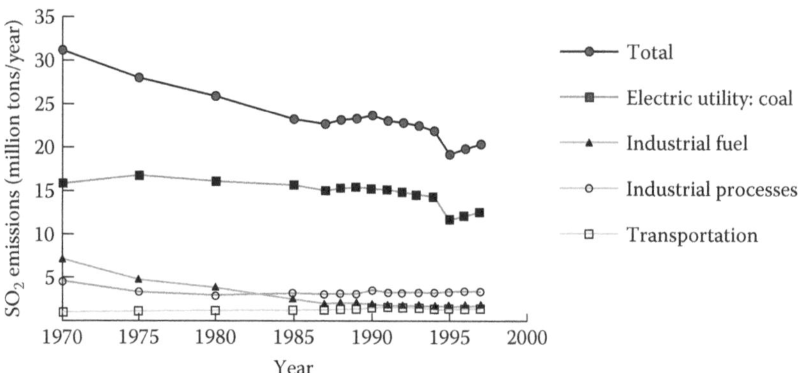

FIGURE 18.2 SO_2 emission trend.

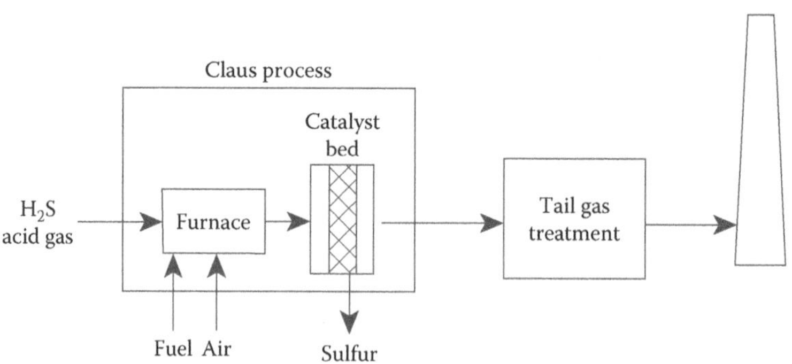

FIGURE 18.3 Oil and gas processing sulfur recovery units.

A 93%–97% sulfur recovery can be obtained with the Claus process. However, the exhaust gas from the process, called *tail gas*, may contain small amounts of unreacted H_2S, SO_2, as well as carbon disulfide (CS_2), carbonyl sulfide (COS), and sulfur vapor (S_8). Some facilities simply incinerate the tail gas to convert the sulfur species to SO_2 before exhausting to the atmosphere. But a variety of approaches are used to eliminate the SO_2 emissions. The sulfur species may be hydrogenated to H_2S, and then an H_2S removal process treats the tail gas. Alternatively, the H_2S and SO_2 may be absorbed and a liquid phase reaction used to produce sulfur. Another alternative is to use an SO_2 removal system for the SO_2 in the tail gas.

18.2 SO_2 (AND HCL) REMOVAL

A variety of processes are available for control systems to treat SO_2 and HCl emissions.[1,2] Selection of the best process for each facility must, of course, include a traditional assessment of capital versus operating costs. Additional considerations include operability, maintainability, plus site-specific preferences for the handling of slurries, aqueous solutions, and dry powder.

SO_2 control processes are used for coal-fired industrial boilers, coke calciners, and catalyst regenerators. SO_2 and HCl controls are required for hazardous and municipal solid waste combustors. HCl control is used for aluminum furnaces fluxed with chlorine and regenerators for chlorinated solvent drums. Flue gas desulfurization is well known to the coal-fired electric utility industry. Many facilities in that industry use wet limestone scrubbers that have a relatively high capital cost in order to utilize inexpensive limestone reagent, although other systems sometimes are used. Smaller, industrial-scale facilities typically use more expensive reagents in systems with lower equipment costs.

18.2.1 Reagents

Reagent selection is a key decision that affects both process design and operating cost. Common reagents and approximate reagent costs are listed in Table 18.1.

18.2.1.1 Calcium-Based Reactions

Limestone is an inexpensive rock that is quarried and crushed. It can be used directly as a reagent either in an aqueous slurry or by injection into a furnace, where the heat decarbonates the limestone. Quicklime is produced by calcining limestone, so quicklime is more expensive than limestone because of the high energy requirement for calcination.

$$CaCO_3 + heat \rightarrow CaO + CO_2 \quad (18.2)$$

Slaking or hydrating quicklime produces more expensive slaked or hydrated lime:

$$CaO + H_2O \rightarrow Ca(OH)_2 \quad (18.3)$$

Wet limestone scrubbing produces calcium sulfite and calcium sulfate reaction products. Oxidation of sulfite to sulfate can be inhibited by adding emulsified sulfur, which forms the thiosulfite ion. Alternatively, oxidation can be enhanced by blowing air into the slurry in the reactor holding tank.

TABLE 18.1
Reagent Chemicals

Common Name or Mineral	Chemical	Formula	Cost ($/ton)
Limestone	Calcium carbonate	$CaCO_3$	$25[a]
Lime, quicklime	Calcium oxide	CaO	$57[b]
Lime, slaked lime, hydrated lime	Calcium hydroxide	$Ca(OH)_2$	$70[b]
Soda ash	Sodium carbonate	Na_2CO_3	$105[c]
Caustic soda	Sodium hydroxide	NaOH	$200[d]
Nahcolite[e]	Sodium bicarbonate	$NaHCO_3$	$260[c]
Trona[f]	Sodium sesquicarbonate	$NaHCO_3\ Na_2CO_3\ 2H_2O$	$65[c]

[a] Delivered. Cost will be lower for very large quantities delivered by barge.
[b] Free on board, supplier's plant. Many locations throughout the United States.
[c] Free on board, Wyoming.
[d] Freight equalized (to be competitive, supplier offers same freight as closer competitor). Based on dry weight of NaOH in 50% solution.
[e] Nahcolite is naturally occurring sodium bicarbonate that contains impurities. Sodium bicarbonate can be refined to remove impurities.
[f] Trona is naturally occurring sodium sesquicarbonate that contains impurities. Sodium sesquicarbonate can be refined to remove impurities.

$$CaCO_3 + SO_2 + \tfrac{1}{2} H_2O \rightarrow CaSO_3 \bullet \tfrac{1}{2} H_2O + CO_2 \tag{18.4}$$

$$2CaSO_3 \bullet \tfrac{1}{2} H_2O + 3H_2O + O_2 \rightarrow 2CaSO_4 \bullet 2H_2O \tag{18.5}$$

Wet lime scrubbing, lime spray drying, and hydrated lime processes also form a mixture of calcium sulfite and sulfate reaction products:

$$Ca(OH)_2 + SO_2 \rightarrow CaSO_3 \bullet \tfrac{1}{2} H_2O + \tfrac{1}{2} H_2O \tag{18.6}$$

$$Ca(OH)_2 + SO_2 + \tfrac{1}{2} O_2 + H_2O \rightarrow CaSO_4 \bullet 2H_2O \tag{18.7}$$

18.2.1.2 Calcium-Based Reaction Products

Calcium sulfite is the reaction product from the inhibited wet limestone scrubber process. It forms needle-like crystals that are hard to de-water because of their high surface area. It has no market value, and it has a chemical oxidation demand if ponded. It can be landfilled, but must be stabilized (typically with fly ash) to improve load-bearing properties.

On the other hand, calcium sulfate product from the forced oxidation process has some desirable features. It forms block-like crystals that are easy to de-water. Also, pure calcium sulfate is gypsum, which can be used to manufacture wallboard.

The dry solid waste product from dry calcium-based systems typically contains about 75% calcium sulfite and 25% calcium sulfate. It can be landfilled, but any unreacted reagent in the waste product will result in a leachate with a high pH.

18.2.1.3 Sodium-Based Reactions

18.2.1.3.1 Wet Sodium-Based Scrubbers

Either soda ash or caustic soda may be used as reagents. They produce a clear scrubber liquor solution with a high pH. The two reagents often are cost competitive with each other on a weight basis, although soda ash tends to be a little less expensive. Caustic soda and soda ash solutions can be used interchangeably in most scrubbers. Caustic soda typically is sold as a 50% solution with pricing on a dry ton NaOH basis. A separate solids storage silo, solids feeder, and dissolution vessel are required to prepare soda ash solution. The market price for caustic soda tends to fluctuate more than other reagents due to market demand. To be competitive, caustic suppliers frequently offer shipping *freight equalized.* This means they will offer the same freight cost as a closer competitor despite being further away. Soda ash is dissolved in water to make the scrubber liquor solution:

$$Na_2CO_3\ (s) + H_2O \rightarrow 2Na^+\ (aq) + CO_3^=\ (aq) + H_2O \tag{18.8}$$

$$CO_3^=(aq) + H_2O \leftrightarrow HCO_3^-(aq) + OH^- \tag{18.9}$$

Caustic soda typically is sold as a 50% solution, which eliminates the need for solid-handling equipment.

$$NaOH\ (s) + H_2O \rightarrow Na^+\ (aq) + OH^- + H_2O \tag{18.10}$$

The first step in alkaline scrubbers is absorption of SO_2 into the aqueous solution. The alkalinity keeps the following equilibrium reactions progressing to the right, which prevents buildup of sulfurous acid from limiting the solubility of gaseous SO_2:

$$SO_2(g) + H_2O \leftrightarrow H_2SO_3\,(aq) \tag{18.11}$$

Sulfurous acid dissociates to form bisulfite or sulfite, depending on the pH:

$$H_2SO_3\ (aq) \leftrightarrow HSO_3^- + H^+ \leftrightarrow SO_3^= + 2H^+ \tag{18.12}$$

Therefore, the overall reactions with SO_2 produce a mixture of sodium sulfite, sodium sulfate, and sodium bisulfite. The exact proportions of the sulfur species depend on the pH and degree of oxidation. The simplified overall reactions are as follows:

$$2Na^+ + 2OH^- + SO_2 \rightarrow Na_2SO_3(aq) + H_2O \tag{18.13}$$

$$2Na^+ + 2OH^- + SO_2 + {}^1\!/\!_2\,O_2 \rightarrow Na_2SO_4\ (aq) + H_2O \tag{18.14}$$

$$Na^+ + OH^- + SO_2 \rightarrow NaHSO_3\ (aq) \tag{18.15}$$

pH is an important variable affecting the solubility of SO_2 gas, as well as reagent consumption. It takes two moles of sodium to remove one mole of SO_2 as sodium sulfite or sulfate, but only one mole of sodium to remove one mole of SO_2 as sodium bisulfite. However, a significantly higher scrubbing liquor flow rate may be required to overcome the lower equilibrium concentration of total dissolved sulfur species at the lower pH, where bisulfite is predominant. Also, if aeration is required to remove the chemical oxidation demand of sulfite and bisulfite in discharged wastewater, the final waste product is sodium sulfate, which requires two moles of sodium despite the pH.

18.2.1.3.2 Dry Sodium-Based Systems

When exposed to heat, dry sodium bicarbonate decomposes to produce dry, high-surface-area soda ash. The soda ash in turn reacts with SO_2. This can be achieved at moderate temperatures of 300°F–600°F, which is in the range of typical exhaust temperatures for many processes.

$$2NaHCO_3 + heat \rightarrow Na_2CO_3 + CO_2 + H_2O \tag{18.16}$$

$$Na_2CO_3 + SO_2 \rightarrow Na_2SO_3 + CO_2 \tag{18.17}$$

Sodium bicarbonate may be purchased as a refined mineral, or as nahcolite, a naturally occurring mineral. Trona, which is mined as naturally occurring sodium sesquicarbonate, also can be used as a dry reagent. Its composition, shown in Table 18.1, is half sodium bicarbonate and half sodium carbonate.

18.2.1.4 Sodium-Based Reaction Products

Sodium-based processes produce sodium bisulfite ($NaHSO_3$), sodium sulfite (Na_2SO_3), and sodium sulfate (Na_2SO_4). These sodium salts are water soluble, which may create a disposal problem in some locations. The liquid waste from these wet scrubber systems cannot be discharged to fresh receiving water. However, in locations near the ocean, wastewater disposal from wet systems is simplified because the wastewater can be discharged directly after pH adjustment and aeration.

Because they are water soluble, dry reaction products from sodium-based powder injection systems produce leachate with high dissolved solids content.

18.2.2 Capital versus Operating Costs

The classic tradeoff of capital versus operating costs requires a good estimate of total installed cost for the system. Capital cost comparisons have been published for large, coal-fired power plants. However, these estimates may not scale down accurately for smaller industrial-scale systems. Industrial systems cover a wide range of sizes, and a traditional capital cost estimate should be developed during process selection.

18.2.2.1 Operating Costs

The major operating cost is the reagent, which can be estimated for a wide range of system sizes in terms of dollars per ton SO_2 and HCl removed. Power and waste product disposal also are significant operating costs that depend upon the process.

The wet limestone scrubber, reagent, and power and waste product costs contribute about 33% each to the total operating cost. For the lime spray dryer or circulating lime systems, each of these components contributes approximately 60%, 15%, and 25%, respectively, to the operating costs. Also, operating labor and maintenance costs should be included if significant differences between processes can be estimated.

The quantity of reagent added to the process is quantified in terms of the reagent ratio. This is defined as the ratio of the amount of reagent added to the stoichiometric amount of reagent required to react with the pollutants in the gas stream.

$$\text{Calcium reagent ratio} = \frac{\text{Moles calcium added}}{\text{Moles Ca}^{++}\text{ required by stoichiometry to remove 100\% SO}_2\text{ and HCl in gas stream}} \qquad (18.18)$$

Note that this definition of reagent ratio is used by the dry and semi-dry acid gas scrubbing industry. It is different from the usual definition of reagent ratio used for wet limestone scrubbers, which is the ratio of moles calcium added per mole SO_2 removed. To be consistent, reagent ratio in this chapter is based on moles pollutant in the incoming gas stream.

$$\text{Sodium reagent ratio} = \frac{\text{Moles Na}^{+}\text{ added}}{\text{Moles Na}^{+}\text{ required by stoichiometry to remove 100\% SO}_2\text{ and HCl in gas stream}} \qquad (18.19)$$

The reagent cost per ton of pollutant removed is

$$\frac{\$}{\text{Ton pollutant remove}} = \text{Reagent ratio} \times \frac{\text{Mole Ca}^{++}\text{ or Na}^{+}\text{ required}}{\text{Mole pollutant}}$$

$$\times \frac{\text{Mole reagent}}{\text{Mole Ca}^{++}\text{ or Na}^{+}} \times \frac{MW_{\text{reagent}}}{MW_{\text{pollutant}}} \times \frac{\text{\% reagent purity}}{100\%} \qquad (18.20)$$

$$\times \frac{100\%}{\text{\% pollutant removal}} \times \frac{\$\text{reagent} + \text{freight}}{\text{Ton reagent}}$$

18.2.3 SO_2 Removal Processes

Several commercially available and demonstrated processes for SO_2 removal are summarized in Table 18.2 and are discussed further below.

18.2.3.1 Wet Limestone

The process flow sheet for the wet limestone scrubbing process is shown in Figure 18.4. Equipment typically includes limestone crusher, storage silo, corrosion resistant absorber tower with mist eliminator, holding tank, slurry pumps, settler, vacuum filter, corrosion resistant ductwork, induced draft (ID) fan, and reheat or corrosion resistant stack. Additional blowers are required for the forced oxidation process. Particulate removal is typically done upstream by electrostatic precipitator or baghouse, but can be done downstream or as an integral part of the absorber.

TABLE 18.2
Summary of Common Processes for Acid Gas Control

Process	Efficiency	Capital Cost	Reagent Cost	Complexity	Comments
Wet limestone	High	High	Low	High	Low reagent cost offsets high cost of operation and maintenance in very large systems.
Wet soda ash/ caustic	High	Moderate	High	Moderate	No slurry or solids handling. Very effective for smaller systems.
Lime spray dryer	Moderate	Moderate	Moderate	Moderate	Mature technology widely used for industrial applications. Subject to deposits accumulation during upset conditions.
Circulating lime reactor	Moderate to high	Moderate	Moderate	Moderate	High solids circulation rate prevents deposits accumulation, enabling slightly higher reactivity or lower reagent cost than lime spray dryer.
Sodium bicarbonate/ trona injection	Moderate	Low	High	Low	Lower cost of trona an advantage in HCl applications, but mostly offset by lower reactivity in SO_2 applications.

For example, a venturi scrubber for particulate control could double as the absorber when limestone slurry is used as the liquid.

Wet limestone scrubbing is the workhorse process for coal-fired electric utility power plants. The high capital cost and the cost of operating and maintaining a complex system is offset by the low cost of limestone used to remove very large quantities of SO_2.

Originally, wet limestone scrubbers were difficult to operate because they were plagued by scaling problems. Partial oxidation of calcium sulfite to calcium sulfate causes sulfate crystals to build up on vessels, pipe walls, and appurtenances. This problem has been largely solved with the development of inhibited oxidation and forced oxidation technologies. Today, wet limestone scrubbers are much more reliable. Single absorbers are considered for a boiler unit at significant cost savings over multiple absorbers.

Inhibited oxidation uses emulsified sulfur as an additive to produce the thiosulfate ion, which inhibits the oxidation of sulfite to sulfate.

$$S + SO_3^{=} \rightarrow S_2O_3^{=} \tag{18.21}$$

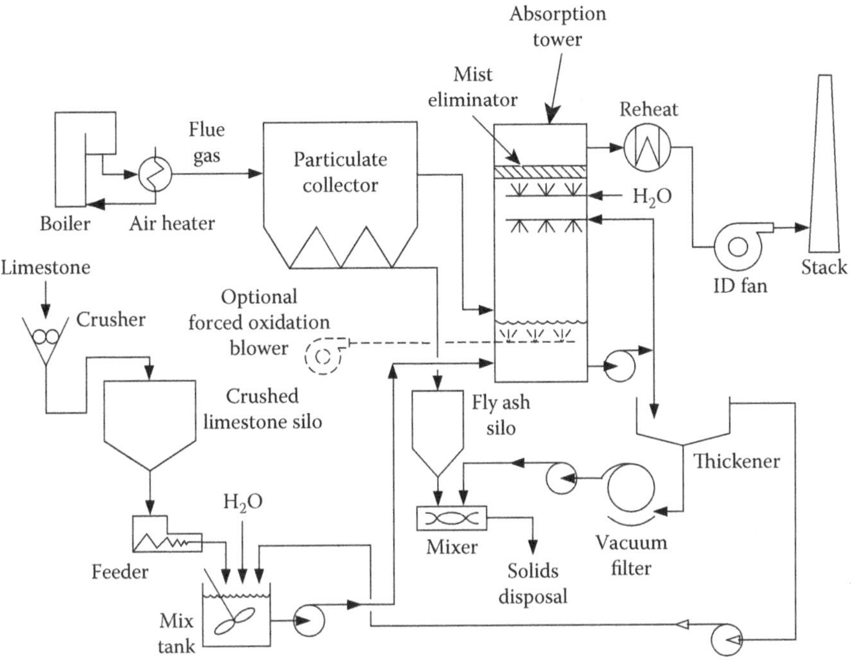

FIGURE 18.4 Simplified wet limestone process flow diagram. (Reproduced with permission of the American Institute of Chemical Engineers. Copyright 1998 AIChE. All rights reserved.)

Thiosulfate stays in solution and is recycled back to the absorber tower. The reaction product of the inhibited oxidation process is calcium sulfite.

In the forced oxidation process, air is injected into the scrubber liquor to oxidize all of the sulfite to sulfate. This produces a large amount of seed crystals. As calcium sulfate precipitates, the crystals grow instead of forming scale. The byproduct of forced oxidation is calcium sulfate. The disadvantage of forced oxidation is the equipment and operating cost associated with blowers for air injection.

SO_2 and HCl removal efficiencies greater than 95% are attainable with wet limestone scrubbers using a reagent ratio of about 1.1. The performance is a function of the absorber configuration, liquid surface area for gas–liquid contact in the absorber, and slurry alkalinity.

18.2.3.2 Wet Soda Ash or Caustic Soda

A simplified process flow sheet for a wet soda ash or caustic soda scrubber is shown in Figure 18.5. Equipment includes soda ash storage silo or caustic tank, corrosion resistant absorber with mist eliminator, holding tank, recirculation pumps, corrosion resistant ductwork, ID fan, and reheat or corrosion resistant stack. Particulate removal is typically done upstream of the absorption tower.

Wet soda ash or caustic scrubbing is simpler than wet limestone scrubbing, since the reagent and the waste products are water soluble. There are no slurries or solid

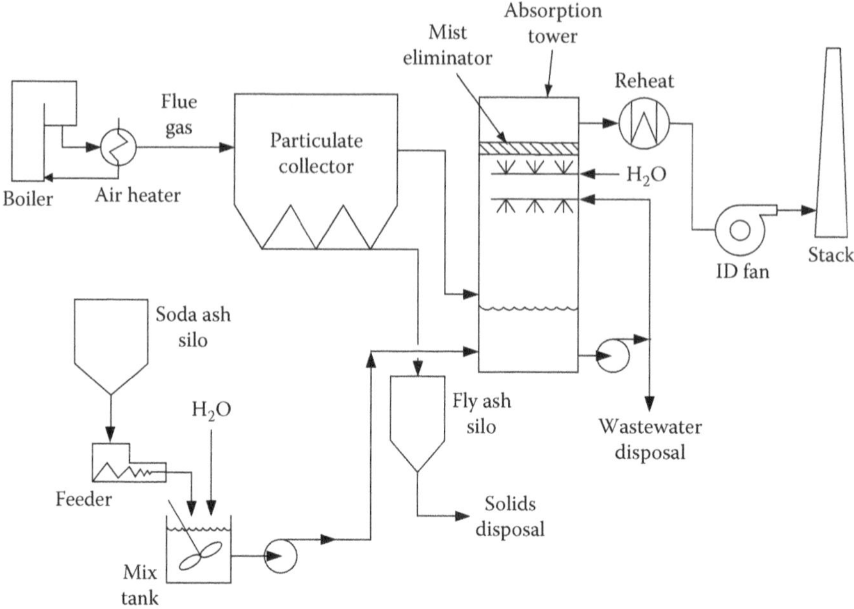

FIGURE 18.5 Soda ash wet scrubber. (Reproduced with permission of the American Institute of Chemical Engineers. Copyright 1998 AIChE. All rights reserved.)

waste products to handle. At the same time, very high acid gas removal efficiency can be obtained. A common application for wet soda ash or caustic scrubbing is catalyst regeneration for fluid catalytic crackers in the refinery industry.

One disadvantage of this process is the relatively high reagent cost, which is tolerable only for smaller systems. Also, the use of aqueous solutions instead of dry powders may be considered a disadvantage by some people, although others dislike handling powders so much that this is considered an advantage. A potential problem that can occur if SO_3 is present when saturating the gas stream with water is the formation of submicron sulfuric acid mist. Once formed, submicron acid mist can pass through the scrubber and pose a difficult particulate collection or emission problem.

Some midsize soda ash/caustic systems have been installed because they were simple and effective. However, they may have reagent costs that are higher than necessary. It is possible to reduce reagent cost significantly by retrofitting lime treatment. Calcium hydroxide regenerates the sodium ion by ion exchange. Some additional equipment is required, including a lime storage silo, solids feeder, slaker, contact tank, pump, settler, and vacuum filter. This modified system is called *dual alkali* because both sodium- and calcium-based materials are used.

18.2.3.3 Lime Spray Drying

The simplified lime spray dryer process flow sheet is shown in Figure 18.6. Widespread use has resulted in spray drying as a mature, developed technology for smaller and low-sulfur coal-fired boilers and for solid waste incinerators. Equipment

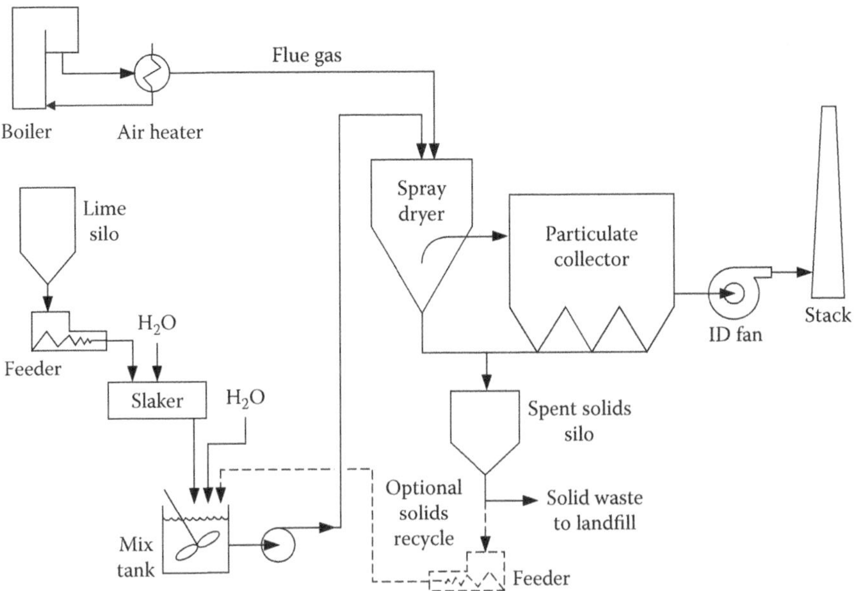

FIGURE 18.6 Simplified spray dryer flow diagram. (Reproduced with permission of the American Institute of Chemical Engineers. Copyright 1998 AIChE. All rights reserved.)

includes carbon steel spray dryer vessel, rotary atomizer or two-fluid abrasion resistant nozzles requiring compressed air, lime storage silo, solids feeder, slaker, mix tank, slurry pumps, particulate collector, carbon steel ductwork, ID fan, stack, optional recycle mix tank and slurry pumps, and optional reheat.

Spray drying with slaked lime slurry is relatively simple compared to a wet scrubber. Carbon steel vessels and ductwork can be used because the flue gas is kept above its saturation temperature. However, the process does require slaking and slurry handling equipment, plus careful control of operating conditions. SO_2 removal performance is enhanced by the presence of surface moisture on the solid lime particles, but excess moisture can result in accumulation of deposits in the spray dryer vessel. The optimum temperature is maintained by control of the approach-to-adiabatic saturation (approach). This is the difference between the spray dryer exit temperature and the adiabatic saturation temperature. For air/water systems, the adiabatic saturation temperature can be estimated by measuring the wet bulb temperature.

Operating conditions have a marked effect on SO_2 removal performance. The approach temperature and reagent ratio are the two major factors. Other variables include SO_2 concentration and inlet gas temperature. The use of a baghouse instead of an electrostatic precipitator can enhance SO_2 removal performance by forcing intimate contact between the gas stream and unreacted reagent in the dust cake. SO_2 removal efficiency of 90% is attainable with a lime spray dryer under good operating conditions. Typical good operating conditions to achieve this level of performance require a reagent ratio of 1.3 at an approach of 20°F. Very high inlet temperature or high inlet SO_2 concentration will reduce the level of performance. Performance is strongly affected by both approach temperature and reagent ratio at ratios below about 2.0 moles Ca/mole SO_2. Increasing the reagent ratio above this level has a diminishing effect on improving performance.

Calcium chloride plays a major role in selecting the desired approach temperature. It is a deliquescent salt, which means that it absorbs moisture then dissolves in the absorbed moisture. Spray dryer solids that contain calcium chloride will be sticky when cool. While this improves acid gas removal performance, it can lead to accumulation of deposits. To avoid this problem, municipal solid waste incinerator systems that scrub both SO_2 and HCl operate at an approach of 50°F–100°F above adiabatic saturation, while spray dryers for coal-fired boilers typically operate at an approach of 20°F–50°F. Small changes in flue gas chloride content have a pronounced effect on SO_2 removal performance. A change in residual chloride content in coal from 0.05% to 0.25% can increase SO_2 removal performance by 8–10 percentage points in a coal-fired power plant application.[3]

HCl is removed in a spray dryer more readily than SO_2. Therefore, SO_2 removal usually is the limiting criteria for performance. The reagent requirement should consider HCl to be completely removed first, followed by the quantity required for SO_2 removal.

It is possible to run a spray dryer system using soda ash solution instead of slaked lime slurry as the reagent. Soda ash precludes the need for a slaker and slurry handling. Deposit buildup can still be an issue, however, and soda ash is significantly more expensive than lime. Therefore, this option in not often used since the technology provides only moderate performance. The expense of soda ash is accepted in wet scrubbers that are capable of high SO_2 removal efficiency.

18.2.3.4 Circulating Lime Reactor

Simplified process flow sheets for circulating lime reactors using quicklime and hydrated lime are shown in Figures 18.7 and 18.8, respectively. Equipment includes a carbon steel reactor, lime storage silo, particulate precollector, final particulate collector, carbon steel ductwork, ID fan, and stack. Either a hydrator or a slaker is required to use less expensive quicklime reagent. Hydrated lime can be used with less equipment, but a hydrator or slaker usually can pay for itself quickly even in small systems.

The circulating lime reactor process is similar to the spray drying process, but has one significant difference: in this system, a much larger quantity of solids can be recycled from the particulate collectors to the reactor because the solids recycle rate is not limited by the water balance and maximum pumpable slurry concentration. The recycle of larger solids produces two advantages. First, it enhances reagent utilization by providing additional gas–solid contact time with additional passes of solids through the reactor. Second, the recycled solids provide surface area to absorb moisture and scour the walls of the reactor vessel. The second advantage is the more significant of the two. It allows the reactor to operate at a lower approach temperature than a spray dryer, without experiencing deposits accumulation. Results show that, under otherwise identical conditions, 90% SO_2 removal can be obtained at a reagent

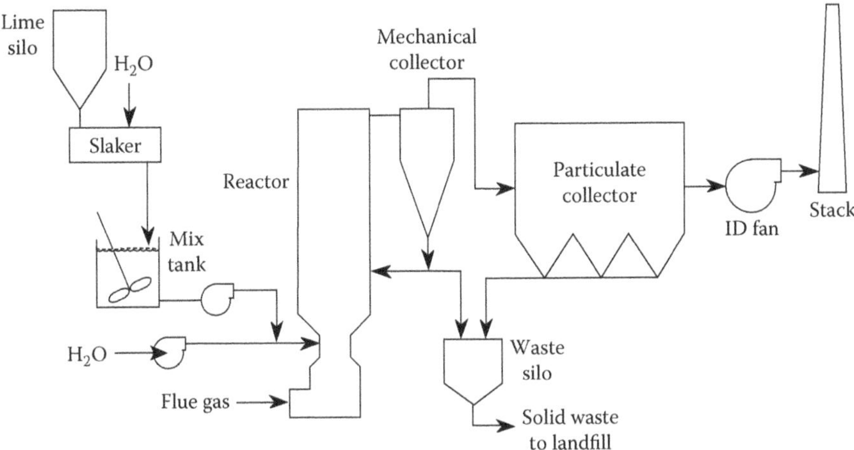

FIGURE 18.7 Simplified hydrated lime reactor flow diagram: lime slurry. (Reproduced with permission of the American Institute of Chemical Engineers. Copyright 1998 AIChE. All rights reserved.)

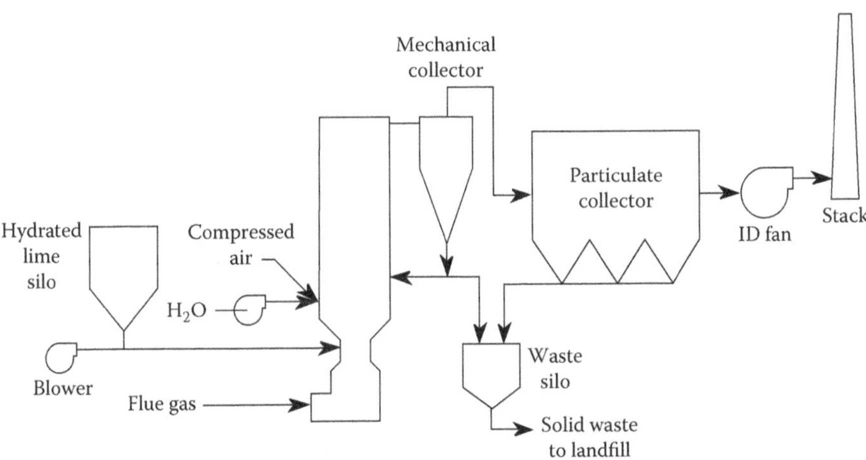

FIGURE 18.8 Hydrated lime reactor flow diagram: humidification and hydrated lime. (Reproduced with permission of the American Institute of Chemical Engineers. Copyright 1998 AIChE. All rights reserved.)

ratio of 1.2 and approach of 12°F with a circulating lime reactor.[4] This compares to a reagent ratio of 1.3 at an approach of 18°F for a spray dryer. At the same operating conditions, there is little difference in the performance between the circulating lime reactor and the spray dryer with optional solids recycle in the slurry.[5] However, high removal efficiency performance levels of 98% can be achieved with the proper combination of high reagent ratio (>1.5), low approach temperature, and sufficient solids chloride content.[6]

Cooling is required to promote the reaction between lime and SO_2 in the presence of surface moisture on lime particles. Vendors of circulating lime reactor systems use different methods of cooling, including prehumidification, humidification in the reactor, slurry spray (with quicklime reagent), and heat exchange. Air dilution can also be used, although it increases the gas flow and requires a larger baghouse and ID fan.

Prehumidification of acid gases must be done in a properly designed humidification chamber to avoid corrosion. Two-fluid nozzles are used to atomize the water spray. In larger systems, the cost of compressed air is significant.

Humidification within the reactor vessel avoids the corrosion problem because the alkalinity of the reagent neutralizes any acid. Simpler water nozzles can be used because the solids absorb moisture and a very fine spray is not required. Both prehumidification and humidification systems use hydrated lime as the reagent. Hydrated lime powder can be produced from quicklime using a hydrator. Small users may purchase hydrated lime, which is more expensive than quicklime.

Slaked lime slurry can be used to humidify and inject lime in one step, but requires a slaker, slurry pumps, and an abrasion-resistant slurry spray nozzle. It also requires compressed air to atomize the slurry.

Cooling with an air-to-gas heat exchanger avoids the use of water, but adds the cost of the exchanger. This is an attractive approach in applications where water is scarce or there is a use for the recovered low-grade heat.

Because the circulating lime process can operate at a lower approach temperature than a lime spray dryer, the circulating lime process has slightly better reagent utilization. This gives a reagent cost advantage to the circulating lime process when quicklime is used.

18.2.3.5 Sodium Bicarbonate/Sodium Sesquicarbonate Injection

A process flow sheet for the sodium bicarbonate process is shown in Figure 18.9. The basic flowsheet for sodium sesquicarbonate injection is the same, although the reagent reactivity and costs are different. Equipment includes reagent storage, solids feeder, blower for pneumatic injection, particulate collector, carbon steel ductwork, ID fan, and stack. This is a very simple system that has minimal operation and maintenance requirements.

Sodium bicarbonate decarbonates in the temperature range of 300°F–500°F. This is a typical range for exhaust gases from combustion systems with air heaters or heat recovery. The decarbonation process causes particles to *popcorn*, creating high surface area reagent and exposing fresh reagent that was covered by reaction products. While the decarbonated reagent is Na_2CO_3, injecting less expensive Na_2CO_3 does not work well because the popcorn effect is important to reactivity.

Ninety percent SO_2 removal performance can be achieved with sodium bicarbonate at a reagent ratio of approximately 1.2 moles Na/mole SO_2.[7] Trona, which is mined as naturally occurring sodium sesquicarbonate, is an alternate reagent. Its use can reduce reagent cost, particularly in HCl applications. In SO_2 applications, the performance of trona is lower than sodium bicarbonate, which reduces its cost advantage. To achieve 90% SO_2 removal performance with trona, a relatively high reagent ratio of about 1.5–1.9 moles Na/mole SO_2 is required.[7,8] In HCl applications,

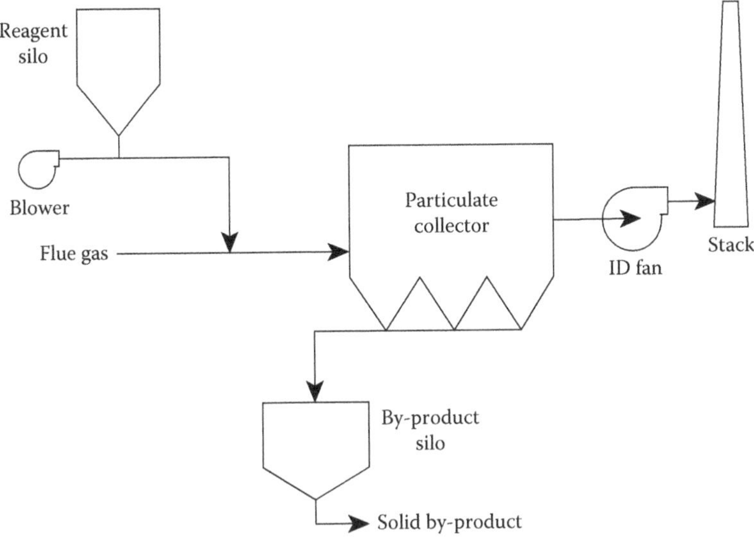

FIGURE 18.9 Sodium bicarbonate flow diagram. (Reproduced with permission of the American Institute of Chemical Engineers. Copyright 1998 AIChE. All rights reserved.)

however, the performance of trona is nearly equal to sodium bicarbonate at the same reagent ratio.[8] The estimated reagent ratio to achieve 90% HCl removal using trona is approximately 1.2 moles Na/mole HCl.

A disadvantage of sodium bicarbonate injection for SO_2 applications is the potential for formation of a brown, NO_2 plume. A NO_2 plume can develop under certain combinations of SO_2, NO, moisture, and temperature. The necessary combination is typical of flue gas from coal-fired power plants. Overall NO_x removal of about 20% can be obtained with the process, but some NO is oxidized to NO_2, resulting in increased NO_2 concentrations despite overall NO_x removal. NO_2 emissions can be controlled with additives such as ammonia, ammonium bicarbonate, and urea.[7] The plume problem does not arise at elevated temperatures (>400°F), at high moisture concentrations, or in HCl removal applications where SO_2 is not present.

The low cost, simple sodium bicarbonate system comes with a very high reagent cost of $260/ton plus freight from Wyoming. That is the only location in the United States where sodium bicarbonate is mined commercially. Freight makes up a substantial portion of the reagent cost and must be included in a site-specific cost analysis. The high reagent cost makes economic sense only when the quantity of pollutant is low.

Trona, at $65/ton plus freight from Wyoming, is much less expensive than sodium bicarbonate. Since the reactivity for HCl removal is almost the same as sodium bicarbonate, the cost advantage for HCl removal is huge. Much of the cost advantage is lost in SO_2 removal applications, however, due to lower reactivity with SO_2. A reagent ratio of 1.9 may be required to achieve 90% removal in typical SO_2 applications.

18.2.3.6 Other SO_2 Removal Processes

Several other processes, such as duct injection, duct spray drying, furnace injection, economizer injection, ADVAnced siliCATE (ADVACATE), E-SO_x, and HYbrid Pollution Abatement System (HYPAS) have been piloted or developed, but are not widely used for industrial systems. However, a basic understanding of the reagents and reaction mechanisms and the inherent advantages and disadvantages of the processes described above will enable one to evaluate modifications of these processes, as well as other new processes that may be offered.

18.2.4 Example Evaluation

An industrial facility has a 100,000 acfm exhaust gas stream at 400°F that contains 1000 ppm SO_2 and 200 ppm HCl. Each pollutant requires 90% removal efficiency. This facility wants a five-year payback on any capital investment that reduces operating costs, using a 10% interest rate. The facility is located next to the ocean, so dissolved sodium salts could be discharged after aeration. The operators have expressed a dislike for slurry handling systems. Which system is recommended?

An evaluation of capital and operating costs, summarized in Table 18.3, shows that all of the sodium-based reagents have very high reagent costs for this application. The cost savings of calcium-based systems are significantly greater that the extra cost of landfilling the waste product. Also, the high capital cost of a wet limestone scrubber makes it expensive to recover the small savings in reagent cost. Given the choice between a lime spray dryer and a circulating lime reactor at the same capital cost, a small savings in reagent requirement for the circulating lime reactor, and the operators' preference for avoiding slurry handling, then a circulating lime system that uses water instead of slurry for humidification is recommended for this hypothetical case.

18.3 SO_3 AND SULFURIC ACID

18.3.1 SO_3 and H_2SO_4 Formation

An interesting aspect of sulfur oxide emissions is the oxidation of a small fraction of SO_2 to SO_3. Oxidation is promoted by the presence of metals in contact with the flue gas that catalyze the reaction. Some catalysis may occur from iron on boiler tubes. Fly ash from coal combustion may contain metals that catalyze the oxidation of SO_2. Petroleum coke typically has a relatively high vanadium content, which comes from crude oil, so petroleum coke combustion and calcining can have a high oxidation rate. Also, fluid catalytic cracker regenerator flue gas contains entrained catalyst.

SO_3 combines readily with water vapor in the temperature range of 400°F–900°F, producing sulfuric acid vapor. The equilibrium relationship between SO_3, H_2O, and H_2SO_4 is given by Gmitro and Vermeulen[9]:

$$\ln K_p = -6.71464 \ln\left(\frac{298}{T}\right) - \frac{81016.1}{T^2} - \frac{9643.04}{T} + 14.74965 - 0.009458T + 0.00000219062T^2 \tag{18.22}$$

TABLE 18.3
Example Evaluation

System	Wet Limestone Scrubber	Lime Spray Dryer	Circulating Lime Reactor	Soda Ash Scrubber	Caustic Scrubber	Sodium Bicarbonate Injection	Trona Injection
Total installed cost ($)	7,000,000	4,000,000	4,000,000	6,000,000	6,000,000	2,500,000	2,500,000
Capital recovery cost ($)	1,847,000	1,055,000	1,055,000	1,583,000	1,583,000	659,000	659,000
Reagent ratio for SO_2	1.0	1.3	1.2	1.0	1.0	1.2	1.7
Reagent ratio for HCl	1.0	1.0	1.0	1.0	1.0	1.2	1.2
Reagent use (tons/year)	4,600	3,280	3,040	4,880	3,680	9,270	11,420
Reagent cost ($)	115,000	187,000	174,000	512,000	736,000	2,967,000	1,427,000
Power cost ($)	115,000	47,000	47,000	115,000	115,000	47,000	47,000
Landfill cost ($)	115,000	78,000	78,000	0	0	0	0
Total annual cost ($)	2,192,000	1,367,000	1,354,000	2,210,000	2,434,000	3,673,000	2,133,000

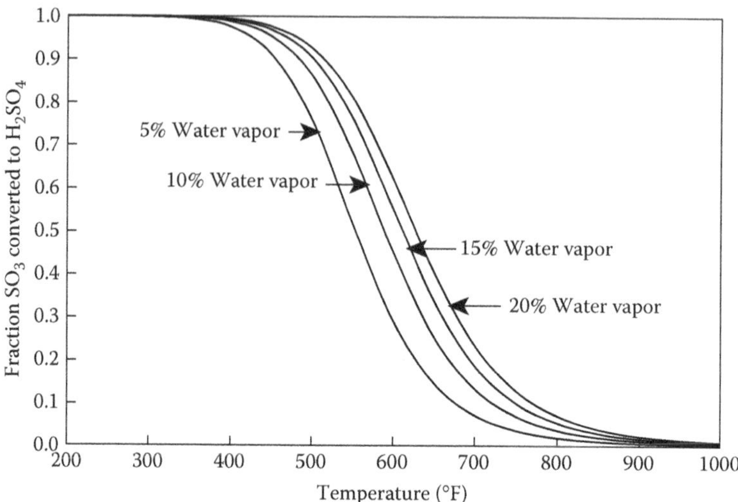

FIGURE 18.10 SO_3/H_2SO_4 equilibrium.

where:

$$K_p = \frac{\overline{P}_{SO_3}\overline{P}_{H_2O}}{\overline{P}_{H_2SO_4}}, \text{ atm}$$

T is the *temperature*, °K

Using this equilibrium relationship, the fraction of SO_3 converted to H_2SO_4 as a function of temperature and water vapor concentration in flue gas is plotted in Figure 18.10.

Condensed sulfuric acid is extremely corrosive, and, as noted in the discussion of wet limestone scrubbing, requires the use of corrosion-resistant materials, where the flue gas temperature is below the acid dew point. When the temperature is below the acid dew point, sulfuric acid vapor condenses into liquid aerosol droplets. Haase and Borgmann[10] proposed a simple correlation for the acid dew point based on experimental data:

$$T_{dp,C} = 255 + 27.6 \log \overline{P}_{H_2O} + 18.7 \log \overline{P}_{SO_3} \tag{18.23}$$

where:

$T_{dp,C}$ is the sulfuric acid dew point, °C
$\overline{P}$ is the partial pressure, atmospheres

Although useful because of its simplicity, it does not predict dew point accurately at high temperatures. Other correlations do not show a linear relation between dew point and the logarithm of the SO_3 partial pressure. Verhoff and Banchero[11] proposed another equation for sulfuric acid dew point:

$$\frac{1}{T_{dp,K}} = 0.002276 - 0.00002943 \ln P_{H_2O} - 0.0000858 \ln P_{SO_3} + 0.0000062\left(\ln P_{H_2O}\right)\left(\ln P_{SO_3}\right) \quad (18.24)$$

where:

$T_{dp,K}$ is the sulfuric acid dew point, °K
P is the partial pressure, mm Hg

Okkes and Badger[12] proposed a correlation that is based on the simplicity of the correlation of Haase and Borgmann,[10] but is adjusted to curve-fit data:

$$T_{a,C} = 203.25 + 27.6 \log P_{H_2O} + 10.83 \log P_{SO_3} + 1.06\left(\log P_{SO_3+8}\right)^{2.19} \quad (18.25)$$

where:

$T_{a,C}$ is the sulfuric acid dew point, °C
P is the partial pressure, atmospheres

One problem with using the acid dew point correlations is determining the partial pressure of SO_3. The oxidation of SO_2 to SO_3 is affected by a number of variables, including oxygen and SO_2 concentrations, temperature, and the presence of catalysts that may be in entrained particulate or on metal boiler tubes. Neglecting the influence of catalysts, thermodynamic equilibrium may be used to estimate the SO_3 concentration at temperatures above 1000°C, where kinetics do not hinder the reaction rate by using the relationship:

$$K_p = \frac{\bar{P}_{SO_3}}{\bar{P}_{SO_2}\left(\bar{P}_{O_2}\right)^{0.5}} \quad (18.26)$$

The equilibrium constant, K_p, may be calculated from:

$$\ln K_p = \left(\frac{12120}{T}\right)\left[1 - (0.000942T) + \left(7.02\times10^{-8}T^2\right) - \left(1.08\times10^{-5}T\ln T\right) - \frac{3.1}{T}\right] \quad (18.27)$$

where:

T is the temperature, °K

The SO_3 concentration in cooled flue gases can be estimated using the relationship in Equation 18.25 at a temperature of 1000°C by assuming slow kinetics *freezes* the concentration at that value.

An alternative to calculating acid dew point from equations, assumptions, and empirical correlations is to measure it with a commercially available acid dew point analyzer. Typical coal-fired boilers have acid dew points of approximately 225°F–275°F.

Formation of sulfuric acid mist presents another unique problem to wet SO_2 absorbers. Submicron mist can form immediately at the entrance of a wet absorber as the flue gas is saturated with water and before gas phase SO_3 and H_2SO_4 diffuses to large droplets of absorber liquor solution. Once the submicron mist is formed, it is difficult to remove with a wet scrubber because the mechanisms of interaction and interception are weak for removing submicron particulate (droplets of sulfuric acid mist). This can result in significant sulfuric acid mist emissions, which can form a visible plume. This has been an aggravating problem for coal-fired utility boilers that retrofitted wet SO_2 absorbers downstream of existing electrostatic precipitators to meet SO_2 emission control requirements. While succeeding in removing substantial amounts of SO_2, some of them went from clear stacks to a stack that emitted a bluish-gray haze.[13]

18.3.2 Toxic Release Inventory

Until recent years, sulfuric acid mist generated by combustion processes was sometimes overlooked and unreported for the annual Superfund Amendments and Reauthorization Act (SARA) 313 Toxic Release Inventory. Also until recently, coal-fired utilities were exempt from SARA reporting. But when reported, the quantity of sulfuric acid mist generated by combustion processes is a very large number. Interestingly, the reporting requirements for sulfuric acid mist exclude SO_3, but include H_2SO_4 in both the vapor and liquid droplet forms.[14] This is one of those many oddities that can be found in the regulations for reporting. But it means that knowing the conversion of SO_3 to H_2SO_4 is important for accurate reporting.

19 Fundamentals of Particulate Control

To understand particulate control, one must first understand some fundamental concepts and properties of particulate. Figure 19.1 shows common terms associated with particulate material as a function of particle diameter. Particles of concern to air pollution are typically measured in microns (i.e., micrometers, or 1×10^{-6} m). Large or coarse particles are those considered to be well above 10 μm. Note for perspective that the diameter of a human hair is approximately 50–110 μm. Very small or fine particles are considered to be those less than 1 μm, or submicron. This size range includes smoke and fumes. Also notice that the size range considered *lung-damaging dust* ranges from approximately 0.7 to 7 μm. Dust in this size range gets into lungs and is the most difficult dust to collect with particulate control equipment for reasons that will be discussed later in this chapter.

19.1 PARTICLE SIZE DISTRIBUTION

Particles frequently are described by their diameter. The amount of the particulate of a given diameter is very informative. But pay attention to the number distribution and the mass distribution. Table 19.1 illustrates the distinction. Given 100,000 particles of 1 μm diameter and assigning each particle a unit mass of one gives a total mass of 100,000 units. Adding 1000 particles with a diameter of 10 μm plus 10 particles with a diameter of 100 μm provides a size distribution consisting of 1, 10, and 100 μm particles. There are 101,010 particles, of which 99% are 1 μm in diameter. Because the volume, and therefore the mass, of each particle varies with the cube of the diameter, the mass of 1-μm particles is only 9% of the total mass of all particles. Size distributions commonly are expressed in terms of the mass distribution. It is always best to be very clear whether the distribution is a mass or a number distribution.

Sample particle size distribution data are listed in Table 19.2. Data are divided into six size bins, the first being 0–2 μm, and the last being greater than 25 μm. Note that the last size bin is open ended. The mass fraction of particulate within each size range is provided. The next column gives the cumulative weight percent less than the top size listed in each size range.

Cumulative size distributions are commonly plotted on a probability paper. The probability paper is based upon the normal distribution, a bell-shaped curve, with the highest density in the center and the lowest densities at each extreme. The *x*-axis on the probability paper is different from that in a logarithmic graph paper, where the highest density would be at the extreme top or right. The *y*-axis on the probability paper may be either linear or logarithmic. The example data listed in Table 19.2 are plotted on a linear probability paper in Figure 19.2 and on a log-probability paper in Figure 19.3. On the probability paper with a linear *y*-axis, a normal distribution would

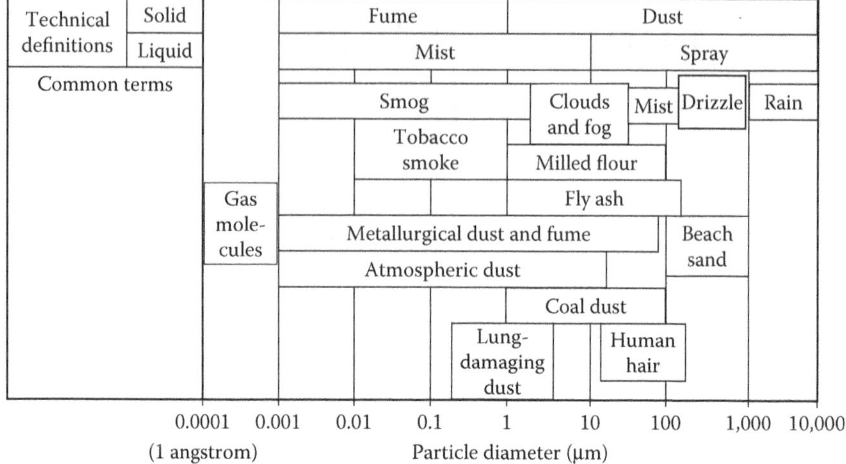

FIGURE 19.1 Common particulate terms and size ranges.

TABLE 19.1
Number Fraction versus Mass Fraction

D_j (μm)	Number	Number Fraction	Mass of Particle	Total Mass	Mass Fraction
1	100,000	0.990000	1	10^5	0.009
10	1,000	0.009900	1000	10^6	0.090
100	10	0.000099	10^6	10^7	0.900
Total	101,010	1.0		1.11×10^7	1.0

TABLE 19.2
Sample Particle Size Distribution Data

Size Range (μm)	Mass Fraction in Size Range (m_j)	Cumulative % Less than Top Size
0–2	0.005	0.5
2–5	0.195	20.0
5–9	0.400	60.0
9–15	0.300	90.0
15–25	0.080	98.0
>25	0.020	100.0

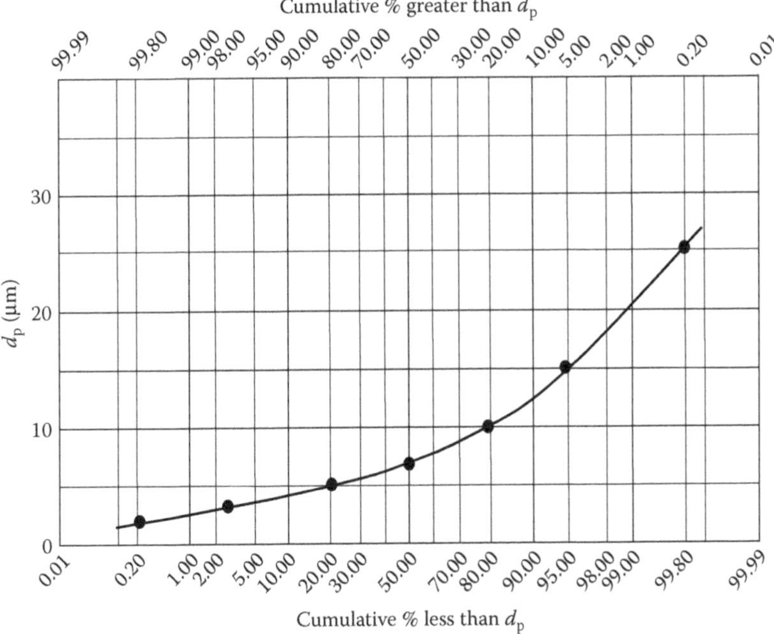

FIGURE 19.2 Sample particle size distribution on a linear-probability plot.

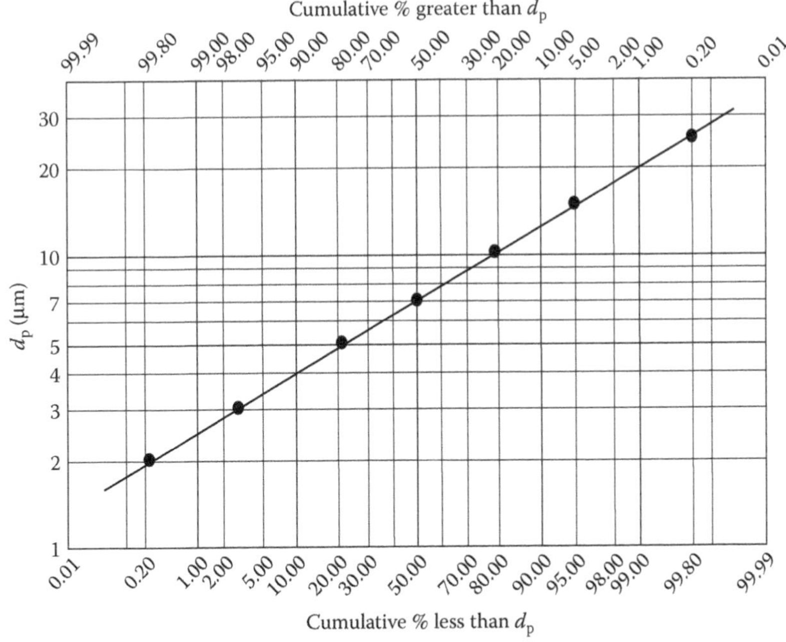

FIGURE 19.3 Sample particle size distribution on a log-probability plot.

be plotted as a straight line. On the log-probability paper, a log-normal distribution would be plotted as a straight line. The sample particle size distribution in Table 19.2 plots as a straight line on the log-probability paper, so it is a log-normal distribution.

These data can be plotted on a linear graph paper, giving the mass fraction in equal micron size range bins, which would show a skewed distribution, as shown in Figure 19.4. The same distribution, but with the mass fraction bins plotted on a log scale, gives the distribution as shown in Figure 19.5. The distribution produces a bell-shaped curve when plotted on the log scale, indicative of a log-normal distribution.

When the same data are plotted as a cumulative distribution, that is, as the total percentage less than the specified particle size, an S-shaped curve is produced.

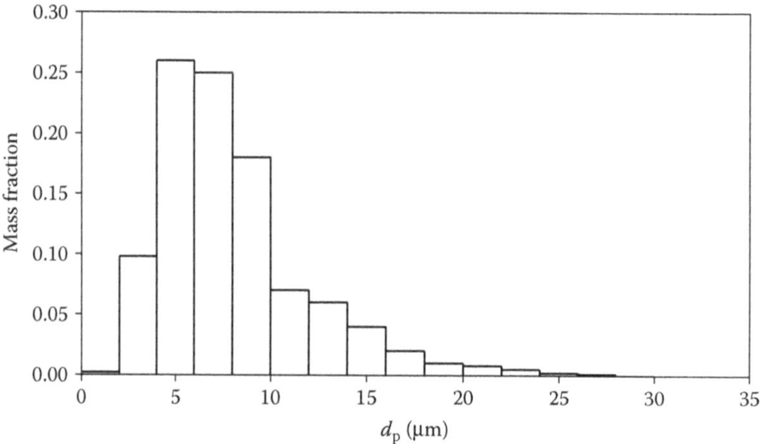

FIGURE 19.4 Particle size bins plotted on a linear graph paper.

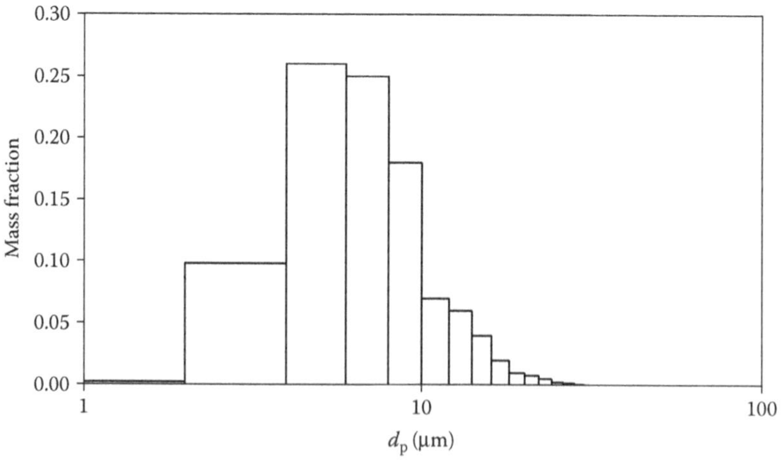

FIGURE 19.5 Particle size bins plotted on a semi-logarithmic graph paper.

The curve for a log-normal distribution is skewed when plotted on the linear graph paper, as shown in Figure 19.6. But when plotted on a semi-logarithmic paper, the curve is symmetrical, as shown in Figure 19.7.

Log-normal distributions are common for many particulate emission sources because a large amount of mass is concentrated in a very few large particles. While common, other distributions are certainly found. Bimodal distributions, those with a large mass of particles in two different size fractions, are sometimes found. They may be indicative of two different mechanisms for generating the particles. For example, larger particles in the distribution may have formed from a grinding process, while smaller particles may have formed from a condensation process.

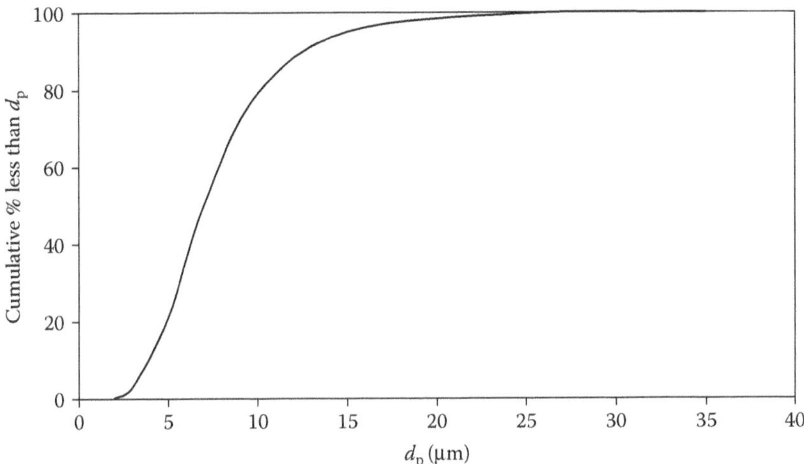

FIGURE 19.6 Sample cumulative size distribution on a linear paper.

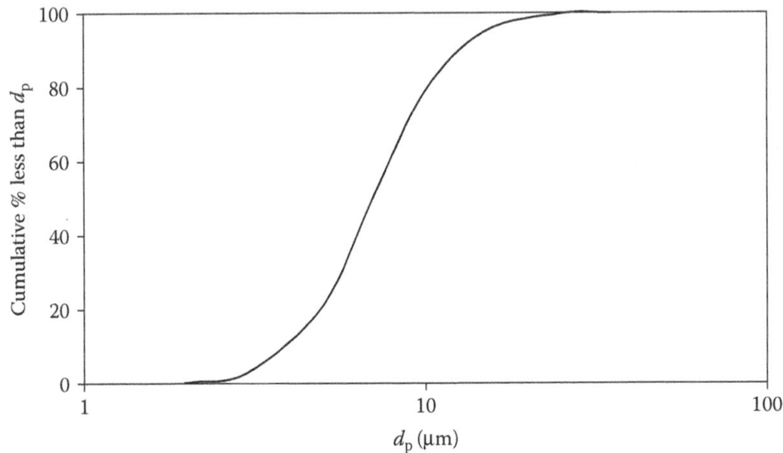

FIGURE 19.7 Sample cumulative size distribution on a semi-logarithmic paper.

19.2 AERODYNAMIC DIAMETER

The diameter of the particle is a common descriptor. But there are two diameters that are commonly used as descriptors: physical diameter and aerodynamic diameter. The physical diameter, of course, is the actual diameter of a spherical or nearly spherical particle. The aerodynamic diameter is the diameter of a spherical particle with the density of 1 g/cm^3 that behaves aerodynamically in the same manner as the subject particle. That is, the actual particle and the spherical particle with density of 1 g/cm^3 have the same momentum and drag characteristics. As a practical example, they have the same terminal settling velocity, which is an aerodynamic property. Remember that terminal settling velocity results from simple force balance between the force of gravity and resisting drag force.

19.3 CUNNINGHAM SLIP CORRECTION

When airborne particles are so small that the particle size approaches the mean free path of gas molecules, less than 5 μm, drag on the particles tends to be reduced. Drag is created as gas molecules impact a moving particle. But when the mean free path is approached, a moving particle will tend to *slip* between the gas molecules with less resistance. This phenomenon is of importance to particulate collection devices and the term *Cunningham slip correction factor* will be found in many small particle correlations. The following empirical correlation for the Cunningham slip correction factor was developed by Davies[1]:

$$C' = 1 + 2\frac{\lambda}{d_p}\left[1.257 + 0.4\exp\left(-0.55\frac{d_p}{\lambda}\right)\right] \tag{19.1}$$

where:

C' is the Cunningham slip correction factor
λ is the mean free path, meters
d_p is the particle diameter, microns

The mean free path, λ, is given by

$$\lambda = \frac{\mu}{0.499\,\rho_g u_m} \tag{19.2}$$

where:

μ is the gas viscosity, kg/m-s
ρ_g is the gas density, kg/m^3
u_m is the mean molecular speed, m/s

The mean molecular speed, u_m, is given by

$$u_m = \sqrt{\frac{8RT}{\pi\,\mathrm{MW}}} \tag{19.3}$$

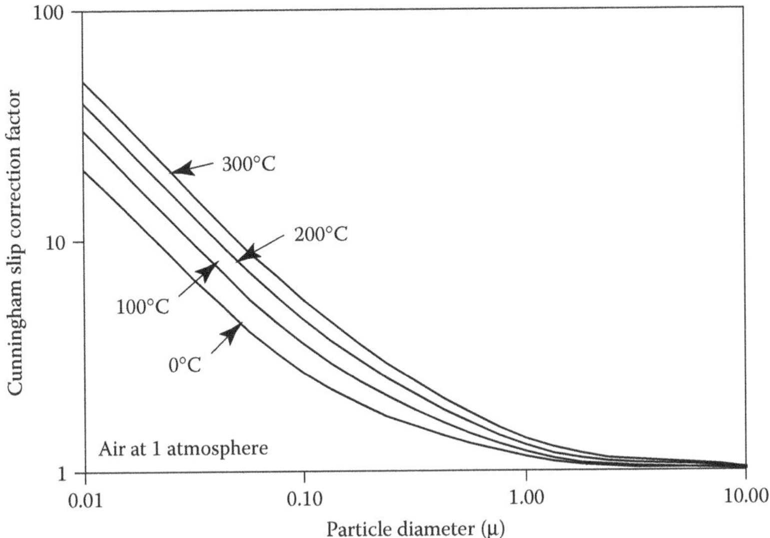

FIGURE 19.8 Cunningham slip correction factor for air at 1 atm.

where:

MW is the molecular weight of the gas
R is the universal gas constant, 8315 gm²/gmole-K-s²
T is the absolute temperature, °K

By the ideal gas law, density is

$$\rho_g = \frac{m}{v} = \frac{PMW}{RT} \tag{19.4}$$

Therefore, the gas mean free path can also be expressed as

$$\lambda = \frac{\mu}{0.499P\sqrt{(8MW/\pi RT)}} \tag{19.5}$$

where:

P is the pressure, Pa

The Cunningham slip correction factor for air at 1.0 atm is plotted in Figure 19.8. It can be seen that the correction factor has a major effect when the particle diameter is less than 1 μm.

19.4 COLLECTION MECHANISMS

19.4.1 Basic Mechanisms: Impaction, Interception, and Diffusion

Consider a particle in a gas stream moving toward or being carried toward a target. If the particle touches the target, it will likely stick to the target due to intersurface forces.

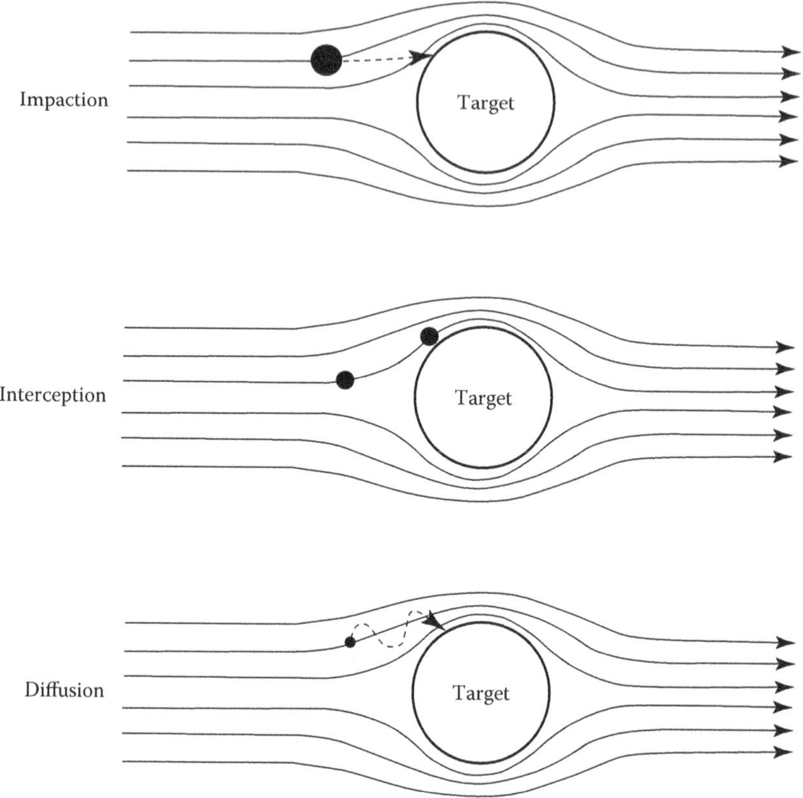

FIGURE 19.9 Basic particle collection mechanisms.

The target may be a liquid droplet, as in the case of wet scrubbers, or a fiber, as in a fabric filter baghouse. The three mechanisms by which the particle touches the target are illustrated in Figure 19.9. Small-, medium-, and large-size particles are depicted as being carried by the gas stream toward round targets. The gas flow streamlines are shown as diverging as they approach the target, then moving around the target.

In each of these mechanisms, a large number of targets will increase the probability that a particle will touch a target. Therefore, having abundant targets enhances collection efficiency.

19.4.1.1 Impaction

In the mechanism called *impaction*, large particles moving toward the target have mass, and therefore momentum, which causes each particle to travel in a straight line toward the target. The particle leaves the streamline as the streamline bends to move around the target. The greater the mass of the particle, the more likely that it will travel in a straight line. Also, as the velocity difference between the particle and the target increases, the particle will have increased momentum and will be more likely to be carried into the target.

The radius of curvature of the bend in the streamline has a very important effect on the probability that a particle will be carried into the target. The smaller the radius of curvature, the less likely that a particle will follow the streamline. Therefore, small targets are more likely to be impacted than large targets.

19.4.1.2 Interception

Interception is the mechanism by which particles of roughly 0.1–1 μm diameter are carried by the gas streamline sufficiently close to the surface of the target that the particle touches the target. These particles have insufficient inertia to leave the gas streamline and are carried with the streamline. Some gas will flow very close to the particle.

Interception is a relatively weak mechanism for particle collection compared to impaction and, as discussed in Section 19.4.1.3. It is coincidental that the path of the streamline and the particle happens to be close to the target. It is for this reason that particles in this size range are difficult to collect compared to larger and smaller particles. For the same reason, particles in this size range are not collected by natural cleaning mechanisms in nasal and tracheobronchial passages, and enter the lungs where they can lodge in the alveoli.

19.4.1.3 Diffusion

Diffusion of extremely small, submicron particles is a result of Brownian motion. These particles are so small that the mass of the particles is very small and the number of collisions with air molecules is low. Therefore, random collisions with air molecules cause the particle to bounce around. They are moved from one gas streamline to the next by random motion. If sufficient time is allowed, and if the distance to the target is small, then diffusion can be an effective collection mechanism. This is why fabric filter baghouses can be effective for collecting submicron particles, and why it is difficult for wet venturi scrubbers to collect these particles.

19.4.2 Other Mechanisms

19.4.2.1 Electrostatic Attraction

If particles acquire a charge and are placed in an electric field, the electrostatic force will move the particles across gas flow streamlines. Electrostatic forces on small particles can be quite large, making this a very effective mechanism for particle collection. This mechanism is utilized in electrostatic precipitators.

19.4.2.2 Gravity

The force of gravity is sufficient to pull very large particles out of a gas stream. Some mechanical separators are designed to slow a gas stream to allow particles to settle. However, gravity is a weak mechanism for all but the heaviest particles.

19.4.2.3 Centrifugal Force

Centrifugal force is the basis for cyclonic separation as a dusty gas is spun into a circle. Cyclones are discussed in Chapter 21.

19.4.2.4 Thermophoresis

When a temperature gradient exists across a gas space, there will be a small temperature difference from one side of a particle to the other side. Gas molecules on the high-temperature side of the particle collide with the particle with more energy than gas molecules on the cooler side. This causes the particle to move slightly toward the cold side. Thermophoresis is a relatively weak mechanism for particle collection, but it can have a small effect on collection efficiency.

19.4.2.5 Diffusiophoresis

Diffusiophoresis can be illustrated by considering the example of water vapor in a gas stream condensing on a cold target. As water vapor molecules are removed from the gas stream by condensation in the vicinity of the target, the concentration of water molecules, and hence the partial pressure of water vapor, is decreased. The concentration gradient causes water molecules from the bulk gas space move toward the cold target. The moving water vapor molecules collide with particles, causing them to be driven slightly toward the cold target. Again, this is a relatively weak mechanism that can have a small effect on particulate collection efficiency.

20 Hood and Ductwork Design

20.1 INTRODUCTION

The design of hoods and ductwork is often a very important part of air pollution control. Hoods and the air exhausted must be adequate to prevent escape of contaminants to the atmosphere. On the other hand, the air exhausted through a hood must usually be treated to remove the contaminant. Therefore, to keep capital and operating costs of control equipment to a minimum, no more air should be exhausted than is necessary for complete capture. By designing the hood with the minimum openings necessary, the air quantity to be exhausted can often be decreased.

When an exhaust system must exhaust gases from a number of points, it is important that the ductwork sizing and layout be carefully engineered for the proper flow and velocity in all branches, for conveying both velocity and pressure drop. It is axiomatic that the pressure drop through each branch at its design flow must be the same from its intake to the point of common junction. If it is not, the flow will redistribute itself in operation to create the same pressure drop in each branch.

When handling abrasive dusts, duct wear can be quite severe. Situations are known to have occurred where ¼-in.-thick elbows have been worn through in two weeks time in poorly designed exhaust systems. Therefore, when handling abrasive dusts, the duct routing should be as simple and direct as possible. Bends, where necessary, should be gradual with long radii. Velocity should be kept as low as possible, consistent with keeping the dust particles in suspension. Where bends in the ductwork are necessary and rapid wear is encountered, designs can be developed, which employ easily replaceable wear pads on the bends. These pads can be thick metal, made of wear-resistant steel, castings, hard-surfaced with special abrasion-resistant alloys such as stellite, or have thick rubber-coated linings.

A number of good reference sources are available for the design of hoods and ductwork and should be consulted before design of an elaborate dust-collecting network of ducts. Among the better general references recommended are Dalla Valle,[1] Hemeon,[2] American Conference of Governmental Industrial Hygienists,[3] *ASHRAE Handbook*,[4] and Goodfellow.[5]

A recent article by King[6] describes the proper installation of fans for optimal system performance. A concise summary of exhaust system design was prepared by Brandt.[7] While it is an excellent article on hood and duct design, it was written in 1945. Thus, it was written more from the standpoint of industrial ventilation rather than pollution control. The information presented is still accurate and useful. However, it must be understood that statements concerning control velocity, which relate to capturing sufficient contaminants to prevent a worker health hazard, must now be interpreted as meaning capture of contaminants sufficient to prevent pollution of air.

The principle in design of hoods and ducts is to choose a velocity that fits the situation. For both hoods and ducts, the velocity must be sufficient to overcome the pressure drop. For hoods, the material must be drawn up and into the duct. For ducts, the velocity must be sufficient to keep any particles suspended. Once the operating velocity is set, the flow is governed by the following law for volumetric flow:

$$Q = AV \tag{20.1}$$

Here Q is the volumetric flow rate, V is the velocity chosen, and A is the required area of flow. Once a velocity is chosen for a hood, the area required to cover the source from which the pollutant must be removed determines the flow rate. The flow rate needed in a duct sets the area of the duct.

20.2 HOOD DESIGN

The purpose of a hood is to collect contaminants from a workplace. Significant amounts of air are also drawn into the hood. The air flow is set by the distance between the source and the hood and by the pressure loss created by the air entering the duct. A sophisticated approach is given by Goodfellow.[5] Brandt[7] lists the following four major groups of hoods and designates the relationship governing the flow into the hood:

1. Enclosing hoods
2. Rectangular or round hoods
3. Slot hoods
4. Canopy hoods

In each case, the minimum control velocity must first be selected. Table 20.1 is a guide for the selection of this control velocity.

TABLE 20.1
Minimum Recommended Control Velocities

Condition of Release of Contaminant	Example of Process or Operations	Minimum Control Velocity (ft/min)
Released with no significant velocity into quiet air	Evaporation from open vessels	100
Released with low initial velocity into moderately quiet air	Spray paint booths, welding, dumping of dry materials into containers	100–200
Released with considerable velocity or into zone of rapid air movement	Spray painting in small booths with high pressure, active barrel or container filling, conveyor loading	200–500
Released with high velocity or into zone of rapid air movement	Grinding, abrasive blasting, and surfacing operations on rock	500–2000

20.2.1 Flow Relationship for Various Types of Hoods

20.2.1.1 Enclosing Hoods

An enclosing hood completely shuts off the ineffective outside area. Paint booths and laboratory hoods are typical examples of an enclosing hood. Equation 20.1 describes this type of hood, where the area, A, is the area of any opening into the hood. This opening is necessary to assure air flow. However, the area should be kept to a minimum for good performance.

20.2.1.2 Rectangular or Round Hoods

This type of hood is used for welding, stone surfacing, cleaning and degreasing, and drilling. The shape of the velocity pattern in front of these hoods determines the velocity into the hood. The total air flow entering the hood is determined by a point at x distance from an imagined suction point. Thus, the air velocity V is measured at a distance x from this suction point. The area A is essentially a spherical surface with area = $4\pi x^2 = 12.57x^2$. Equation 20.1 then becomes

$$Q = 12.57x^2V \tag{20.2}$$

Experimentally, Dalla Valle[1] noted that the contour shape actually changes and flattens slightly in front of the hood, and thus, he modified this theoretical relationship to the following:

$$Q = \left(10x^2 + a\right)V \tag{20.3}$$

where:

- x is the distance in feet along hood centerline from the face of the hood to the point, where the air velocity is V in ft/min
- a is the area in ft^2 of the hood opening

This equation is applied to the centerline or axial velocity and not to the velocity at any point. Consequently, Dalla Valle[1] or Brandt[7] should be consulted before applying this equation.

20.2.1.3 Slot Hoods

Slot hoods are lateral hoods recommended for ventilation at tanks used for degreasing, pickling, or plating. These hoods are an extreme form of rectangular hood with 50 ft or more of hood opening and are very narrow, with as little as 4 in. of height. The opening can be visualized as cylindrical surface with an area of $2\pi rL$. Hence, Equation 20.1 becomes

$$Q = 2\pi rL = 6.28rL \tag{20.4}$$

where:

- r is the radius in feet or distance from source of suction to point where the air velocity is V
- L is the length of cylinder in feet

Brandt[7] modifies this equation to account for the actual noncylindrical form of the cylinders. The modified expression is

$$Q = kLWV \tag{20.5}$$

where:

L is the length of hood in feet
W is the tank width in feet (distance from hood to remotest source of contamination)
k is a constant
$k = 6.28$, for a freely suspended slot hood with cylindrical contours
$k = 3.27$, for freely suspended slot hoods
$k = 2.8$, for slot hoods adjacent to tank tops
$k = 1.5$, if slot hood has flanges or other restricting surfaces at right angles to each other, so that air can enter from only one quadrant

20.2.1.4 Canopy Hoods

Canopy hoods are a class of rectangular or round hoods. They are used for tanks and furnaces. They are more effective if the contaminated air is warmer than the surrounding air. In this case, Brandt[7] recommends that the area A be replaced by PY where P is the perimeter of the hood face in feet and Y is the perpendicular distance in feet from the hood face to the top of the tank
Therefore,

$$Q = PYV \tag{20.6}$$

where:

V is the average velocity through the opening between the hood edge and the tank

Della Valle[1] found that for canopy hoods located between 3.5 and 4 ft above the source of contamination, the velocity at the top edge of the tank was about 0.7 of the average velocity. Equation 20.6 was then modified to

$$Q = 1.4\ PDV \tag{20.7}$$

where:

V is now the minimum control velocity given in Table 20.1

Note that the constant could be as low as 1.0 if the canopy hood is close to the surface from which this contaminant source is coming. Also, the constant could be greater than 1.4 if the distance between the hood and the contaminant source is much greater than 4 ft.

20.3 DUCT DESIGN

Ducts have the purpose of carrying the contaminated air from a hood, a piece of process equipment, or another piece of control equipment to another piece of equipment or to the discharge stack. Ducts can be water cooled, refractory lined, or made of stainless steel or plain carbon steel. For corrosive materials in the air or temperatures in the range of 1150°F–1500°F, stainless steel is used. At lower temperature and noncorrosive materials in the air, plain carbon steel may be used.

20.3.1 Selection of Minimum Duct Velocity

The ductwork, if carrying particulates, must be designed to keep the particulates in suspension. This means that carrying velocities must be sufficiently high to prevent settling of the largest particles being conveyed. An empirical formula recommended by Brandt[7] in use for estimating duct velocity required to prevent settling is

$$V = 15{,}700\left(\frac{S}{S+1}\right)\sqrt{dP} \tag{20.8}$$

where:

V is the duct velocity in feet/minute
S is the specific gravity of the particle
d is the diameter in inches of the largest particle to be conveyed

The above equation has been developed for use with ambient air. While it considers the effect of density of the particle, it ignores the density of the conveying gas. Where the gas density is considerably different from sea level ambient air, the need to alter the equation could be anticipated. Although the velocity chosen by Equation 20.8 is to convey particulates, it is generally desirable in exhaust ductwork to avoid long horizontal runs, where possible and to provide some slope to the essentially horizontal portions of the ductwork. In addition, moist and sticky particulates can produce duct buildup, and the velocities predicted by the above equation are not adequate to prevent duct-wall caking in such situations. Higher duct velocities, providing frequent duct cleanouts, and fluorocarbon-sheet lining of the duct are practices employed in such situations.

Table 20.2 should be consulted for determining the minimum duct velocity. The area depends on the source of the air flow. If the duct work originates from a hood, the flow rate will be determined from that hood as suggested in Section 20.2.1 If the duct work originates from a piece of process equipment or another piece of control equipment, that equipment will set the flow rate. Knowing the flow rate from both the sources and the desired velocity estimated from Table 20.2, for example, the area for flow can be determined from Equation 20.1. Then the duct work can be designed. Fluid flow in the duct is described by the mechanical energy balance presented in Section 20.3.2.

TABLE 20.2
Minimum Recommended Duct Velocities

Nature of Contaminant	Examples	Minimum Control Velocity (ft/min)
Vapors, gases, smoke, fumes, very light dusts	VOC, all smoke, and acid gases	2000
Medium density dry dusts	Cotton, jute lint, wood, grain, rubber, and polymers	3000
Average industrial dust	Wool, wood, sand blast, and wood shavings	4000
Heavy dust	Lead, foundry emissions, and metal turnings	5000
Large particles of heavy moist materials	Foundry dust and wet lead	5000 and over

20.3.2 Mechanical Energy Balance

The following equation comes from Chapter 8. Here the Δ represents the difference between the output and input value of the variable.

$$M\left(\Delta H+\frac{\Delta \overline{U}^2}{2g_c}+\Delta zg/g_c\right)=Q+W_s$$

In this equation, written for turbulent flow, where $\alpha = 1.0$, the fundamental equation for enthalpy with no phase change or chemical reaction can be written as follows:

$$dH = TdS + vdP$$

where:
 S is the entropy
 v is the specific volume

Then from the definition of entropy, this equation can be rewritten as

$$dH = dQ + vdP$$

then

$$\Delta H = Q + \int_{P_1}^{P_2} vdP \tag{20.9}$$

And then the mechanical energy balance results from substituting Equation 20.9 into Equation 8.7.

$$W_S = \int_{P_1}^{P_2} vdP + \frac{\Delta \overline{U}^2}{2g_c} + g/g_c\,\Delta z \tag{20.10}$$

This equation applies strictly to a reversible idealized process. Because mechanical energy is dissipated into heat through friction, a term is added for friction. The equation then can be used to describe real situations of fluid flow. Equation 20.10 now becomes

$$W_S = \int_{P1}^{P2} vdP + \frac{\Delta \overline{U}^2}{2g_c} + g/g_c\,\Delta z + F \tag{20.11}$$

where:
 F is the added friction term

Furthermore, for incompressible fluids,

$$\int_{P_1}^{P_2} vdP = v\Delta P = \frac{\Delta P}{\rho} \tag{20.12}$$

where:

ρ is the fluid density

Since W_S is the shaft work done on a system, the equation can be rewritten to include the efficiency of the fan and motor used in this case. If η is the efficiency, then

$$W_S = \eta W \tag{20.13}$$

where:

W is the work done on the system by the fan whose efficiency is η

Now the mechanical energy balance becomes

$$\eta W = \frac{\Delta P}{\rho} + \frac{\Delta \overline{U}^2}{2g_c} + g/g_c \, \Delta z + F \tag{20.14}$$

The usual practice is to consider the velocity head $\overline{U}^2$, and the friction head, F, when making a calculation for the total pressure for a fan to encounter, discounting the potential energy loss due to change in elevation Δz because it will be so small and the change in pressure ΔP will be small since the flow is nearly incompressible.

20.3.2.1 Velocity Head

For the average velocity, $\overline{U} = V$, the velocity head H_V is

$$H_V = \frac{V^2}{2g} \text{(in feet of fluid)} \tag{20.15}$$

To convert to inches of H_2O for the standard conditions of 70°F, 50% air humidity, and 1 atm pressure, this equation becomes

$$\text{VP} = \left(\frac{V}{4005}\right)^2 \tag{20.16}$$

The velocity head is now designated as the velocity pressure (VP) in inches of H_2O.

20.3.2.2 Friction Head

The static pressure is sometimes called the *friction pressure* or *friction head*. In ducts, the friction pressure is due to skin friction generated by flow and energy losses. It is also generated due to turbulence in bends, fittings, obstructions, and sudden expansion and contractions. The friction loss in smooth circular pipes and ducts can be calculated from

$$H_f = 4f\left(\frac{L}{D_c}\right)\left(\frac{V}{2g}\right) \tag{20.17}$$

where:

f is the friction factor

D_c is the duct diameter

McCabe et al.[8] report that the friction factor, f, can be calculated from the von Karmen equation.

$$\frac{1}{\sqrt{f}}=4.07\log_{10}\left(N_{\mathrm{Re}}\sqrt{f}\right)-0.60 \tag{20.18}$$

where:

N_{Re} is the Reynolds number

$$N_{\mathrm{Re}}=\frac{D_c V\rho}{\mu} \tag{20.19}$$

A nomograph based on this type of equation is given in Figure 20.1, friction losses for air in circular ducts. For noncircular, square ducts, it is possible to use the hydraulic radius concept.

$$r_{\mathrm{H}}\equiv\frac{S}{L_{\mathrm{P}}} \tag{20.20}$$

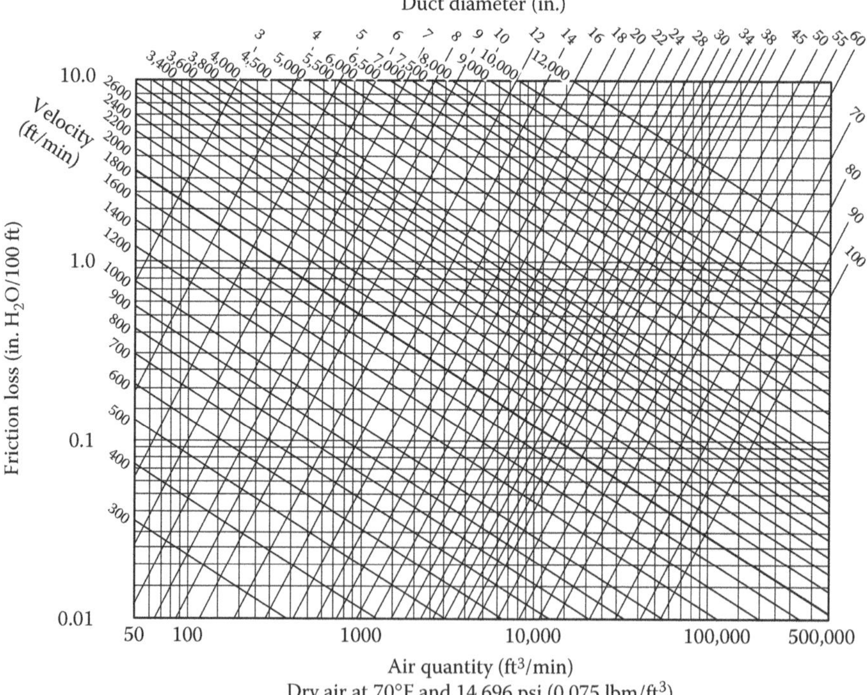

FIGURE 20.1 Friction losses for air in circular ducts—U.S. customary units. (Reprinted by permission from the American Society of Heating, Refrigerating, and Air-Conditioning Engineers, *ASHRAE 1985 Handbook—Fundamentals*, www.ashrae.org, Copyright 1985.)

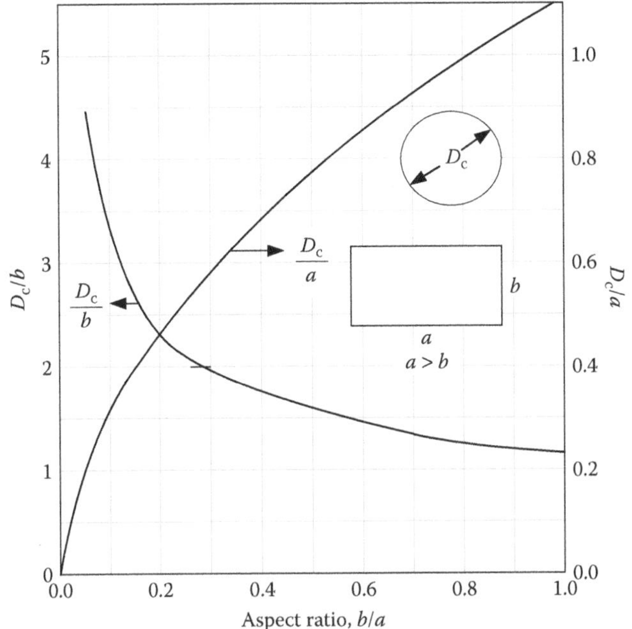

FIGURE 20.2 Equivalent rectangular and circular ducts having equal pressure drop and flow rate. (From Crawford, M., *Air Pollution Control Theory*, McGraw-Hill, New York, 1976. With permission.)

where:

r_H is the hydraulic radius
S is the cross-sectional area of the channel
L_P is the perimeter of the channel in contact with the fluid

The diameter used in the Reynolds number calculation is taken as four times the hydraulic radius. This concept is especially good for square ducts. It is better to use Figure 20.2, equivalent rectangular and circular ducts having equal pressure drop (Crawford[9]), for nonsquare, rectangular ducts.

The effect of bends, fittings, obstructions, and sudden expansion and contractions can be accounted through a relationship, where the head loss is proportional to the velocity in the pipe section squared.

$$H_{fx} = K_x \frac{V^2}{2g_c} \tag{20.21}$$

where:

K_x is the proportionality constant

McCabe et al.[8] list the following formulas for head loss due to a contraction, H_{fC},

TABLE 20.3
Head Loss Constant for Fittings and Branches

Fitting	K_{fF}
Tee	2.0
90° elbow	0.9
60° elbow	0.6
45° elbow	0.45
Branch into duct	
30° angle	0.2
45° angle	0.3

$$K_C = 0.4\left(1 - \frac{S_D}{S_U}\right) \tag{20.22}$$

and the head loss due to an expansion, H_{fE},

$$K_E = \left(1 - \frac{S_U}{S_D}\right)^2 \tag{20.23}$$

where:
S_U is the upstream cross-sectional area
S_D is the downstream cross-sectional area

Cooper and Alley[10] have adapted the fitting coefficients listed in Table 20.3 from *Industrial Ventilation*,[3] for head loss, H_{fF}, in fittings and branches.

For each of these cases where the friction head has been calculated, the head is reported in feet of fluid flowing. The conversion factor to inches of H_2O is accomplished by making the following substitution with V in ft/min, which now becomes the velocity pressure VP(std) at standard conditions.

$$\text{VP(std)} = \frac{V^2}{2g} = \left(\frac{V}{4005}\right)^2 \tag{20.24}$$

This VP(std) is for air flowing at 70°F and 1.0 atm, where the air density is $\rho = 0.075$ lbm/ft^3. A correction can be made for other temperatures as follows:

$$\frac{\text{VP(act)}}{\text{VP(std)}} = \frac{\rho(\text{act})}{\rho(\text{std})} \tag{20.25}$$

If the pressure is nearly atmospheric, which it most generally is, then the ideal gas law can be applied, and the density ratio can be replaced by the inverse temperature ratio.

$$\frac{\text{VP(act)}}{\text{VP(std)}} = \frac{\rho(\text{act})}{\rho(\text{std})} = \frac{530}{\text{T(act)}} \tag{20.26}$$

TABLE 20.4
Entrance Loss Coefficients for Hoods

Type of Entrance	K_{EH}
Square entrance	0.7–1.25
Round entrance	0.5–0.9
Slot with bend	1.6
Canopy	0.5

20.4 EFFECT OF ENTRANCE INTO A HOOD

The hood static pressure, SP_H, measured as a short distance from the hood, is a direct measurement of the energy required to accelerate the fluid from rest to the duct velocity and to account for the turbulence losses due to the shape of the hood structure. The hood entry losses, SP_{EH}, can be expressed as a function of the VP as with the fittings. Thus,

$$SP_{EH} = K_{EH} VP \tag{20.27}$$

Goodfellow[5] made a detailed presentation of this situation. Table 20.4 has been adapted from his work to apply to the hood types described above.

The total effect is then the sum of the entry loss, SP_{HE}, and the acceleration to get the air up to the duct velocity, which is 1.0 VP. The total static pressure due to the entrance at the hood is then

$$SP_H = (1 + K_{HE}) VP(\text{std}) \tag{20.28}$$

20.5 TOTAL ENERGY LOSS

The static pressure is the same as the friction head. The total pressure, TP, is then the sum of the static pressure, SP, and the VP. This can be expressed as the sum of duct, hood, and fittings losses.

$$TP = \left[f\left(\frac{D}{V}\right) + (1 + K_H) + \Sigma K_x \right] \tag{20.29}$$

Here, $f(D/V)$ for the duct is found from Figure 20.1.

20.6 FAN POWER

In the case of ductwork requiring a fan, the operating cost is mostly related to the cost of operating the fan. The work of the fan can be calculated from

$$W_f = \frac{K\, Q\, \Delta P_f}{\eta} \tag{20.30}$$

where:

k is the constant dependent on the units of the other parameters

η is the mechanical efficiency

For Q in acfm, ΔP_f in inches of H_2O, $k = 0.0001575 \text{ HP}/\left(\text{ft}^3/\text{in. of } H_2O\right)$ and W_f is in HP. For Q in m^3/s and ΔP_f in Pa, $k = 1.0$ and W_f is in J/s or watts. The fan static pressure is then

$$\Delta P_f = \text{TP}_{\text{outlet}} - \text{TP}_{\text{inlet}} \tag{20.31}$$

where:

$$\text{TP}_i = \text{SP}_i + \text{VP}_i \tag{20.32}$$

20.7 HOOD–DUCT EXAMPLE

Figure 20.3 is a schematic of a hood–duct system to be installed in a chemical plant. One branch of the ductwork begins in a canopy hood. The other branch of the duct work begins in a flanged slot hood. The ductwork ends in an air-pollution-control apparatus. The pressure is 20 in. of H_2O at this end of the ductwork due to the pressure drop through the control apparatus. The air is flowing at standard conditions of 1.0 atm and 70°F. The fan, which is to be installed, has a mechanical efficiency of 83%. Determine the horsepower of the motor required for the fan.

Hood information

Hood 1:

$$V = 100 \text{ fpm}, Y = 3.0 \text{ ft}, P = 40 \text{ ft}, K_H = 0.5$$

$$Q = PYV = 40 \times 3 \times 100 = 12{,}000 \text{ scfm}$$

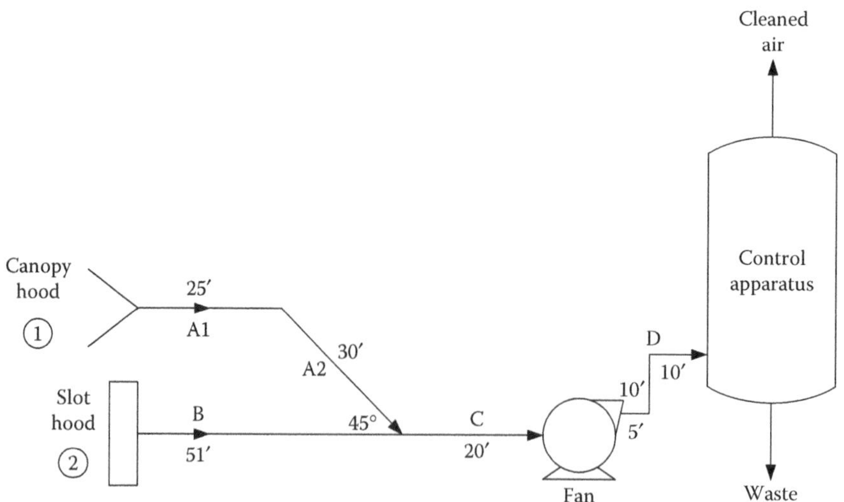

FIGURE 20.3 A schematic of a hood–duct system.

Hood 2: $V = 175 \text{ fpm}, L = 18 \text{ ft}, W = 5.33 \text{ ft}, k = 1.5, K_H = 1.0$

$$Q = kLWV = 1.5 \times 18 \times 8 \times 175 = 37{,}800 \text{ scfm}$$

Duct information

$$\text{Duct area}: A = \frac{Q}{V}, \quad \text{duct diameter}: D = \left(\frac{4A}{\pi}\right)^{1/2}$$

Determine duct velocity and calculate duct area. The ducts will all be sized as round ducts. Then calculate duct diameter. Round off diameter to the next highest or lowest whole number.

Duct	Flow (scfm)	Velocity (*V* in ft/min)	Area (*A* in ft²)	*D* (ft)	D_{act} (ft)	A_{act} (ft²)	V_{act} (ft/min)
A	12,000	2000	6.0	2.76	3.0	19.64	1698
B	37,800	2000	18.9	4.91	5.0	19.64	1925
C	49,800	2000	24.9	5.63	5.0	19.64	2536
D	49,800	1000	49.8	7.96	8.0	50.26	990

$$\text{TP} = \left[f\left(\frac{D}{V}\right) + (1 + K_H) + \Sigma K_x\right]$$

Duct A: V = 1698 ft/min, L = 55 ft

$$V(\text{std}) = \left(\frac{1698}{4005}\right)^2 = 0.18$$

$$f(D/V) = (2\,\text{in.}/100\,\text{ft}) \times 55/100 = 1.10 \text{ in. of } H_2O$$

$$1 + K_H = 1 + 0.5 = 1.5$$

$$K_x = \underset{\text{elbow}}{0.45} + \underset{\text{branch}}{0.30} = 0.75$$

$$\text{TP} = (1.10 + 1.5 + 0.75) \times 0.18 = 0.60 \text{ in. of } H_2O$$

Duct B: V = 1925 ft/min; L = 51 ft

$$V(\text{std}) = \left(\frac{1925}{4005}\right)^2 = 0.23$$

$$f(D/V) = (1.4\,\text{in.}/100\,\text{ft}) \times 51/100 = 0.71 \text{ in. of } H_2O$$

$$1 + K_H = 1 + 1.0 = 2.0$$

$$K_x = 0.0$$

$$\text{TP} = (0.71 + 2.0 + 0.0) \times 0.23 = 0.62 \text{ in. of } H_2O$$

The two branches are nearly balanced, and therefore, no more calculations are required.

Duct C: $V = 2536$ ft/min, $L = 20$ ft

$$V(\text{std})=\left(\frac{2536}{4005}\right)^2=0.40$$

$$f(D/V)=(2.6\,\text{in.}/100\,\text{ft})\times 20/100=0.52\,\text{in. of } H_2O$$

$$1+K_H=0.9\,(\text{no hood})$$

$$K_x=0.0\,(\text{no fittings})$$

$$\text{TP}=(0.52+0.0+0.0)\times 0.40=0.21\,\text{in. of } H_2O$$

$$\text{Total TP UP to fan}: \text{TP}=-(0.62+0.21)=-0.83\ \text{in. of } H_2O$$

Duct D: $V = 990$ ft/min, $L = 25$ ft

$$V(\text{std})=\left(\frac{990}{4005}\right)^2=0.06\,\text{ft}$$

$$f(D/V)=(0.22\,\text{in.}/100\,\text{ft})\times 25/100=0.055\,\text{in. of } H_2O$$

$$1+K_H=0.0\,(\text{no hood})$$

$$K_x=\text{expansion}=(1-8/12)^2=0.11$$

$$\text{TP}=\underbrace{(0.055+0.0+0.11)\times 0.06}_{\text{negligible}}+\underbrace{20.0}_{\text{at entrance to control equipment}}=20.0\,\text{in. of } H_2O$$

Fan: $Q = 49{,}800$ scfm, $\Delta P = 20 - (-0.83) \approx 21$ in. of H_2O

$$\eta=0.83$$

$$W_f=\frac{0.0001575\times 49{,}800\times 21}{0.83}\approx 200\,\text{HP}$$

21 Cyclone Design

Cyclones are very common particulate control devices used in many applications, especially those where relatively large particles need to be collected. They are not very efficient for collecting small particles because small particles have little mass that can generate a centrifugal force. Cyclones are very simple devices that use centrifugal force to separate particles from a gas stream. They are commonly constructed of sheet metal, although other materials can be used. They have a low capital cost, small space requirement, and no moving parts. Of course, an external device, such as a blower or other source of pressure, is required to move the gas stream. Cyclones are able to handle very heavy dust loading, and they can be used in high-temperature gas streams. Sometimes, they are lined with castable refractory material to resist abrasion and to insulate the metal body from high-temperature gas.

A typical cyclone is illustrated in Figure 21.1. It has a tangential inlet to a cylindrical body, causing the gas stream to be swirled around. Particles are thrown toward the wall of the cyclone body. As the particles reach the stagnant boundary layer at the wall, they leave the flowing gas stream and presumably slide down the wall, although some particles may be re-entrained as they bounce off of the wall back into the gas stream. As the gas loses energy in the swirling vortex, it starts spinning inside the vortex and exits at the top.

The vortex finder tube does not create the vortex or the swirling flow. Its function is to prevent short-circuiting from the inlet directly to the outlet. Cyclones will work without a vortex finder, although the efficiency will be reduced.

21.1 COLLECTION EFFICIENCY

When a particle moves at a constant speed in a circular direction, the velocity vector changes continuously in direction, although not in magnitude. This creates an acceleration resulting from a change in direction of the velocity, which is just as real and just as much an acceleration as that arising from the change in the magnitude of velocity. By definition, acceleration is the time rate of change of velocity, and velocity, being a vector, can change in direction, as well as magnitude. Force, of course, is defined by Newton's second law ($F = ma$). Centrifugal force is given by

$$F = \frac{mV^2}{r} \tag{21.1}$$

where:

F is the centrifugal force
m is the mass of the particle
V is the velocity of the particle, assumed to equal inlet gas velocity
r is the radius of cyclone body

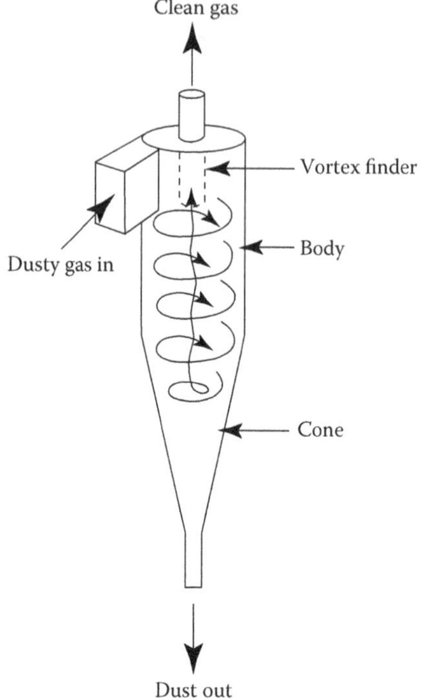

FIGURE 21.1 A schematic of the standard cyclone.

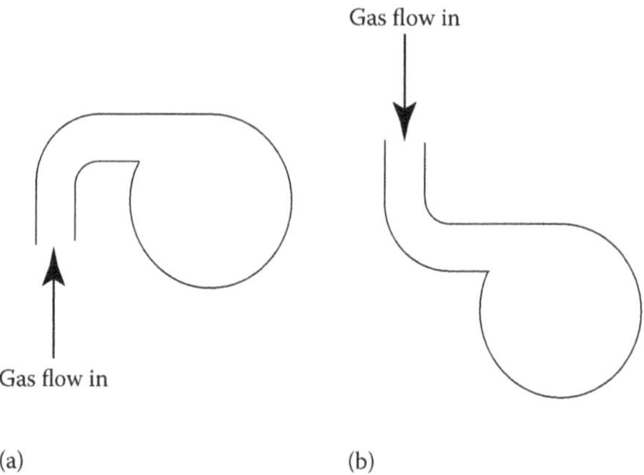

FIGURE 21.2 Inlet piping configuration: (a) right and (b) wrong.

Because the operating principle of a cyclone is based on using centrifugal force to move particles to the cyclone wall, a simple mistake in the piping configuration, shown in Figure 21.2a, reduces efficiency. Ensure that particles are given a head start in the right direction by using the configuration shown in Figure 21.2b.

21.1.1 Factors Affecting Collection Efficiency

Several factors that affect collection efficiency can be predicted. Increasing the inlet velocity increases the centrifugal force, and therefore the efficiency, but it also increases the pressure drop. Decreasing the cyclone diameter also increases centrifugal force, efficiency, and pressure drop. Increasing the gas flow rate through a given cyclone has the effect of efficiency as shown in Equation 21.2:

$$\frac{\mathrm{Pt}_2}{\mathrm{Pt}_1} = \left(\frac{Q_1}{Q_2}\right)^{0.5} \tag{21.2}$$

where:

Pt is the penetration (Pt = 1 − η)
η is the particle removal efficiency
Q is the volumetric gas flow

Interestingly, decreasing the gas viscosity improves efficiency, because drag force is reduced. Centrifugal force drives the particle toward the wall of the cyclone, while drag opposes the centrifugal force. The terminal velocity of the particle toward the wall is the result of the force balance between the centrifugal and drag forces. Increasing gas to particle density difference affects penetration as shown in Equation 21.3:

$$\frac{\mathrm{Pt}_2}{\mathrm{Pt}_1} = \left(\frac{\mu_2}{\mu_1}\right)^{0.5} \tag{21.3}$$

where:

μ is the gas viscosity

Note that decreasing the gas temperature increases the gas density, but contrary to intuition, decreases the gas viscosity, which reduces drag force and results in a small efficiency improvement. However, decreasing the gas temperature also decreases the volumetric flow rate, which affects efficiency, as described above in Equation 21.2.

Finally, particle loading also affects efficiency. High dust loading causes particles to bounce into each other as they move toward the wall, driving more particles toward the wall and their removal.

$$\frac{\mathrm{Pt}_2}{\mathrm{Pt}_1} = \left(\frac{L_1}{L_2}\right)^{0.18} \tag{21.4}$$

where:

L is the inlet particle concentration (loading)

Figure 21.3 shows generalized efficiency relationships for high-efficiency conventional and high-throughput cyclones. It simply demonstrates that the dimensions of the cyclones can be tuned to the application. Figure 21.4 and Table 21.1 illustrate typical cyclone dimensions. Relative dimensions are based upon the diameter of the body of the cyclones. High-efficiency cyclones tend to have long, narrow bodies, while high-throughput cyclones generate less pressure drop with fat bodies.

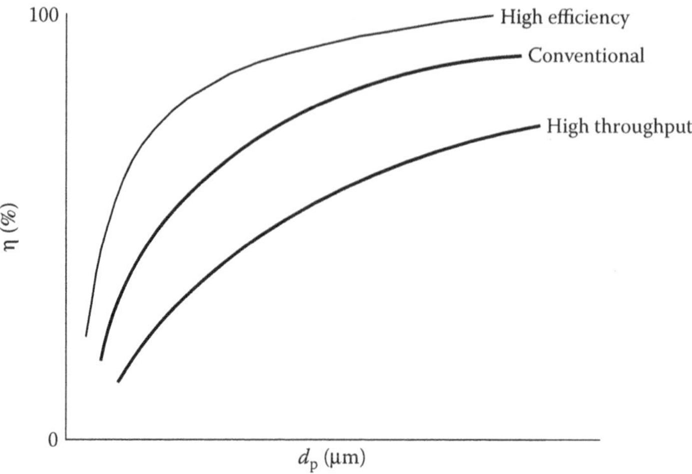

FIGURE 21.3 Generalized efficiency relationships.

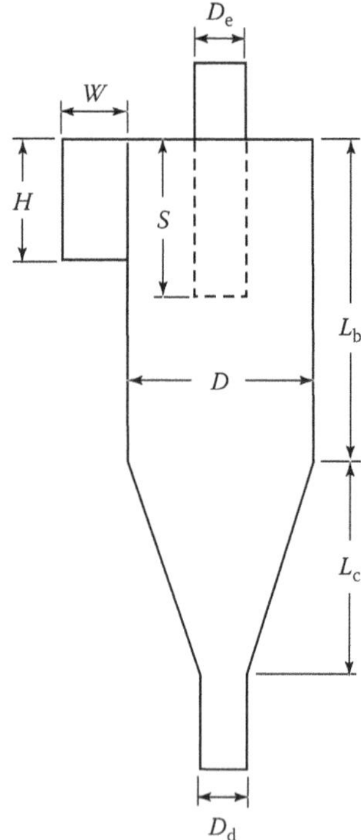

FIGURE 21.4 Cyclone dimensions.

TABLE 21.1
Typical Cyclone Dimensions

Dimension	Fractional Ratio	High Efficiency	Standard	High Throughput
Inlet height	H/D	0.44	0.5	0.8
Inlet width	W/D	0.21	0.25	0.35
Gas exit diameter	D_e/D	0.4	0.5	0.75
Body length	L_b/D	1.4	1.75	1.7
Cone length	L_c/D	2.5	2.0	2.0
Vortex finder	S/D	0.5	0.6	0.85
Dust outlet diameter	D_d/D	0.4	0.4	0.4

21.1.2 Theoretical Collection Efficiency

The force balance between centrifugal and drag forces determines the velocity of the particles toward the wall. Resident time of particles in the cyclone, which allows time for particles to move toward the wall, is determined by the number of effective turns that the gas path makes within the cyclone body. An empirical relationship for the number of effective turns is provided in Equation 21.5:

$$N_e = \frac{1}{H}\left(L_b + \frac{L_c}{2}\right) \tag{21.5}$$

where:

N_e is the number of effective turns
H is the height of the tangential inlet
L_b is the length of cyclone body
L_c is the length of cyclone lower cone

The theoretical efficiency of a cyclone can be calculated by balancing the terminal velocity with the residence time resulting from a distance traveled in the cyclone. This force and time balance results in Equation 21.6:

$$d_{px} = \left[\frac{x}{100}\frac{9\mu W}{\pi N_e V_i\left(\rho_p - \rho_g\right)}\right]^{0.5} \tag{21.6}$$

where:

d_{px} is the diameter of a particle with x% removal efficiency
μ is the viscosity
W is the inlet width
N_e is the number of effective turns
V_i is the inlet velocity
ρ_p is the density of particle
ρ_g is the density of gas

21.1.3 Lapple's Efficiency Correlation

Unfortunately, the theoretical efficiency relationship derived above does not correlate well with real data. The relationship works reasonably well for determining the 50% cut diameter (the diameter of the particle that is collected with 50% efficiency). To better match data with reasonable accuracy, the efficiency of other particle diameters can be determined from Lapple's empirical efficiency correlation,[1] which is shown in Figure 21.5. This correlation can be set up for automated calculations using the algebraic fit given by Equation 21.7:

$$\eta_j = \frac{1}{1+\left(d_{p50}/d_{pj}\right)^2} \tag{21.7}$$

where:

η_j is the collection efficiency of particle with diameter j
d_{p50} is the diameter of particles with 50% collection efficiency
d_{pj} is the diameter of particle j

Lapple's efficiency curve was developed from measured data for cyclones with the *standard* dimensions shown in Table 21.1. The efficiency curve can be tailored

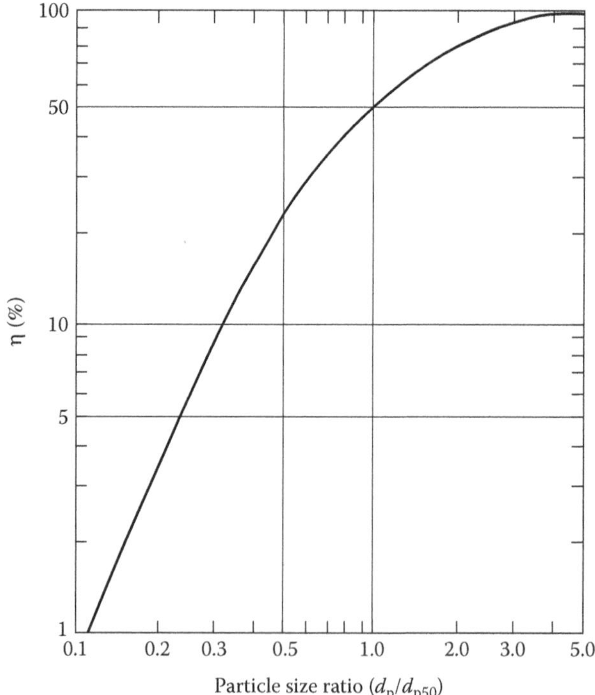

FIGURE 21.5 Lapple's efficiency curve.

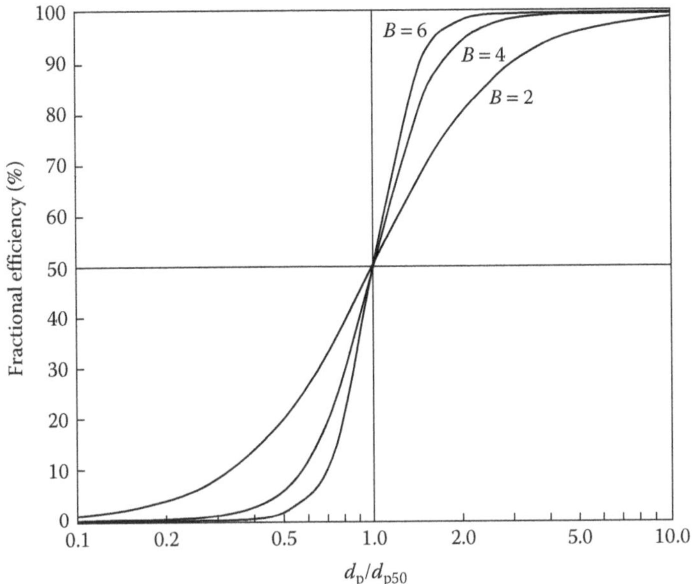

FIGURE 21.6 Effect of slope parameter, *B*.

for different industrial cyclone dimensions by adding a slope parameter, *B*, to the correlation:

$$\eta_j = \frac{1}{1 + \left(d_{p50}/d_{pj}\right)^B} \tag{21.8}$$

where:

B is the slope parameter, typically ranging from 2 to 6

Figure 21.6 illustrates the effect of the slope parameter, *B*. Note that the larger value for *B* results in a sharper cut. Since more mass is associated with larger particles, the sharper cut results in higher overall mass removal efficiency.

21.1.4 Leith and Licht Efficiency Model

Other models have been developed to predict cyclone performance. One is the Leith and Licht model[2] shown in Equation 21.9:

$$\eta = 1 - \exp\left(-\Psi d_p^M\right) \tag{21.9}$$

$$M = \frac{1}{m+1} \tag{21.10}$$

TABLE 21.2
Geometric Configuration Parameter

Geometric Parameter	Fractional Ratio	Standard	Stairmand	Swift
Inlet height	H/D	0.5	0.5	0.44
Inlet width	W/D	0.25	0.2	0.21
Gas exit diameter	D_e/D	0.5	0.5	0.4
Body length	L_b/D	2.0	1.5	1.4
Cone length	L_c/D	2.0	2.5	2.5
Vortex finder	S/D	0.625	0.5	0.5
Dust outlet diameter	D_d/D	0.25	0.375	0.4
Geometric configuration parameter	K	402.9	551.3	699.2

$$m = 1 - \left[\left(1 - 0.67 D_c^{0.14}\right)\left(\frac{T}{283}\right)^{0.3}\right] \tag{21.11}$$

$$\Psi = 2\left[\frac{KQ\rho_p C'(m+1)}{18\mu D_c^3}\right]^{\frac{M}{2}} \tag{21.12}$$

where:

d_p is the particle diameter in meters
D_c is the cyclone body diameter in meters
T is the gas temperature, °K
K is the dimensional geometric configuration parameter
Q is the volumetric gas flow
ρ_p is the particle density
C' is the Cunningham slip correction factor
μ is the gas viscosity

The geometric configuration parameter is estimated based on the cyclone configuration. Table 21.2 shows relative dimensions for three types of cyclones: the standard cyclone, the Stairmand design,[3] and the Swift design.[4] Note that the Stairmand and the Swift cyclones have smaller inlet openings than those in the standard design, which means a higher inlet velocity for the same size body. This results in more centrifugal force and increased efficiency. In the Leith and Licht model, a larger geometric configuration parameter results in a higher predicted efficiency.

21.1.5 Comparison of Efficiency Model Results

Efficiency models are adequate for getting a fair idea of performance, but there can be a rather wide variation in model predictions. Part, but not all, of the variation can be explained by empirical factors for the cyclone configuration. Figure 21.7 shows cyclone efficiency curves as a function of particle diameter based on several sources.

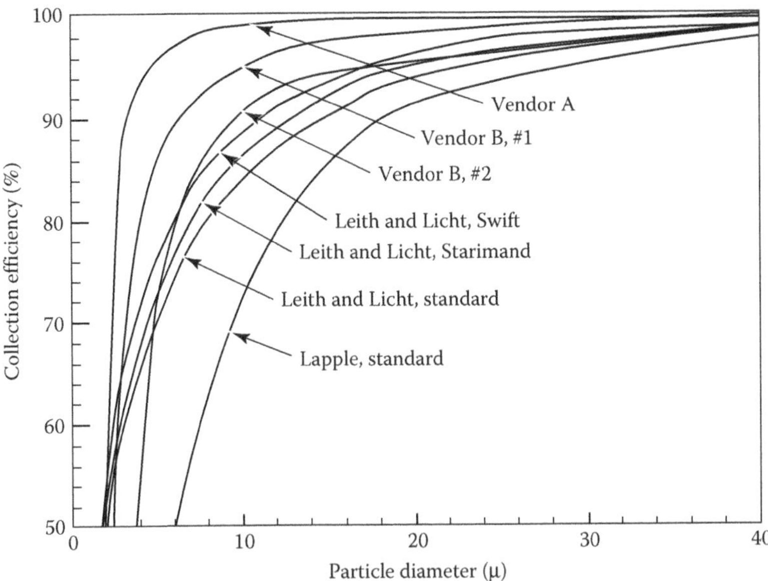

FIGURE 21.7 Cyclone efficiency curves.

Each curve is based upon the same gas flow and gas and particle conditions. The lowest efficiency is predicted by Lapple's curve for a standard cyclone. Interestingly, the Leith and Licht model for the same standard cyclone predicts a significantly higher efficiency. The Leith and Licht model for the higher efficiency Stairmand and Swift cyclone designs shows incremental improvement over the standard design. Vendor data also were collected for the same set of gas and particle conditions, with significant predicted performance improvement. Perhaps the vendors were being overoptimistic about their designs, or perhaps there have been significant improvements in cyclone design over the years. It does point out that performance guarantees for cyclones must be written with specific information about the gas and particle properties, including the particle size distribution, to ensure that vendor guarantees can be measured and substantiated after installation.

21.2 PRESSURE DROP

Pressure drop provides the driving force that generates gas velocity and centrifugal force within a cyclone. Several attempts have been made to calculate pressure drop from fundamentals, but none of them has been very satisfying. Most correlations are based on the number of inlet velocity heads as shown in Equation 21.13:

$$\Delta P = \frac{1}{2g_c} \rho_g V_i^2 N_H \tag{21.13}$$

where:

ΔP is the pressure drop

ρ_g is the gas density

V_i is the inlet gas velocity
N_H is the pressure drop expressed as number of the inlet velocity heads

One of the correlations for number of inlet velocity heads is by Miller and Lissman[5]:

$$N_H = K_{\Delta P1}\left(\frac{D}{D_e}\right)^2 \tag{21.14}$$

where:
$K_{\Delta P1}$ is the constant based on the cyclone configuration and operating conditions
D is the diameter of the cyclone body
D_e is the diameter of the exit tube

A typical value for $K_{\Delta P}$ in the Miller and Lissman correlation is 3.2. For the standard cyclone configuration described above, the Miller and Lissman correlation results in 12.8 inlet velocity heads.

Another correlation for number of inlet velocity heads is by Shepherd and Lapple[6]:

$$N_H = K_{\Delta P2}\frac{HW}{D_e^2} \tag{21.15}$$

where:
$K_{\Delta P2}$ is the constant for cyclone configuration and operating conditions
H is the height of the inlet opening
W is the width of the inlet opening
D_e is the diameter of the exit tube

The value for $K_{\Delta P}$ in the Shepherd and Lapple correlation is different, typically ranging from 12 to 18. The Shepherd and Lapple correlation results in 8 inlet velocity heads for the standard cyclone dimensions, 6.4 inlet velocity heads for the Stairmand cyclone design, and 9.24 inlet velocity heads for the Swift cyclone design. As can be seen, there is a substantial difference among the correlations. Again, it is best to rely upon vendors' experience when your own experience is lacking; however, to enforce a performance guarantee, ensure that the specification is well written and can be documented for the expected conditions.

21.3 SALTATION

The previous discussion of efficiency and pressure drop leaves the impression that continually increasing the inlet gas velocity can give incrementally increasing efficiency. However, the concept of *saltation* by Kalen and Zenz[7] indicates that more than just diminishing return with increased velocity, collection efficiency actually decreases with excess velocity. At velocities greater than the saltation velocity, particles are not removed when they reach the cyclone wall, but are kept in suspension as the high velocity causes the fluid boundary layer to be very thin. A correlation for the saltation velocity was given by Koch and Licht[8]:

$$V_s = 2.055 D^{0.067} V_i^{0.667} \left[4 g \mu \frac{(\rho_p - \rho_g)}{3\rho_g^2} \right]^{0.333} \left\{ \frac{(W/D)^{0.4}}{\left[1 - (W/D)\right]^{0.333}} \right\} \tag{21.16}$$

where:

V_s is the saltation velocity, ft/s
D is the cyclone diameter, ft
V_i is the inlet velocity, ft/s
g is the acceleration of gravity, 32.2 ft/s^2
μ is the gas viscosity, lbm/ft-s
ρ_p is the particle density, lbm/ft^3
ρ_g is the gas density, lbm/ft^3
W is the width of inlet opening, ft

The maximum collection efficiency occurs at $V_i = 1.25V_s$, which typically is between 50 and 100 ft/s.

22 Design and Application of Wet Scrubbers

22.1 INTRODUCTION

In wet scrubbing, an atomized liquid, usually water, is used to capture particulate dust or to increase the size of aerosols. Increasing size facilitates separation of the particulate from the carrier gas. Wet scrubbing can effectively remove fine particles in the range from 0.1 to 20 μm. The particles may be caught first by the liquid, or first on the scrubber structure, and then washed off by the liquid. Because most conventional scrubbers depend upon some form of inertial collection of particulates as the primary mechanism of capture, scrubbers when used in a conventional way have a limited capacity for controlling fine particulates. Unfortunately, inertial forces become insignificantly small as particle size decreases, and collection efficiency decreases rapidly as particle size decreases. As a result, it becomes necessary to greatly increase the energy input to a wet scrubber to significantly improve the efficiency of collection of fine particles. Even with great energy inputs, wet scrubber collection efficiencies are not high with particles less than 1.0 μm in size.

Wet scrubbers have some unique characteristics useful for fine particulate control. Since the captured particles are trapped in a liquid, re-entrainment is avoided, and the trapped particles can be easily removed from the collection device. Wet scrubbers can be used with high-temperature gases, where cooling of the gas is acceptable and also with potentially explosive gases. Scrubbers are relatively inexpensive when removal of fine particulates is not critical. Also, scrubbers are operated more easily than other sophisticated types of particulate removal equipment.

Wet scrubbers can be employed for the dual purpose of absorbing gaseous pollutants while removing particulates. Both horizontal and vertical spray towers have been used extensively to control gaseous emissions when particulates are present. Cyclonic spray towers may provide slightly better particulate collection, as well as higher mass transfer coefficients and more transfer units per tower than other designs. Although there is theoretically no limit to the number of transfer units that can be built into a vertical countercurrent packed tower or plate column, if it is made tall enough, there are definite limits to the number of transfer units that can be designed into a single vertical spray tower. As tower height and gas velocities are increased, more spray particles are entrained upward from lower levels, resulting in a loss of true countercurrency. Achievable limits have not been clearly defined in the literature, but some experimental results have been provided.[1] There have been reports of 5.8 transfer units in a single vertical spray tower and 3.5 transfer units in horizontal spray chambers. Researchers have attained seven transfer units in a single commercial cyclonic spray tower. Theoretical discussion and a design equation for cyclonic spray towers of the Pease-Anthony type are available. Whenever more transfer units are required, spray towers can be used in series.

When heavy particulate loads must be handled or are of submicron size, it is common to use wet particulate collectors that have high particle collection efficiencies along with some capability for gas absorption. The venturi scrubber is one of the more versatile of such devices, but it has absorption limitations because the particles and spray liquid have parallel flow. It has been indicated that venturi scrubbers may be limited to three transfer units for gas absorption.[1] The liquid-sprayed wet electrostatic precipitator is another high-efficiency particulate collector with gas absorption capability. Limited research tests have indicated that the corona discharge enhances mass-transfer absorption rates, but the mechanism for this has not been established.

The disadvantages of wet scrubbers include the necessity of reheating cooled scrubber effluents for discharge up a stack. Furthermore, the water solutions may freeze in winter and become corrosive at other times. In some cases, the resultant liquid sludge discharge may have to be treated for disposal. It should be noted also that operating costs can become excessive due to the high energy requirements to achieve high collection efficiencies for removal of fine particulates.

22.2 COLLECTION MECHANISMS AND EFFICIENCY

In wet scrubbers, collection mechanisms such as inertial impaction, direct interception, Brownian diffusion, and gravity settling apply in the collection process. Most wet scrubbers will use a combination of these mechanisms; therefore, it is difficult to classify a scrubber as predominately using one particular type of collection mechanism. However, inertial impaction and direct interception play major roles in most wet scrubbers. Thus, in order to capture finer particles efficiently, greater energy must be expended on the gas. This energy may be expended primarily in the gas pressure drop or in atomization of large quantities of water. Collection efficiency may be unexpectedly enhanced in a wet scrubber through methods that cause particle growth. Particle growth can be brought about by vapor condensation, high turbulence, or thermal forces in the confines of the narrow passages in the scrubber structure. Condensation, the most common growth mechanism, occurs when a hot gas is cooled or compressed. The condensation will occur preferentially on existing particles rather than producing new nuclei. Thus, the dust particles will grow larger and will be more easily collected. When hydrophobic dust particles must be collected, there is evidence that the addition of small quantities of nonfoaming surfactants may enhance collection. The older literature is contradictory on this point, but careful experiments by Hesketh[2] and others indicate enhancement can definitely occur.

22.3 COLLECTION MECHANISMS AND PARTICLE SIZE

When a gas stream containing particulates flows around a small object such as a water droplet or a sheet of water, the inertia of the particles causes them to move toward the object where some of them will be collected. This phenomenon is known as *inertial impaction*, which customarily describes the effects of small-scale changes in flow direction. Because inertial impaction is effective on particles as small as a few tenths of a micrometer in diameter, it is the most important collection mechanism for wet scrubbers. Since this mechanism depends upon the inertia of the particles, both their size and density are important in determining the efficiency with which they will be collected. All important particle properties may be lumped into one parameter, the aerodynamic impaction diameter. For the basic definition of the aerodynamic impaction diameter see Section 19.2. The aerodynamic impaction diameter can be calculated from the actual particle diameter by the following relationship:

$$d_{ap} = d_p \left(p_p C' \right)^{1/2} \tag{22.1}$$

where:

d_{ap} is the aerodynamic impaction diameter in μm-g/cm³
d_p is the physical diameter in μm
p_p is the density of particle in g/cm³
C' is the Cunningham's correction factor

By a fortunate circumstance, most methods for measuring particle size determine the aerodynamic impaction diameter. The Cunningham correction factor is given by the following formulas:

$$C' = 1 + \frac{2\lambda}{d_p}\left[1.257 + 0.400 \exp - \left(\frac{0.55 d_p}{\lambda}\right)\right] \tag{22.2}$$

$$\lambda = \frac{\mu}{0.499 \rho_g} \sqrt{8RT/\pi \mathrm{MW}} \tag{22.3}$$

where:

λ is the mean free path of the gas in m
d_p is the diameter of particle in m
μ is the gas viscosity in N-s/m² or kg/m-s
MW is the mean molecular weight of the gas
ρ_g is the gas density in kg/m³
R is the universal gas constant (8.3144 J/kg-mol-K)
T is the gas temperature in K

For air at room temperature and pressure, Equation 22.4 is a good approximation of C':

$$C' = 1.0 + \frac{0.16}{d_p} \tag{22.4}$$

Knowing the value of the mean free path of molecules at a given temperature and pressure, the mean free path at other conditions can be calculated from Equation 22.5:

$$\lambda = \lambda_o \left(\frac{\mu}{\mu_o}\right)\left(\frac{T}{T_o}\right)^{1/2}\left(\frac{P_o}{P}\right) \tag{22.5}$$

where:

$\lambda_o = 0.0653$ μm for air at 23°C and 1.0 atm
μ_o, T_o, and P_o are viscosity, temperature, and pressure, respectively, at the same conditions for which λ_o is known

TABLE 22.1
Particle Size Collection Efficiency of Various Wet Scrubbers

Type of Scrubber	Pressure Drop (in Pa)	Minimum Collectible Particle Diameter[a] (in μm)
Gravity spray towers	125–375	10
Cyclonic spray towers	500–2500	2–6
Impingement scrubbers	500–4000	1–5
Packed and moving bed scrubbers	500–4000	1–5
Plate and slot scrubbers	1200–4000	1–3
Fiber bed scrubbers	1000–4000	0.8–1
Water jet scrubbers	–	0.8–2
Dynamic	–	1–3
Venturi	2500–18,000	0.5–1

Source: Porter, H.F. et al., *Chemical Engineers' Handbook,* 6th ed., McGraw-Hill, New York, 20–77 to 20–110, 1984; Sargent, G.D., *Chem. Eng.*, 76, 130–150, 1969; Celenza, G.F., *Chem. Eng. Prog.*, 66(11), 31, 1970.

[a] Minimum particle size collected with approximately 85% efficiency.

22.4 SELECTION AND DESIGN OF SCRUBBERS

Calvert et al.[3] have prepared an extensive report of wet scrubbers from both theoretical considerations and literature data. In considering the types of scrubbers to use for a particular application, the designer must have in mind the required collection efficiency for a particular size emission. The data of Table 22.1 can be used as a rough guide for initial consideration of adequacy of different devices.

22.5 DEVICES FOR WET SCRUBBING

The following material is a compilation of facts and figures for typical wet scrubbers. Table 22.2 serves as a guideline to the general operational characteristics of various types of devices. Figures 22.1 through 22.6 are schematics illustrating the six types of scrubbers listed in Table 22.2. The scrubbers depicted in Figures 22.1 and 22.2 are the same as Figures 11.2 and 11.3. This emphasizes the fact that these devices may be used both to collect particles, primarily by inertial impaction, and to absorb gases, usually with a solvent that also promotes chemical reaction.

22.6 SEMRAU PRINCIPLE AND COLLECTION EFFICIENCY

In order to capture fine particles, greater energy must be expended on the gas. There are two ways to do this:

1. Increase the gas pressure
2. Atomize large quantities of water

TABLE 22.2
Characteristics of Wet Scrubbers

	Spray Tower	Cyclonic Spray Tower	Self-Induced Sprays	Plate Scrubbers	Venturi	Venturi Jet
Efficiency	90%	95%	90%	97	95	92
(d_p-μm)	>8	>5	>2	>5	>0.2	>1.0
Velocity (ft/s)	3–6	150–250 inlet	–	–	150–500 throat	–
Nozzle pressure (psig)	35–50	–	–	–	–	50–150
Pressure drop (inches of H_2O)	1–4	4–8	8–12	8–12	–	2–4 (Draft)
Liquid to gas ratio (gal/1000 ft^3)	10–20	3–6	0.5	2–3	3–10	30–80
Power input (HP/1000 cfm)	0.5–2	1–3.5	2–4	2–5	3–12	–

Collection efficiency may be unexpectedly enhanced in a wet scrubber through methods that will cause particle size growth. Particle growth can be brought about by the following factors:

- Lower temperatures that cause vapor condensation
- Increased flow rates that increase turbulence
- Thermal gradients in the narrow passages of the scrubber, which increase diffusion of particles into the liquid

Condensation is the most common growth mechanism. The hot gases are cooled by the lower temperatures resulting from contact with the scrubbing liquid. The gases may also be compressed in the narrow passages of the scrubber, which would tend to enhance condensation. The condensation occurs on the existing particles rather than the new nuclei. Thus, the dust particles will grow larger and will be more easily collected.

In general, the efficiency of a wet scrubber is directly related to the energy expended to produce the gas–liquid contact. The more the energy expended, the greater the turbulence in the contacting process and the higher the efficiency of collection. Semrau[7,8] defined the *contacting power* as the energy dissipated per unit volume of gas treated. Contacting power should be determined from the friction loss across the wetted portion of the scrubber. Pressure losses due to the gas stream kinetic energy should not be included. However, energy provided by the mechanical devices, along with the energy provided by the gas and liquid, is part of the contacting power. Semrau treated scrubber efficiency by relating *the number of transfer units to the contacting power* as follows:

$$N_t = \alpha P_t^{\beta} \tag{22.6}$$

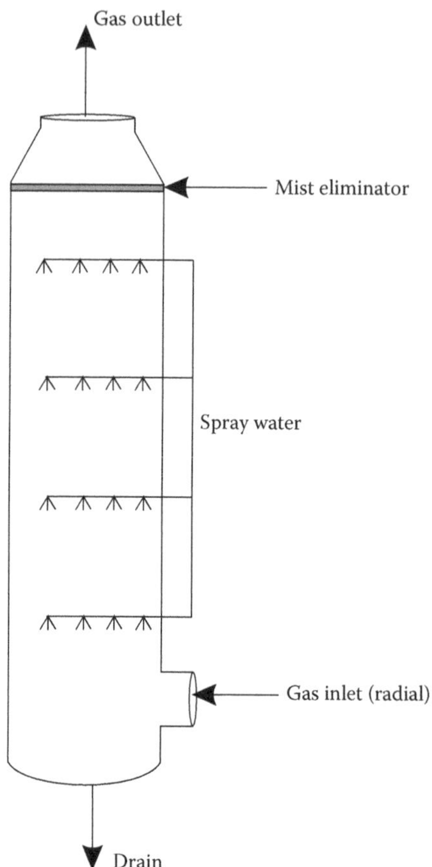

FIGURE 22.1 Spray tower (same as Figure 11.2 for absorption): • A vertical countercurrent tower; • Droplets sufficiently large so that the settling velocity is greater than the upward gas velocity; • Droplet size controlled to optimize particle contact and to provide easy droplet separation.

where:

N_t is the number of transfer units, dimensionless
P_t is the contacting power, hp/1000 cfm or kWh/1000 m^3
α is the coefficient to make N_t dimensionless
β is the dimensionless exponent

The overall efficiency, η_o, of the scrubber can be calculated from the following:

$$\eta_o = 1 - \exp(-N_t) \quad (22.7)$$

The following plot from Semrau, Figure 22.7, is performance curve for a venturi scrubber collecting a metallurgical fume.[8] Table 22.3 reports values of the parameters that could be used in Semrau's equation for the number of transfer units.[7]

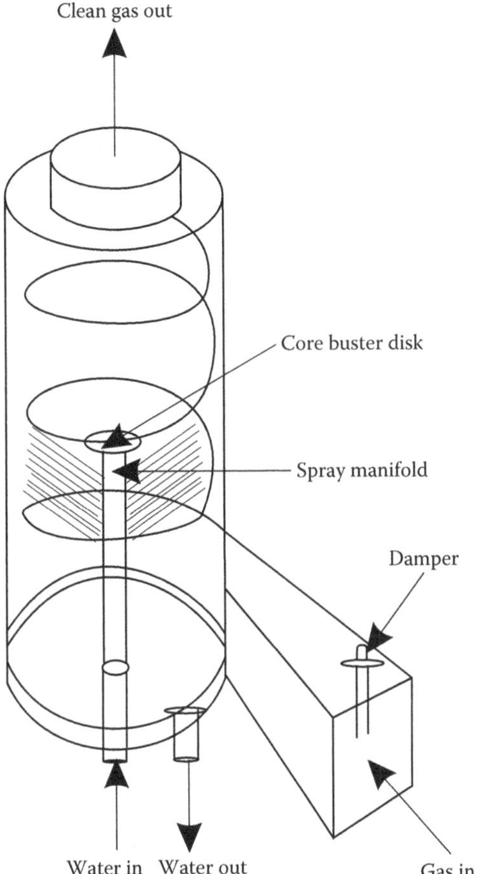

FIGURE 22.2 Cyclonic spray tower (same as Figure 11.3 for absorption): • Gas is introduced tangentially, which increases the forces of collision and relative velocity of the droplets and gas stream; • Well-designed cyclonic spray towers greatly increase the collection of particles smaller than 10 μm when compared to simple countercurrent spray towers; • Droplets are produced by spray nozzles; • Droplets collected by centrifugal force; • Two types of spinning motion are produced—(1) spinning motion imparted by tangential entry and (2) spinning motion produced by fixed vanes.

22.7 MODEL FOR COUNTERCURRENT SPRAY CHAMBERS

Drops are formed by atomizer nozzles and then sprayed into the gas stream. In the countercurrent tower, drops settle vertically against the rising gas stream, which is carrying the particles. Atomization provides a wide variety of droplet size. It is customary to take the Sauter mean drop diameter equivalent to the volume/surface area ratio and defined by the following equation to represent all the droplets:

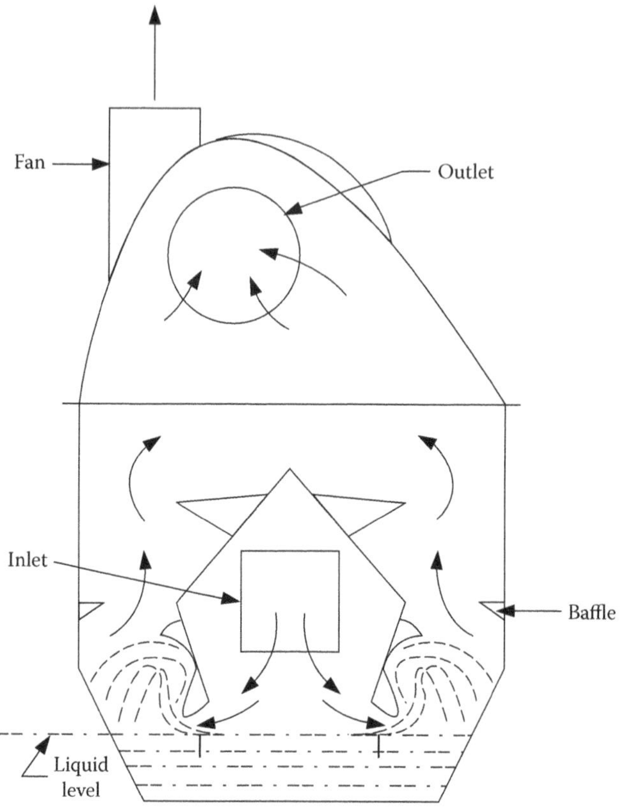

FIGURE 22.3 Self-induced spray tower: • Air impinges on a liquid surface then on a series of baffles; • Particles are initially captured in pool of water by direct interception or inertial impingement; • Some water is atomized into spray droplets which aids collection; • A final change in gas direction, or by baffles, serves as an entrainment separator; • The water circulation rate is low, and water is primarily required to replace evaporation losses; • Droplets are formed by breaking through a sheet of liquid or by impinging on a pool of water; • Droplets are collected by gravity attraction.

$$d_d = \frac{58{,}600}{V_g}\left(\frac{\sigma_L}{\rho_L}\right)^{0.5} + 597\left[\frac{\mu_L}{(\sigma_L\rho_L)^{0.5}}\right]^{0.45}\left[1000\frac{Q_L}{Q_G}\right]^{1.5} \tag{22.8}$$

where:

d_d is the Sauter mean droplet diameter, μm
ρ_L is the density of the liquid, g/cm^3
σ_L is the liquid surface tension, dyne/cm
μ_L is the liquid viscosity, poise
V_g is the superficial gas velocity, cm/s
Q_L is the volumetric liquid flow rate, m^3/s
Q_G is the volumetric gas flow rate, m^3/s

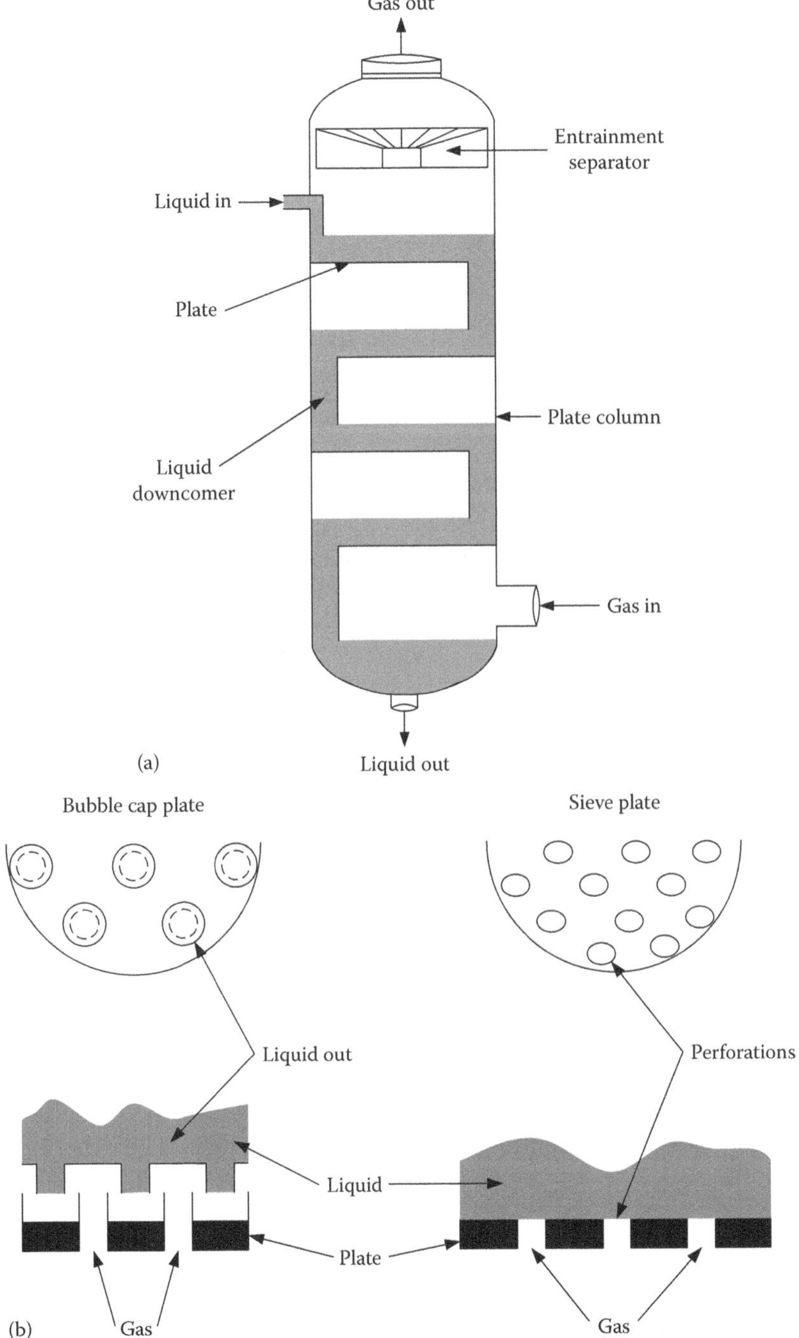

FIGURE 22.4 (a) Impingement plate scrubber. (b) Several plate types: • Plates of all types used—sieve plates, slot plates, valve trays, and bubble cap trays; • One modification, the turbulent contact absorber, uses a layer of fluidized plastic spheres.

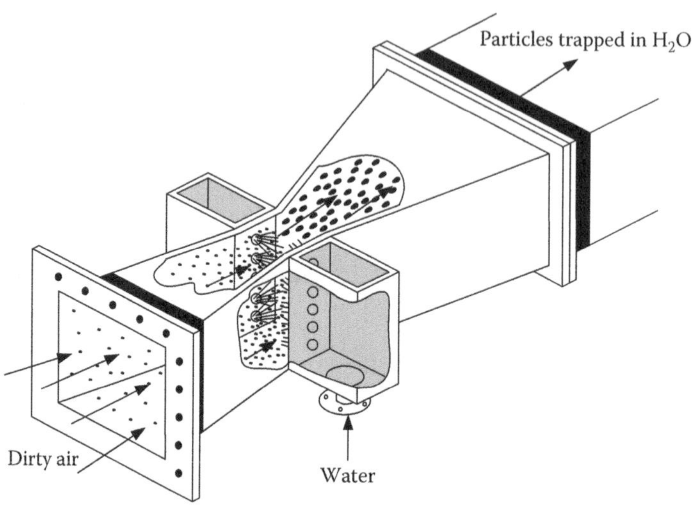

FIGURE 22.5 Venturi scrubber: • The gas is accelerated at the throat; • Atomized water droplets added at throat as a spray or jet, collect particles; • This scrubber can be combined with a cyclonic collector to disengage water droplets from air stream; • This scrubber has a large pressure drop.

Inertial impaction is depicted in Figure 22.8, primary capture mechanism. In this case of inertial impaction, a particle is carried along by the gas stream. Approaching the collecting body, which is a water droplet in the case of a spray scrubber, the particles tend to follow the streamlines. However, for many particles, their inertia will result in the particle separating from the gas stream and striking the water droplet. The result is for the water droplet to collect the particle. The separation number in Figure 22.8, N_{si}, is the same as the inertial impaction parameter, K_p, defined by the following equation:

$$N_{si} = K_p = K_m \rho_p d_p^2 \frac{V_p}{18\mu_g D_b} = \frac{C' \rho_p d_p^2 V_p}{18\mu_g d_d} = \frac{d_{ap}^2 V_p}{18\mu_g d_d} \qquad (22.9)$$

where:

C' is the Cunningham correction factor
ρ_p is the particle density, g/cm^3
d_p is the physical particle diameter, cm
d_{ap} is the aerodynamic impaction diameter in μm-g/cm^3 (See Equation 22.1)
V_p is the particle velocity, cm/s
μ_g is the gas viscosity, poise

Figure 22.9 exhibits single-droplet target efficiency for ribbons, spheres, and cylinders. The curves apply for conditions in which Stokes' law holds for the motion of the particle. Equation 22.10 is an approximate equation for a single-droplet target efficiency, η_d.

$$\eta_d = \left(\frac{K_p}{K_p + 0.35} \right)^2 \qquad (22.10)$$

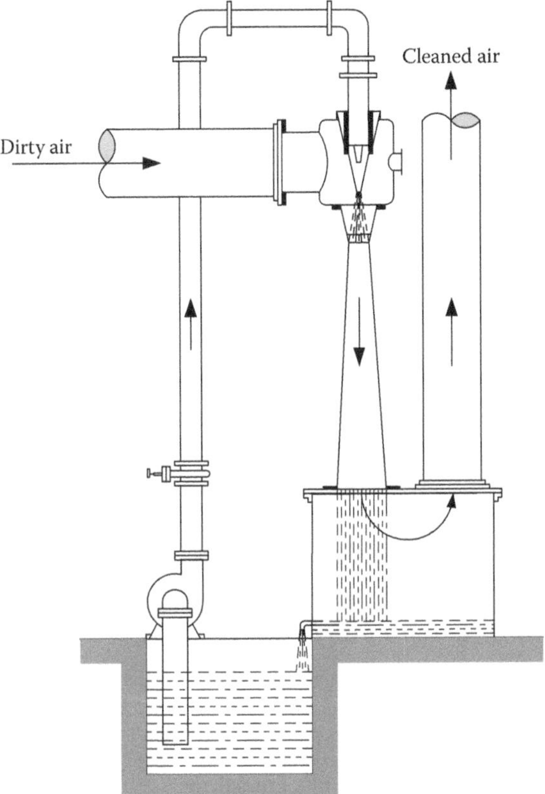

FIGURE 22.6 Venturi jet scrubber: • This scrubber is used for fume scrubbing; • High pressure water atomized from a jet nozzle into a throat of a Venturi, which induces the flow of the gas to be scrubbed.

22.7.1 Application to a Spray Tower

The principles stated above can be used to derive a model for the design of a spray tower. This model can be used along with particle distribution to determine the overall penetration or efficiency of a spray tower making use of the grade efficiency calculation method. The basic assumptions for the model are as follows:

1. Droplets are of uniform size and immediately reach the terminal velocity.
2. Droplets do not channel down to the scrubber walls.

Figure 22.10 depicts the relative velocities of a water droplet and a particle. In this case, we can define the following parameters:

V_{dt} is the droplet terminal velocity, cm/s
V_{pt} is the particle terminal velocity, cm/s
V_g is the superficial gas velocity, cm/s

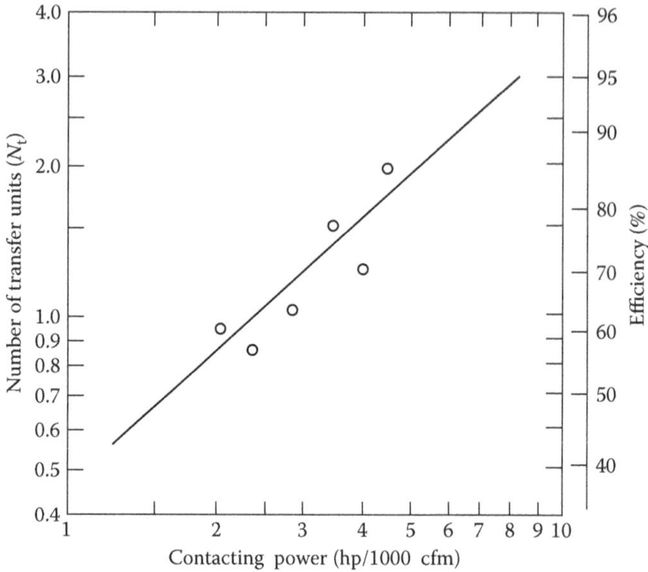

FIGURE 22.7 Performance curve for a venturi scrubber collecting a metallurgical fume. (From Semrau, K. T., *J. Air Poll. Contr. Technol. Assoc.*, 13, 587, 1963. With permission.)

TABLE 22.3
Correlation Parameters for Semrau's Equation

Aerosol	Scrubber	Correlation Parameter A	Correlation Parameter B
Lime kiln dust and fume			
Raw gas (lime dust and soda fume)	Venturi and cyclonic spray	1.47	1.05
Prewashed gas (soda fume)	Venturi, pipeline, and cyclonic spray	0.915	1.05
Talc dust	Venturi	2.97	0.362
	Orifice and pipeline	2.70	0.362
Black liquor recovery furnace fume cold scrubbing water humid gases	Venturi and cyclonic spray	1.75	0.620
Hot fume solution for scrubbing (humid gases)	Venturi, pipeline, and cyclonic spray	0.740	0.861
Hot black liquor for scrubbing (dry gases)	Venturi evaporator	0.522	0.861
Phosphoric acid mist	Venturi	1.33	0.647
Foundry cupola dust	Venturi	1.35	0.621
Open hearth steel furnace fume	Venturi	1.26	0.569
Talc dust	Cyclone	1.16	0.655
Ferrosilicon furnace fume	Venturi and cyclonic spray	0.870	0.459

Source: Semrau, K. T., *J. Air Poll. Contr. Technol. Assoc.*, 10(3), 200, 1960. With permission.

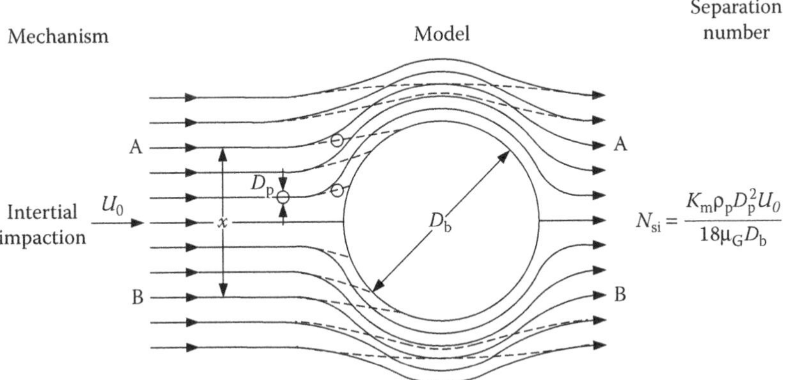

FIGURE 22.8 Primary capture mechanism.

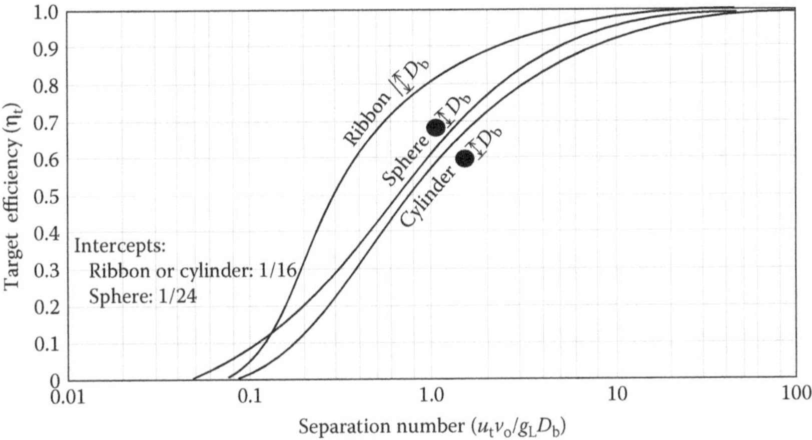

FIGURE 22.9 Single-droplet target efficiency for ribbons, spheres, and cylinders.

$V_g = Q_G/A$
Q_G is the gas flow rate in m^3/s
A is the cross-sectional area of tower, m^2

It is assumed that the particle travels with velocity, V_p, equal to the gas velocity, that is,

$$V_p = V_g$$

A mass balance is made next, based on Figure 22.11, mass balance in tower cross section. The mass balance:

$$\text{Mass in} - \text{mass out} - \text{mass collected} = \text{accumulation}$$

$$C\,Q_G - (C + dC_i)Q_G - M_c = 0$$

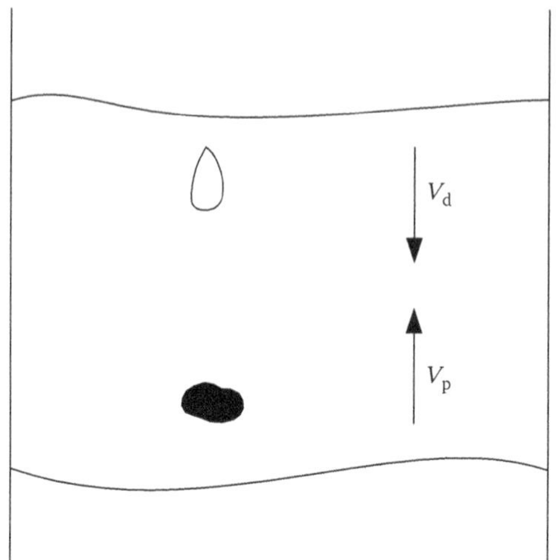

FIGURE 22.10 Relative velocities of a water droplet and a particle.

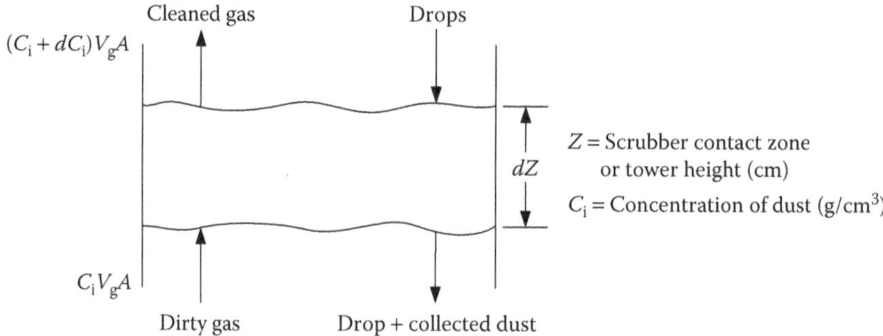

FIGURE 22.11 Mass balance in tower cross section.

The total number of drops entering the cross section per unit of time:

$$\frac{Q_L}{\pi d_d^3/6} \tag{22.11}$$

The mass of particles collected by each drop per unit time:

$$\eta_d \left(V_{dt} - V_{pt}\right) C_i \frac{\pi d_d^2}{4} \tag{22.12}$$

The time for the droplet to pass down through the section dZ,

$$\frac{dZ}{V_{dt} - V_g} \tag{22.13}$$

This is essentially the contact time for the droplet. The mass of particles collected by one droplet during this contact time is then,

$$\left[\eta_d\left(V_{dt}-V_{pt}\right)C_i\frac{\pi d_d^2}{4}\right]\left(\frac{dZ}{V_{dt}-V_g}\right) \tag{22.14}$$

Multiply Equation 22.14 by Equation 22.11, the total amount of drops, and this will be M_c, the total mass collected. The mass balance may then be written as

$$-Q_G dC_i = \frac{3}{2}\eta_d\left(\frac{V_{dt}-V_{pt}}{V_{pt}-V_g}\right)\left(\frac{Q_L}{d_d}\right)C_i\, dZ \tag{22.15}$$

Separate the variables and integrate:

$$-\int_{C_0}^{C}\frac{dC_i}{C_i} = \frac{3}{2}\eta_d\int_{o}^{z}\left(\frac{V_{dt}-V_{pt}}{V_{dt}-V_g}\right)\left(\frac{Q_L}{Q_G}\right)\frac{dZ}{d_d} \tag{22.16}$$

resulting in

$$P_t = \frac{C}{C_o} = \exp\left\{-\left[\frac{3}{2}\eta_d\left(\frac{V_{dt}-V_{pt}}{V_{dt}-V_g}\right)\left(\frac{Q_L}{Q_g}\right)\frac{Z}{d_d}\right]\right\} \tag{22.17}$$

Application of the equation usually involves the following assumptions:

1. $V_{dt} >> V_{pt}$.
2. At Calvert's suggestion, (Q_L/Q_G) is multiplied by 0.2 to correct for wall effects and other loss of effectiveness by the falling droplets.

Then, the final formula is

$$P_t = \exp\left\{-0.30\left[\eta_d\left(\frac{V_{dt}}{V_{dt}-V_g}\right)\left(\frac{Q_L}{Q_g}\right)\frac{Z}{d_d}\right]\right\} \tag{22.18}$$

Figure 22.12 illustrates the effect of the model parameters on penetration. As the penetration decreases, the efficiency increases. Therefore, increasing η_d, Q_L, or Z will increase efficiency, while increasing d_d or decreasing Q_g will decrease efficiency. Note that as d_d becomes smaller, the efficiency should increase. In practice, this is true until the droplet diameter, d_d, becomes so small that the drag coefficient results in a force large enough so that the effect of the water spray is reduced. Thus, the spray becomes less effective in collecting particulates, and the efficiency decreases again. There should, therefore, be an optimum drop size where the efficiency is the maximum. Figure 22.13 presents data to confirm this conclusion.

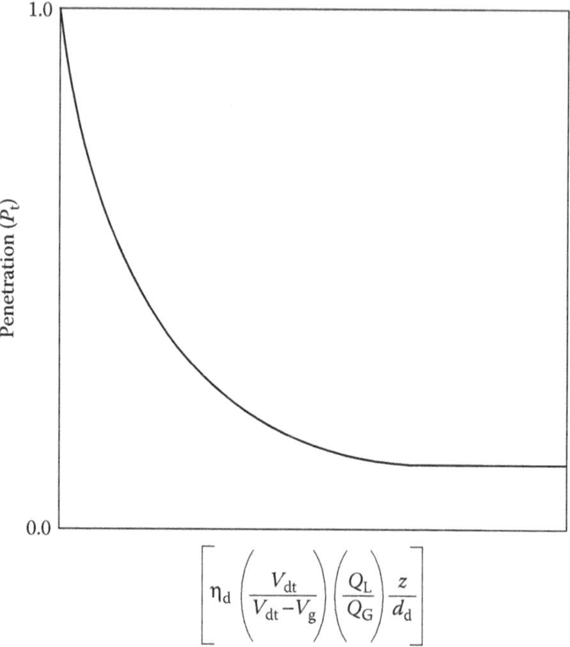

FIGURE 22.12 The effect of the model parameters on penetration.

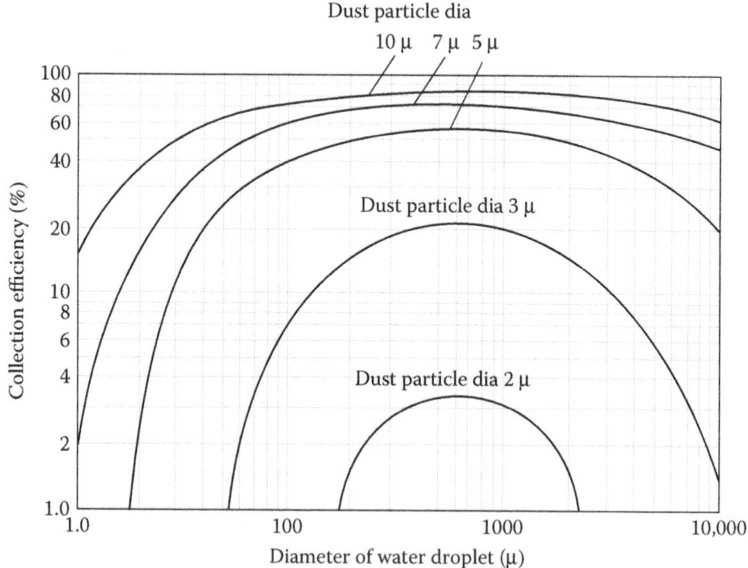

FIGURE 22.13 Optimum droplet size for collection by inertial impaction when droplets are moving in the gravitational field of the earth in a spray tower. (From Stairmand, C. J., *Trans. Inst. Chem. Eng.*, 28, 130, 1950. With permission.)

22.8 A MODEL FOR VENTURI SCRUBBERS

Calvert et al.[3] have developed a model similar to Equation 22.18 for several different types of scrubbers. Following is a model for venturi scrubbers.

$$\mathrm{Pt} = \exp\left[\frac{Q_L V_G \rho_L d_d}{55 Q_G \mu_G}\right] B$$

$$B = \left\{\left[-0.7 - K_p f + 1.4 \ln\left(\frac{K_p f + 0.7}{0.7}\right) + \frac{0.49}{0.7 + K_p f}\right]\frac{1}{K_p}\right\} \quad (22.19)$$

where:

K_p is the inertial impaction parameter at throat entrance

f is the empirical factor ($f = 0.25$ for hydrophobic particles and $f = 0.50$ for hydrophilic particles)

This model works well for $f = 0.50$ for a variety of large-scale venturi scrubbers and other spray scrubbers where the spray is atomized by the gas.

22.9 CALVERT CUT DIAMETER DESIGN TECHNIQUE

The cut diameter method of wet scrubber design was devised by Calvert et al.[3] The method is based on the concept that the most significant single parameter to define both the difficulty of separating particles from the carrier gas and the performance of a scrubber is the particle diameter for which collection efficiency is 50%. As noted previously, aerodynamic diameter defines the particle properties of importance in designing wet scrubbers that depend on inertial impaction as the collection mechanism.

In most cases of particle cleanup, a range of particle sizes is involved. The overall collection efficiency will depend on the amount of each size present and on the efficiency of collection for that size. In order to account for the size distribution effects, the difficulty of separation is defined as the aerodynamic diameter at which collection efficiency (or penetration) must be 50% in order that the required overall efficiency for the entire size distribution be attained. This particle size is defined as the separation cut diameter (d_{RC}). The Calvert design technique is based on a log-normal distribution of particle sizes and relates the separation cut diameter to the overall penetration and the size distribution parameters.

In order to design a wet scrubber, the separation cut diameter is determined from the particle size distribution parameter and the required penetration. A performance cut diameter (d_{PC}) is measured as a function of the varying characteristics of many types of scrubbers. The design process consists of setting the performance cut diameter equal to the separation cut diameter, ($d_{PC} = d_{RC}$) and then looking in a catalog or handbook for the design parameter of the particular type scrubber of interest. In order to be effective, the process obviously requires a series of extensive measurements on many scrubber types. Calvert et al.[3] list performance cut diameter curves for many types of scrubbers. The open literature reports other performance cut diameter curves.[10]

The technique depends upon an exponential relationship for penetration as a function of aerodynamic diameter for inertial impaction collection equipment:

$$\mathrm{Pt_d} = \exp\left(-Ad_{\mathrm{pa}}^{\mathrm{B}}\right) \quad (22.20)$$

This relationship can be simplified and based on actual particle diameter without introducing too much error for larger particles or where the particle distribution is log normal in terms of physical diameter:

$$\mathrm{Pt_d} = \exp\left(-A_{\mathrm{p}}d_{\mathrm{p}}^{\mathrm{B}}\right) \quad (22.21)$$

Packed towers, centrifugal scrubbers, sieve plates, and venturi scrubbers follow the above relationship where B is given in Table 22.4.

From Table 22.4, it can be seen that the last three listings are gas-atomized liquid type of wet scrubbers. Performance of these types of scrubbers fit a value of $B = 2.0$ over a large portion of their operating range. These are scrubbers whose collection mechanism is essentially inertial impaction. Therefore, $B = 2.0$ is representative of scrubbers operating in the inertial impaction range.

With dW/W as the weight fraction of dust in a given size interval, the overall penetration, Pt, of any device on a dust of that type of size distribution will be

$$\overline{\mathrm{Pt}} = \int \mathrm{Pt}\left(\frac{dW}{W}\right) \quad (22.22)$$

Equation 22.22 has been solved by Calvert for a dust, which is log-normally distributed with collection efficiency or penetration defined by Equation 22.20. Figure 22.14 is a plot of penetration, Pt versus the ratio of aerodynamic cut diameter to aerodynamic geometric mean particle diameter, $d_{\mathrm{ap}}^{50}/d_{\mathrm{apg}}$, with $B(\ln \sigma_{\mathrm{g}})$ as a parameter.

d_{ap}^{50} is the *cut diameter* or the aerodynamic particle diameter collected with 50% efficiency
d_{apg} is the aerodynamic geometric mean particle diameter
σ_{g} is the geometric standard deviation

TABLE 22.4
Values of Exponent B in Equation 22.20

Type of Scrubber	Value of B
Centrifugal scrubber	0.67
Packed towers	2.0
Sieve plates	2.0
Venturi scrubbers	2.0

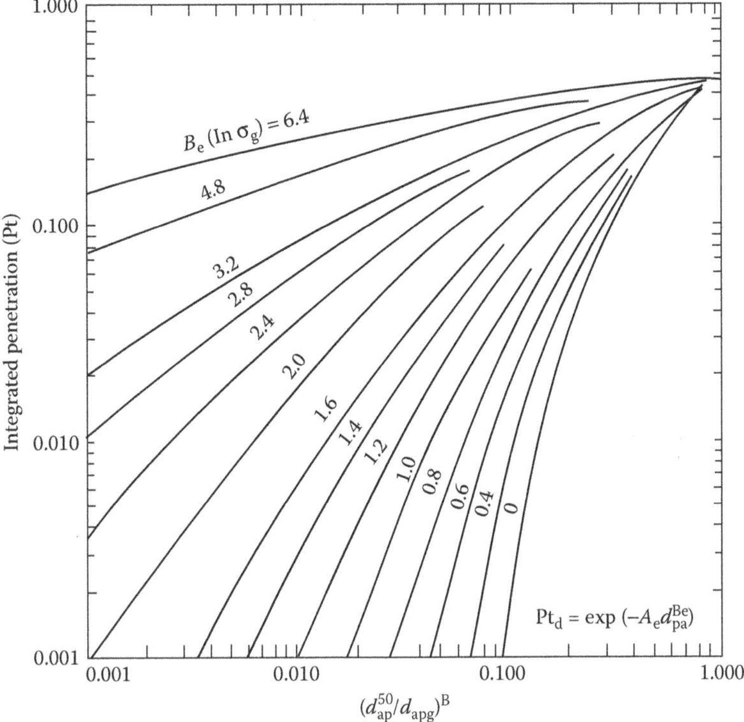

FIGURE 22.14 Penetration, *Pt*, versus the ratio of aerodynamic cut diameter to aerodynamic geometric mean particle diameter, d_{ap}^{50}/d_{apg}, with $B(\ln \sigma_g)$ as a parameter.

It should be noted that

$$\sigma_g = \frac{d_{a84.13}}{d_{apg}} = \frac{d_{apg}}{d_{a15.87}} \tag{22.23}$$

Figure 22.15 presents the special case where $B = 2$ with σ_g as the parameter.

Summarizing the design technique, first determine that the particle size distribution by weight fraction is log normal. If the distribution is obtained from a cascade impactor, the particle size will be in aerodynamic dimensions. From the required efficiency η, the penetration, Pt, can be calculated as $\text{Pt} = 1 - \eta$. The particle distribution will provide the geometric standard deviation σ_g and the value of the geometric mean particle diameter d_{apg}. For a wet scrubber in which the collection action is largely dependent upon inertial impaction, $B = 2$, and the ratio $\left(d_{ap}^{50}/d_{apg}\right)$ can be found from Figure 22.15 knowing the penetration. The separation cut diameter is the aerodynamic particle diameter collected with 50% efficiency, thus $d_{RC} = d_{ap}^{50}$. The design method simply matches the performance cut diameter to the separation cut diameter.

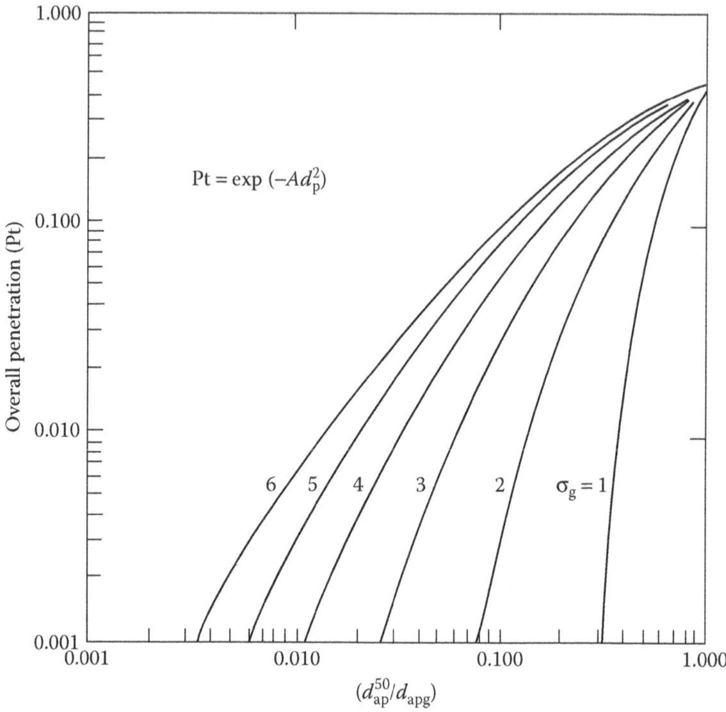

FIGURE 22.15 Penetration, *Pt*, versus the ratio of aerodynamic cut diameter to aerodynamic geometric mean particle diameter, d_{ap}^{50}/d_{apg}, for the special case where $B = 2$ and with σ_g as a parameter.

$$d_{PC} = d_{RC} \tag{22.24}$$

Performance cut diameter plots can then be searched to find feasible scrubber types and their process parameters. The following six curves are presented in Appendix 22A.

22.9.1 Example Calculation

An air stream containing dust must be cleaned to remove 98% of all the particles. The dust is log-normally distributed with $\sigma_g = 3.0$ and $d_{apg} = 20$ μm. The dust density is 3 g/cm³.

$$\overline{Pt} = 1.00 - 0.98 = 0.02 \text{ μm}$$

Determine d_{ap}^{50} from Figure 22.15,

$$\left(\frac{d_{ap}^{50}}{d_{apg}}\right) = 0.85$$

$$d_{ap}^{50} = 0.085 \times 20 - 1.70\ \mu\text{m}$$

$$d_{RC} = d_{ap}^{50} = 1.70\ \mu\text{m}$$

The performance cut diameter is then d_{PC} = 1.70 μm. Performance cut diameter curves for various types of wet scrubbers such as shown in Figures 22A.1 through 22A.6 can now be searched for a suitable scrubber with d_{PC} = 1.70 μm.

22.9.2 Second Example Problem

A vertical countercurrent spray chamber has a 2.0 m contact zone and operates with a liquid-to-gas ratio of 1.01/m³. The average droplet size is 300 μm. The following data were recorded during a test run of the apparatus:

Average operating conditions: 80°F, 1.0 atm
Gas velocity = 0.60 m/s
Inlet loading = 1.5 gr/ft³
Particle density = 2.50 g/cm³

The particle size distribution was obtained by a cascade impactor. From a log-probability plot of the particle size distribution, it was found that

The geometric mean diameter, d_{apg} = 9.7 μm
The geometric standard deviation, σ_g = 2.22

Determine the efficiency of the scrubber by the Calvert cut diameter technique. Examining the plots of Figures 22A.1 through 22A.6, it can be determined that Figure 22A.1 for a vertical countercurrent flow spray tower can be used.

Tower data:

$$Z = 2.0\text{ m},\ \frac{Q_L}{Q_G} = 1.01/\text{m}^3,\ d_d = 300\ \mu\text{m},\ U_G = 0.60\text{ m/s}$$

The performance cut diameter can be read from Figure 22A.1.

$$d_{PC} = 3.3\ \mu\text{m}$$

The separation cut diameter and the cut diameter are therefore

$$d_{ap}^{50} = d_{RC} = d_{PC} = 3.3\ \mu\text{m}$$

To find penetration (Pt), calculate

$$d_{ap}^{50}/d_{apg} = 3.3/9.7 = 0.34$$

then from Figure 22.15 at $\sigma_g = 2.22$

$$\text{Pt} = 0.15 \text{ or } 15.0\%$$

Thus, the overall efficiency for this scrubber would be 85.0%.

22.10 CUT–POWER RELATIONSHIP

Calvert[10] has related the performance cut diameter to gas phase pressure drop or power input to the scrubber. This makes a simple relationship that describes the performance on many scrubbers. Performance tests on many scrubbers combined with Calvert's mathematical model have led to the refined cut–power relationship for several types of scrubbers shown in Figure 22.16.

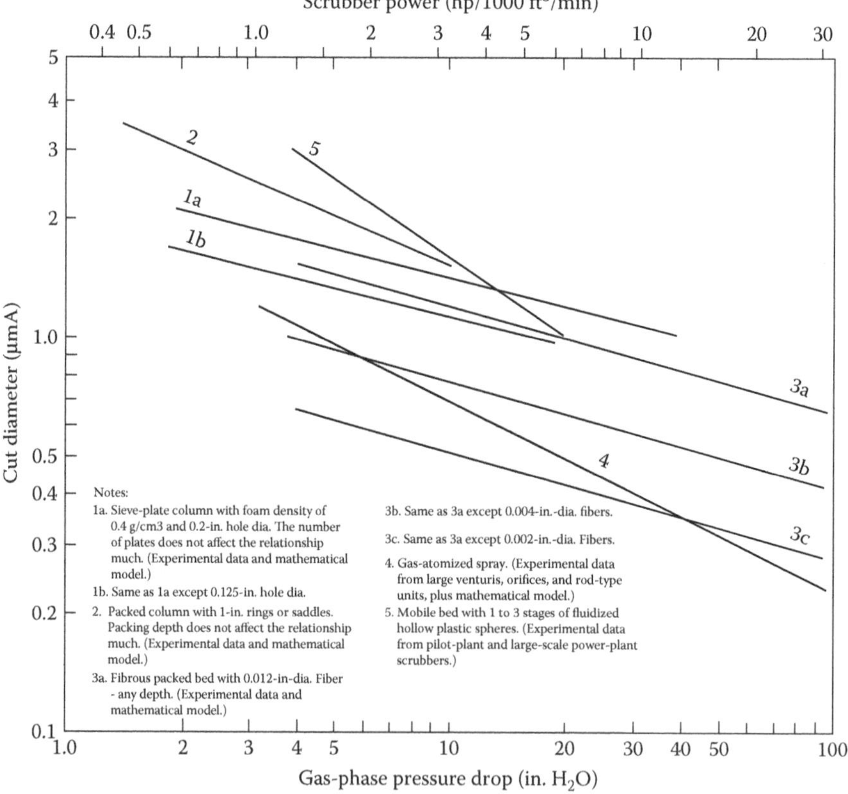

FIGURE 22.16 Cut–power relationship for several scrubber types. (From Calvert, S., *Chem. Eng.*, 29, 54–68, 1977. With permission.)

APPENDIX 22A: CALVERT PERFORMANCE CUT DIAMETER DATA

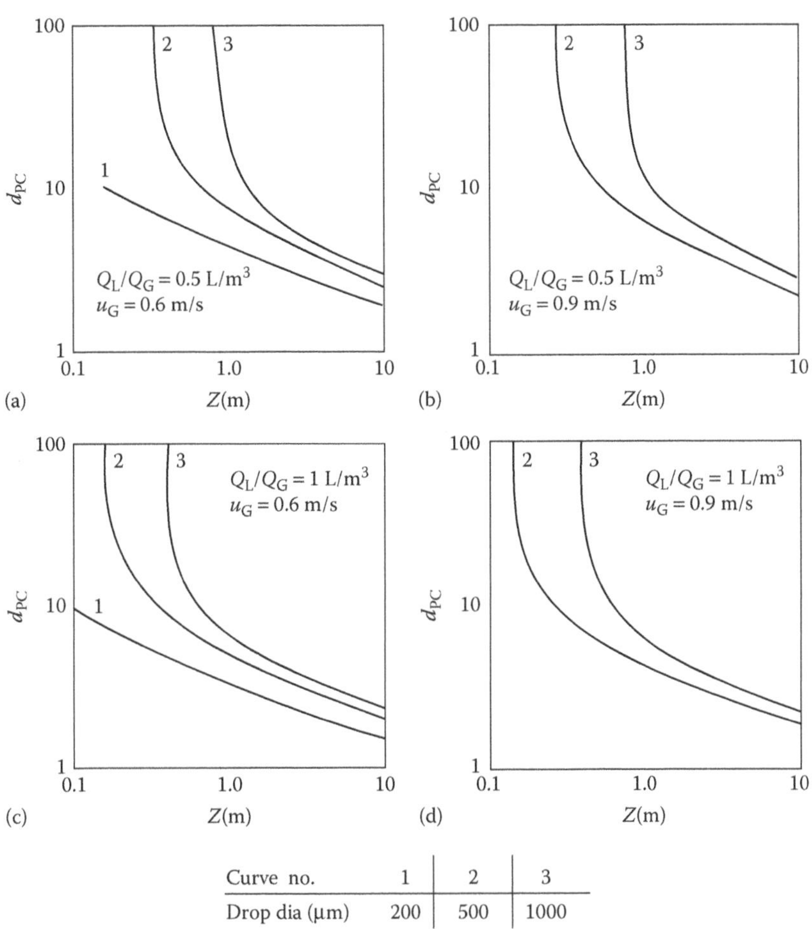

FIGURE 22A.1 (a–d) Vertical countercurrent flow spray towers at various conditions.

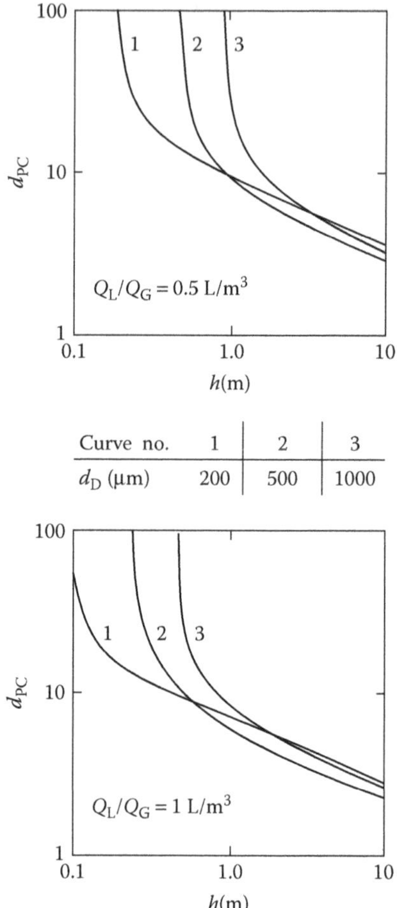

FIGURE 22A.2 Cross-current flow spray towers.

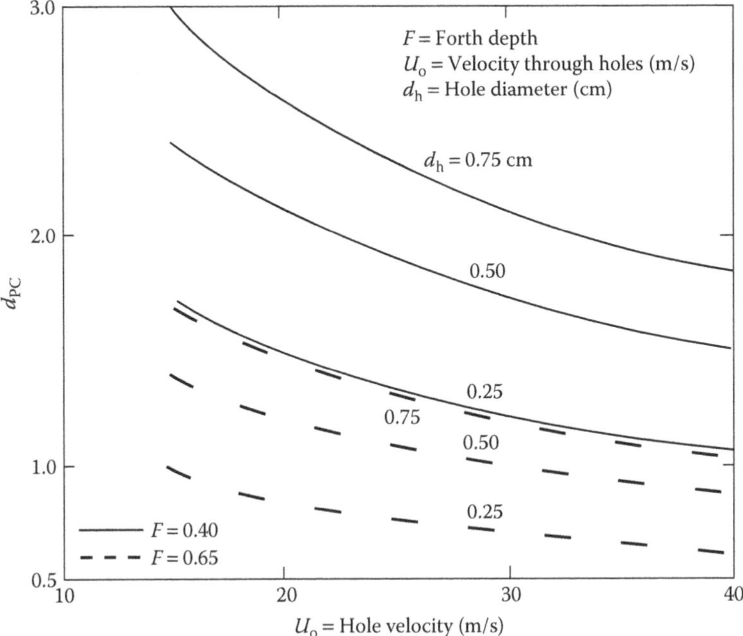

FIGURE 22A.3 Sieve plate column.

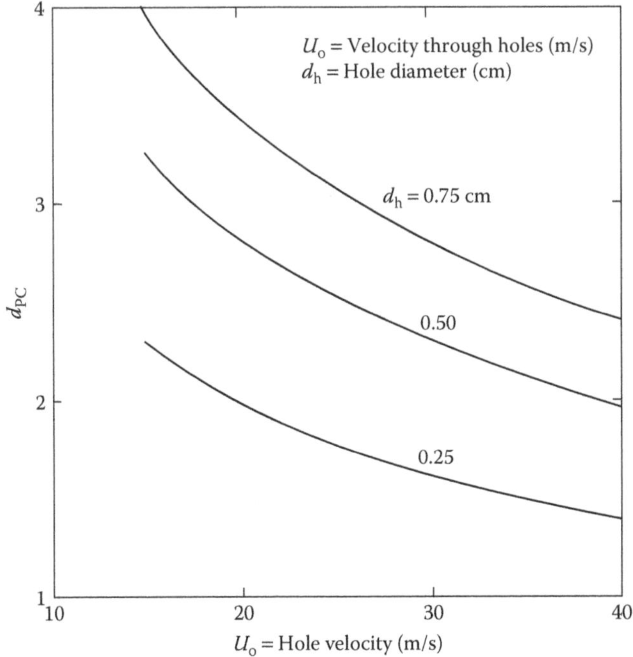

FIGURE 22A.4 Impingement plate column.

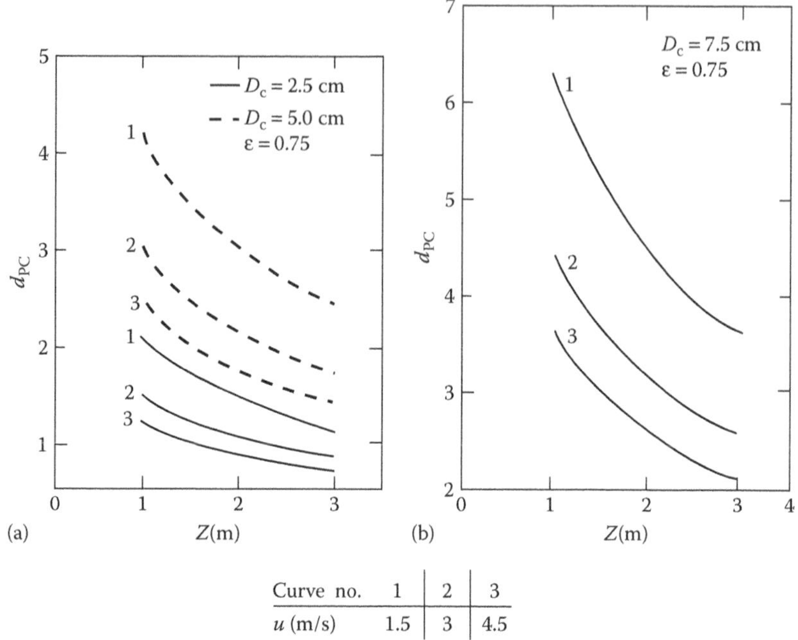

FIGURE 22A.5 (a and b) Packed columns, D_c is the packing diameter, cm and ϵ is the bed porosity at different packing diameters.

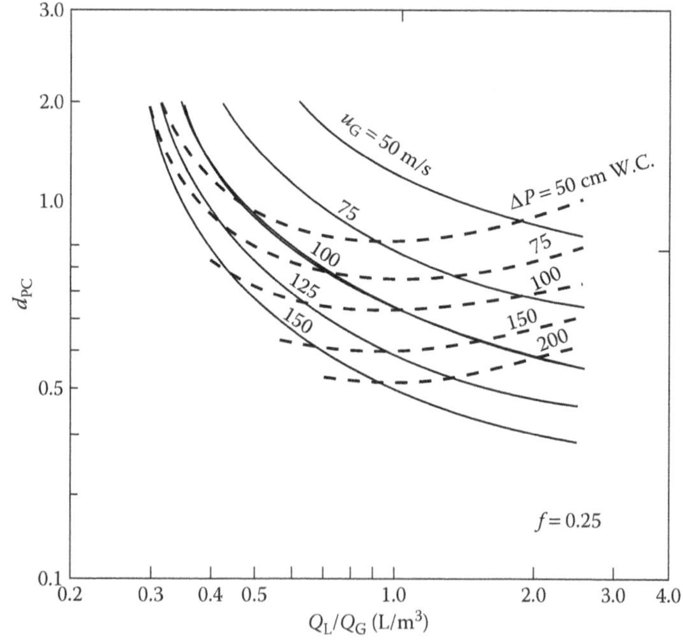

FIGURE 22A.6 Hydrophobic particles in venturi scrubber.

23 Filtration and Baghouses

23.1 INTRODUCTION

Baghouse is a common term for the collection device that uses fabric bags to filter particulate out of a gas stream. The filter bags are mounted on a tubesheet and enclosed in a sheet-metal housing. The housing is visible and the single word *baghouse* is easy to pronounce, but *filtration* is more technically descriptive of the process. Frequently, the term *fabric filtration* is used, partly to be technically accurate and partly to distinguish the technology from water filtration. While filtration of particles from air commonly employs fabric bags as the filter media, porous ceramic candles and paper cartridges also are used to clean gas streams. Finally, a fabric filter *baghouse* system includes the bag cleaning system, dust collection hoppers, and dust removal system, so the total system involves more than just filtration.

It must be understood that the mechanism that achieves filtration of small particles from a gas stream is not simple sieving. The spacing between fabric threads may be on the order of 50–75 μm, yet particles of 1 μm diameter and less are collected efficiently. Indeed, the primary collection mechanisms include impaction, interception, and diffusion, as discussed in Chapter 19. Initially, on a clean filter before any dust accumulates, the fabric threads and fibers are stationary targets for the particles. As soon as a layer of dust (dust cake) accumulates, however, the stationary particles in the dust cake become targets for the particles in the gas stream.

This explains an interesting phenomenon about fabric filtration: emissions from new, clean bags tend to be higher than from used bags. Used bags have been *seasoned*, that is, a number of particles have been lodged in the fabric (cleaning is not perfect and there is always some residual dust after cleaning) that serve as small targets for collection. These embedded particles also tend to fill the gaps between threads, reducing the opening size and increasing the probability for collection by impaction, interception, and diffusion.

Filtration is effective at removing submicron particles because of the diffusion mechanism, especially after a dust cake has been established. The space between target fibers is small, so particles do not have to diffuse a great distance to be collected. And after the dust cake is established, the space between target particles is very small, and the gas path through the dust layer becomes rather long and tortuous.

23.2 DESIGN ISSUES

The basic design parameters for a fabric filter baghouse include the following:

- Cleaning mechanism
    - Shake/deflate

    - Reverse air
    - Pulse jet
- Size
    - Air-to-cloth ratio
    - Can velocity
- Pressure drop
    - Fan power
    - Vacuum/pressure rating
- Fabric
    - Material
    - Weave
- Bag life
    - Cleaning frequency
    - Gas composition
    - Inlet design

23.3 CLEANING MECHANISMS

As dust collects on the fabric, a dust layer builds up, which increases the pressure drop requirement to move gas through the dust cake. Eventually, the dust layer becomes so thick that the pressure drop requirement is exceeded and the dust cake needs to be removed. Three types of dust cake removal (cleaning) systems are used.

23.3.1 Shake/Deflate

This oldest cleaning mechanism is to stop the gas flow through the fabric bags and shake the bags to knock off the dust cake. As shown in Figure 23.1, the dust cake typically collects on the inside of the bags as the gas flows upward through a cell plate or tubesheet near the bottom of the housing and through the bags. The bags are suspended on a rod or frame and the open end at the bottom is clamped onto a thimble on the tubesheet.

Cleaning is accomplished by moving the upper support frame, typically back and forth, up and down, or, if the driver is mounted to an eccentric rod, in a sinusoidal motion. The duration typically is from 30 s to a few minutes and may have a frequency of several times per second, an amplitude of a fraction of an inch to a few inches, and an acceleration of 1–10 g. Typically, the motion is imparted by an electric motor, but may be done using a hand crank on a small baghouse that does not require frequent cleaning.

Stopping the gas flow, or deflating the bags, greatly increases the effectiveness of cleaning. If the process gas flow cannot be interrupted, multiple parallel compartments are required so that one can be isolated. Many shaker baghouses are used in batch applications or noncritical flow applications, where both the process and the baghouse designs are kept very simple intentionally. The dust is allowed to fall to hoppers in the bottom of the baghouse before being immediately re-entrained and re-collected. Also, the bag is looser and moving the top of a bag

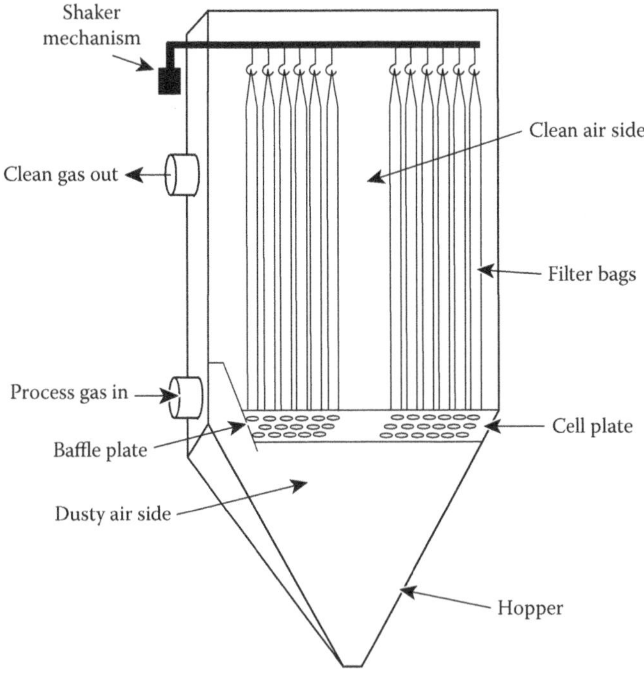

FIGURE 23.1 Shaker baghouse.

imparts more motion to the rest of the bag when it does not have pressure on the inside of the bag.

23.3.2 Reverse Air

Reverse air cleaning involves gently blowing clean gas backward through the fabric bag to dislodge the dust cake. A schematic of the required baghouse arrangement is shown in Figure 23.2. The system requires multiple parallel modules that can be isolated, which typically is accomplished using *poppet* valves, or large disks mounted on a shaft that moves up and down, much like a valve in a car engine. Like the shake/deflate configuration described above, gas flow enters the bag from the bottom and dust collects on the inside of the bag. To clean, a portion of the cleaned gas is withdrawn from the clean-side manifold using a fan. The reverse air valve opens, and low-pressure clean gas is gently blown backward through the fabric. The reverse air fan develops a head of only a few inches of water pressure. The entire cleaning process is sequenced and takes a couple of minutes, including time to close the main discharge valve and let the bag collapse, then open the reverse air valve for 10–30 s, close the reverse air valve and let the dust settle, and finally open the main gas discharge valve.

Blowing gas backward through the bag from the outside to the inside would cause a 20–30 ft long bag to collapse flat. This would not be conducive to allowing a dust

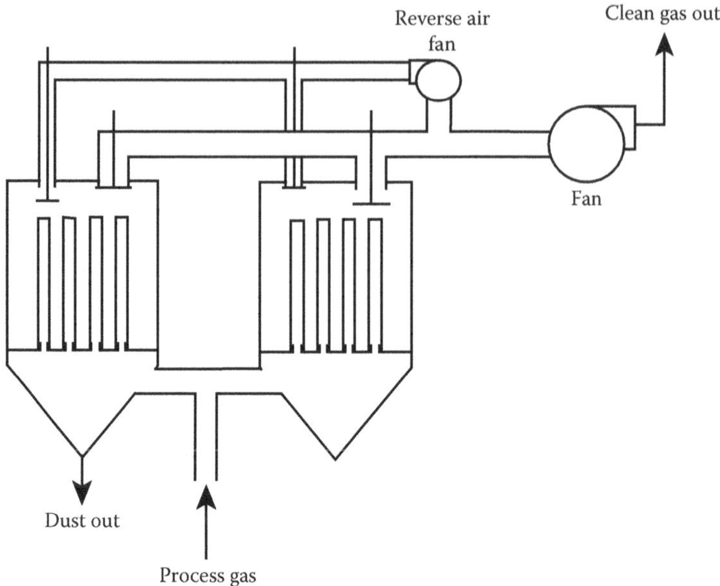

FIGURE 23.2 Reverse air cleaning.

cake to drop. To prevent complete collapse, reverse air bags have stiff anti-collapse rings sewn into them at about 3–4 ft spacing. The partial collapse of the bag flexes the bag, which dislodges the dust cake. The particles are not *blown off* the bag by the gentle reverse air flow.

A primary advantage of reverse air cleaning is that very large bags can be used. The bags can be 12 or more inches in diameter and 30–40 ft long. Thus, a very large reverse air baghouse can have a smaller footprint and fewer bags than a pulse jet baghouse, which is described in Section 20.3.3. Several very large reverse air baghouses at a coal-fired power plant are shown in Figure 23.3.

Another advantage of reverse air cleaning is the potential for increased bag life from the gentle cleaning action. The gentle action minimizes abrasion, which often is the limiting factor for bag life. However, there are several factors that could limit bag life, including buildup of residual dust cake after cleaning, in which case the more vigorous cleaning action of pulse jet could result in longer bag life.

23.3.3 Pulse Jet (High Pressure)

A vigorous and very common cleaning mechanism is high-pressure pulse jet. High-pressure pulse jet cleaning uses a very short blast of compressed air (70–100 psi) to deform the bag and dislodge the dust cake. The common term is simply *pulse jet*, but there are baghouse designs, which also are called as *pulse jet*, that employ air at 7–14 psi and are sufficiently different to warrant a separate discussion in Section 20.3.4.

FIGURE 23.3 Large reverse air baghouse.

A schematic of a pulse jet baghouse is shown in Figure 23.4. Note that in this configuration, the bags are hung from a tubesheet located near the top of the housing, gas flow is from the outside to the inside of the bag, and dust is collected on the outside. To keep the bag from collapsing during normal operation, wire cages are used on the inside of the bags. A compressed air pipe is located over each row of bags, and there is a small hole in the pipe over each bag. A diaphragm valve with a separate solenoid valve operator admits compressed air into the blowpipe, so bags are cleaned one row at a time. A venturi is used at the top of each cage to direct the pulse of compressed air.

The compressed air pulse duration is very short, being about 100–200 msc. The cleaning action often is described as a shock wave or an air bubble that travels down the length of the bag. While conceptually descriptive, this may not be technically accurate. In any case, the pulse distends the bag and dislodges the dust cake.

Pulse jet cleaning can be done online or offline. The obvious advantage to online cleaning is that the gas flow is not interrupted. With small, one-compartment baghouses, this is critical. Disadvantages of online cleaning are that much of the dust is likely to be re-entrained and re-deposited on the bags before falling to the hoppers, and the bags tend to snap back harshly onto the cages at the end of the pulse, aided by the normal gas flow. This can cause excessive bag wear.

Cages typically are constructed with either 10 or 20 vertical wires. Twenty wire cages are more expensive because they have almost twice the wire as 10-wire cages (both require wire circles to hold their shape) and they require more spot welding. The advantage is that with less space between supporting wires, bag flexing is minimized, resulting in less abrasion.

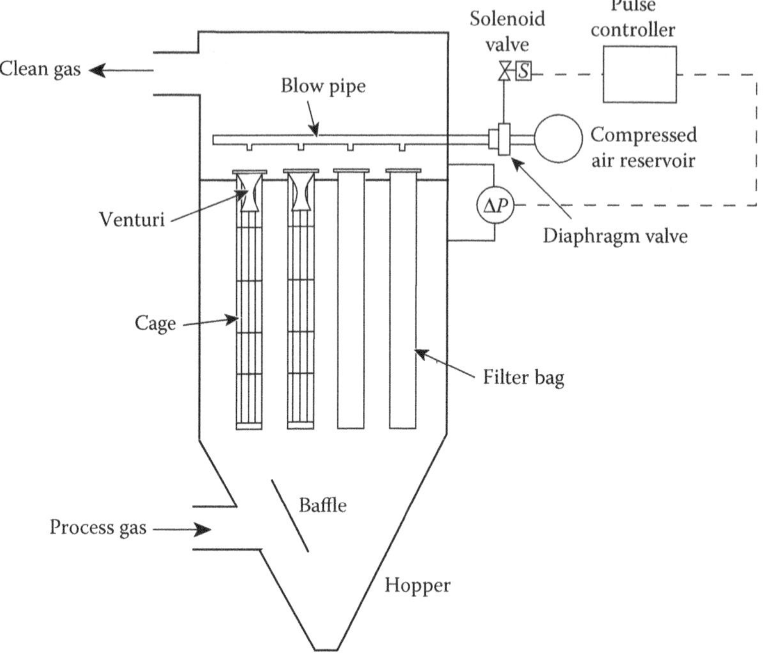

FIGURE 23.4 A schematic of pulse jet baghouse.

23.3.4 Pulse Jet (Low Pressure)

Some baghouses use air at 7–14 psi for cleaning. Although they are called *pulse jet*, the action is different from the high-pressure pulse described above. The bags are distended during cleaning, which dislodges the dust cake. But the pulse of air is longer and less vigorous than a high-pressure pulse. One design uses a single blow pipe, fixed on one end, which travels in a circle over the bags. When a cleaning cycle is triggered, the blow pipe makes one revolution.

Other baghouse designs use intermediate air pressure for cleaning. The air supply is around 40 psi.

23.3.5 Sonic Horns

Sonic or acoustic horns sometimes are used to create low-frequency (150–200 Hz, a very deep bass) sound-wave-induced vibrations to promote cleaning. They are powered by compressed air and generate about an average of 120–140 decibels in a compartment. Sound waves create fluctuations in the static pressure that cause vibrations, which help to loosen particulate deposits. Typically, horns are used to assist reverse air, but can be used as the sole cleaning mechanism in applications where the dust cake releases easily.[1] Sonic horns also are used in baghouse hoppers to prevent accumulated dust from sticking to the sides and bridging across the hopper.

23.4 FABRIC PROPERTIES

Key performance properties for fabric filtration media selection include maximum allowable temperature, chemical resistance, abrasion resistance, weave, weight, and strength. Some of these properties are listed in Table 23.1.

Temperature is a key flue gas property that deserves first consideration in bag material selection. The choice of fabrics that withstand temperatures above 400°F is limited and high-temperature fabrics are expensive. Cooling often is used to reduce the baghouse temperature to an operable and economic value, although each method of cooling has its disadvantages. Adding cold air increases the volume of gas that must be treated by the baghouse and moved by fans. Air-to-gas heat exchangers are expensive. And spraying water into the flue gas must be done carefully to avoid wetting duct walls and causing local corrosion. When the operating temperature exceeds 500°F, ceramic fabrics or candles can be used, although they are very expensive.

Flue gas chemistry is the second key flue gas property that demands consideration in fabric selection. Many applications are in acidic or alkaline environments, and flue gas moisture can promote chemical attack, as well as affect dust cake cohesivity.

TABLE 23.1
Filter Material Properties

	Recommended Max		Chemical Resistance			
Material	**Operating Temp (°F)**	**Excursion Temp (°F)**	**Acid**	**Base**	**Abrasion Resistance**	**Cost (per 8-ft bag)**
Cotton	180	200	Poor	Good	Good	$8
Wool	200	230	Good	Poor	Fair	—
Nylon	200	250	Poor	Good	Excellent	—
Polypropylene	200	200	Excellent	Excellent	Excellent	$8
Polyester	275	300	Good	Fair	Excellent	$9
Acrylic	260	285	Good	Fair	Good	$13
Nomex®	375	400	Fair	Good	Excellent	$22
Ryton®	375	400	Excellent	Excellent	Excellent	—
Teflon®	450	500	Excellent	Excellent	Fair	$26
Fiberglass	500	550	Good	Good	Fair	$24
Coated high-purity silica	900	1050	Good	Good	Fair	$150
Ceramic candle	1650	1830	–	–	–	$1000[a]

[a] 60 mm diameter × 1.5 m.

23.4.1 Woven Bags

Once the material is chosen, the next key selection parameters are the weave and the weight. Woven fabric threads include the warp, which is the thread that runs lengthwise in woven goods, and the fill, which is the thread that interlaces the warp. There are many weave patterns, including plain, twill, and sateen. Selection of the weave pattern is of minor importance compared to the basic choice of woven or felted fabric. Woven fabrics are stronger than felted and can be expected to last longer. Typically, shaker and reverse air baghouses employ woven fabrics. Pulse jet baghouse employs both woven and felted fabrics.

23.4.2 Felted Fabric

Felted fabrics have a woven base scrim to give structure to the cloth, with the scrim filled in with random needle-punched fibers. The felting process is economical because the felting machines can be run at high speed.

From a filtration point of view, the random individual fibers make better targets for the collection mechanisms of impaction and interception, because individual fibers have smaller diameters than woven threads. Felted fabrics also may be thicker than woven fabrics for the same weight, so more time is available for diffusion to be an effective cleaning mechanism. Therefore, new felted fabric can produce higher particulate removal efficiency than new woven fabrics. This advantage does not last long, however. As soon as a dust cake builds up, it becomes the filtering media while the fabric serves merely to support the dust cake. After cleaning to remove the dust cake, some residual particles remain on the fabric, so that older bags never lose their dust cake entirely. This is the reason for the interesting observation that *seasoned* bags often exhibit higher collection efficiency than new bags.

With pulse jet cleaning, the vigorous, high-energy pulse can cause the threads in woven fabrics to separate slightly, resulting in increased bleed-through emissions. This is why felted fabrics sometimes are specified for pulse jet applications, even though the lower strength of felted fabrics might shorten bag life.

23.4.3 Surface Treatment

Surface treatment and finishes commonly are used to modify fabric properties. Fiberglass has become a popular bag material despite its relatively low chemical and abrasion resistance because these weaknesses are overcome with treatment. Silicone, graphite, and fluorocarbon, used alone or in combination, provide lubrication to resist abrasion and protection from acid attack. A new, inorganic, high-temperature coating on high purity silica fibers allows use of woven bags in pulse jet applications up to 900°F.

23.4.4 Weight

Finally, the fabric weight is chosen. It is measured as the weight of one square yard of fabric, or as the denier, which is the weight per unit length. Common fabric weights range from 5 to 26 oz. More fabric adds strength to the fabric and increases the target area for particulate collection. Of course, higher fabric density costs more and adds to the pressure drop of the cloth.

23.4.5 Membrane Fabrics

A unique material, which might be considered to be a surface treatment but changes the concept of fabric filtration sufficiently to warrant a separate discussion, is an expanded polytetrafluoroethylene (PTFE) membrane that is applied to one side of conventional material. Membrane-coated fabrics are commonly known as Gore-Tex®, although the patent for the material has expired and other manufacturers now produce it. The PTFE membrane has extremely fine diameter fibers that are small enough and spaced closely enough together that they act as very efficient primary targets for the impaction and interception collection mechanisms. It is a very thin membrane, so pressure drop is low. Since the membrane serves as the primary target for dust collection, a dust cake is not needed to provide good collection efficiency. In addition, residual dust cake buildup is minimized because dust cake release from PTFE is excellent and little particulate penetrates the membrane. Therefore, membrane bags can operate efficiently with very low pressure drop. The only two disadvantages of this unique material are that it is expensive, and it cannot be used in applications that contain even small amounts of hydrocarbons in the gas stream.

The base material serves as a support for the membrane, not as the target for dust collection. Since PTFE has higher temperature resistance than common bag materials and excellent chemical resistance, the temperature and chemical resistance of the base material limits the material selection. The membrane can be used with most common base materials, including fiberglass.

23.4.6 Catalytic Membranes

A new feature available in fabric filter bags is the addition of catalyst to the felted support fabric of PTFE membrane material. This transforms the filter bag into a multifunctional reactor, where the membrane provides high-efficiency particulate removal and the catalytic support fabric promotes gas phase reactions. This technology is being applied to reduce dioxin/furan emissions from incinerators, metals plants, and crematoria by more than 99%.[2]

23.4.7 Pleated Cartridges

Pleated cartridge filter elements are becoming popular for many applications. Their advantage is much higher collection area per linear foot of element. This allows a more compact baghouse for an original design or allows the air-to-cloth ratio of an older pulse jet baghouse retrofit with cartridges to be decreased. Cartridge filter elements typically are shorter than bags, but the increased area from the pleats more than makes up for the difference in length.

Cartridges originally were available in cellulose or paper, like air filter for a car, for low-temperature nuisance dust applications. They are now available in polyester and Nomex® and can be provided with a PTFE membrane. To make rigid pleats that hold their shape, the materials must be constructed differently from fabric for flexible bags. Polyester can be spun bonded, and Nomex® is impregnated with a resin.

23.4.8 Ceramic Candles

Rigid filter elements that are made in the shape of cylinders are called *candles*. Conceptually, there is no difference in the filtering mechanism between rigid candles and fabric; candles just don't flex when pulsed. The porous media provide the initial targets for particle collection until a dust cake forms; then the dust cake becomes the primary medium for collection by impaction, interception, and diffusion.

Candles are made of ceramic materials, either in a monolithic structure or as composites that contain ceramic fibers. Ceramic candles can be used in very high temperature applications from 1650°F to 2000°F. They are used in extreme services such as pressurized fluid bed combustors, combined cycle combustors, coal gasification, and incinerators, where they are exposed to high temperature and pressure, as well as alkali, sulfur, and water vapor.

Typical monolithic candles may be clay-bonded silicon carbide, silicon nitride, or aluminum oxide particles. In high-temperature applications with alkaline ash, silicon carbide and silicon nitride may oxidize and degrade slowly, while oxide materials will not be further oxidized. A concern with rigid candles is susceptibility to thermal and mechanical shock that can result in a complete failure if cracked. In monolithic materials, the combination of a high elastic modulus and high coefficient of thermal expansion can result in excess thermal stresses.

Candles composed of ceramic fiber composites resist breaking when cracked, which is a significant advantage for this design. One type of composite ceramic fiber material is constructed of continuous ceramic fibers for structural reinforcement and discontinuous ceramic fibers for filtration. The filtration fibers have a small mean diameter of 3.5 μm, which aids in efficient capture. The structural and filtration fibers are bonded with a chemical binder that converts to a stable bond phase with heat treatment.[3]

23.5 BAGHOUSE SIZE

23.5.1 Air-to-Cloth Ratio

The air-to-cloth ratio is simply the gas flow rate divided by the fabric collection area. Volume per unit time divided by area reduces to units of length per unit time, so the air-to-cloth ratio also is called the *superficial velocity.* A high air-to-cloth ratio requires a smaller baghouse, which is less expensive. If the air-to-cloth ratio is too high, the baghouse may experience difficulty in maintaining the desired pressure drop despite frequent cleaning.

A low air-to-cloth ratio provides a large collection area, so dust cake buildup and pressure drop increase at a lower rate than a high air-to-cloth ratio. When clean-on-demand cleaning is used, the overall baghouse pressure drop is set by the pressure drop triggers, so to say that low air-to-cloth ratio reduces pressure drop, can be misleading. Rather, a thicker dust cake can be accumulated and time between cleaning is longer with a low air-to-cloth ratio.

Because air-to-cloth ratio is related to the rate of pressure drop increase, the optimum ratio depends upon the dust permeability (particle size and cohesivity) and dust loading. The long-term rates of blinding and desired bag life also are factors to be considered. Because these factors are difficult to quantify, reasonable ratios generally are chosen based on experience. There are several tables of typical pulse jet air-to-cloth ratios for specific applications available in the literature; however, many of these tables are dated and will not be repeated here to prevent their use in light of experience. Users' experience with a large number of problem baghouses is moving recommended design to lower air-to-cloth ratios as the tradeoffs between initial investment, maintenance cost, troubleshooting cost, and capacity for future growth are weighed. Typical air-to-cloth ratios for shaker and reverse air baghouses are 2.0–2.5 cfm/ft^2. Typical ratios for pulse jet baghouses range from 3 to 10 cfm/ft^2, with recent design practice being close to 4.0 cfm/ft^2 and lower.

23.5.2 Can Velocity

Can velocity is an important sizing criterion for online pulse jet cleaning. As dust is cleaned from the bags, gravity will cause it to fall toward the hopper. But when cleaning online, gas flow is interrupted only momentarily by the pulse. Typical configuration is upward flow with the gas brought in at the bottom. To allow the dust some chance to settle, the upward velocity should be limited. The upward velocity is described as the *can velocity*, which is the air flow rate divided by the horizontal cross-sectional area of the baghouse less the cross-sectional area of the bags. The can velocity should not exceed about 2.5–3.5 ft/s, with the lower value more suitable for fine dust that does not settle easily.

23.6 PRESSURE DROP

Because flow between fibers of fabric cloth and particles in the dust cake is laminar, pressure drop across the filter media will be directly proportional to gas flow, as shown in Equation 23.1.

$$\Delta P = SV \tag{23.1}$$

where:

S is the filter drag, in. H_2O-min/ft
V is the superficial velocity

Note that the superficial velocity is the actual gas flow divided by the filter area; it is not the actual velocity between fibers or dust particles. By this definition, it is frequently called the *gas-to-cloth* or *air-to-cloth* ratio, which has units of volume per time divided by area and reduces to distance per unit time.

Pressure drop across the filter media, which is commonly measured with a local differential pressure gauge for each compartment or module, does not include the pressure drop associated with the inlet and outlet ductwork, which commonly is measured and transmitted to a recorder and is often used to trigger cleaning. Because flow in the ductwork is turbulent, the duct pressure drop varies with the square of the flow. Typically, however, the duct pressure drop is small compared to the drop across the filter media.

The largest contributor to pressure drop typically is the dust cake, although residual pressure drop across plugged fabric may be larger. As illustrated in Figure 23.5, there will be fluctuations in pressure drop as the bags are cleaned and as the gas flow varies. To eliminate gas flow as a variable, the filter drag, S, is defined as

$$S = \frac{\Delta P}{V} \tag{23.2}$$

When plotted as a function of dust loading on the fabric, drag increases linearly with dust loading, as shown in Figure 23.6. Collecting data for a given type of dust with a given fabric as the filter media provides values for clean fabric and dust cake coefficients in the linear drag equation. Drag becomes a useful approach to determine the air-to-cloth ratio required to meet a pressure drop limitation between reasonable cleaning cycle intervals for the dust/fabric combination for which data are available.

$$S = K_e + K_s W \tag{23.3}$$

where:

S is the filter drag, in. of H_2O-min/ft
K_e is the clean cloth filter drag coefficient, in. of H_2O-min/ft
K_s is the dust cake coefficient, in. of H_2O-min/ft
W is the dust loading, lbm/ft^2

Plotting drag data as a function of dust loading after cleaning also provides insight into cleaning effectiveness. Drag and residual dust loading remain high after poor cleaning.

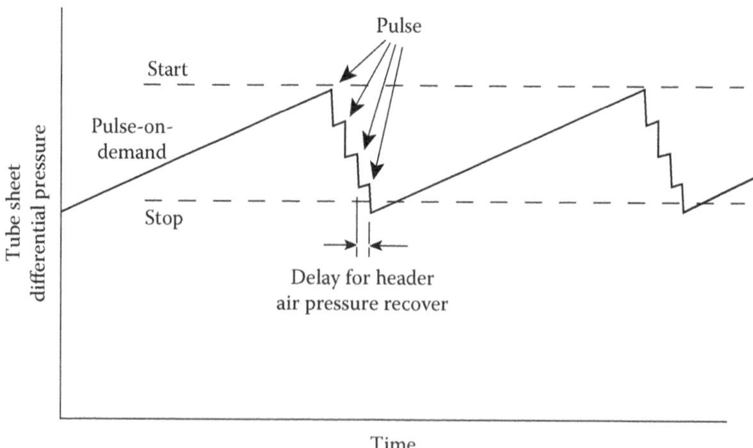

FIGURE 23.5 Overall pressure drop variation with time.

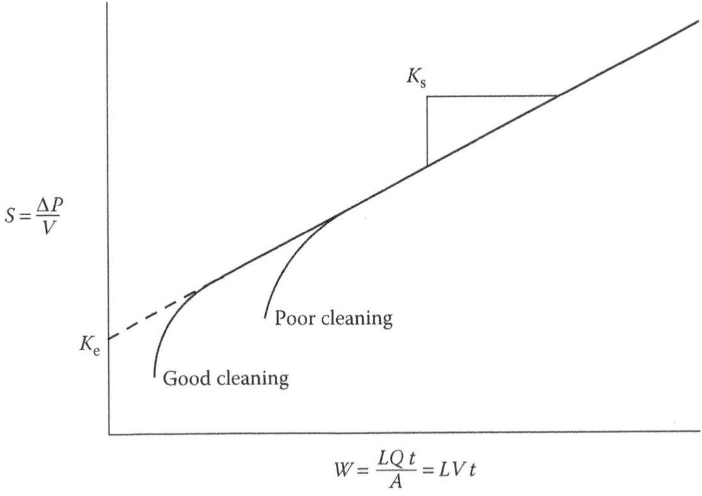

FIGURE 23.6 Determine K_e and K_s from data. Note: title change, letters *k* are capitalized *K* to agree with equations.

Particle size distribution has a dramatic effect on the dust cake coefficient. Small particles pack closely together and leave very small openings for gas flow, resulting in high pressure drop. Dust cohesivity also affects the resistance of gas flow through the dust cake. Particles that have high cohesivity stick to each other and tend to remain on the surface of the fabric, forming a porous dust cake. Low-cohesivity particles can work into the fabric interstices and blind the bag. Particle size, shape, and moisture content can greatly affect the cohesivity of the dust cake.

Additives can condition dust cake properties to improve fabric filter performance. Ammonia and SO_3 are two gaseous additives that are used to change the cohesive properties of some dusts, but they have the disadvantage of having health, safety, and hazardous waste issues associated with them. Also, SO_3 can aggravate acid attack of the fabric. Alternatively, proprietary conditioning agents are being developed to modify dust cake properties and improve fabric filter performance.[4]

23.7 BAG LIFE

23.7.1 Failure Modes

Bag life can be a significant factor in the operating cost of a baghouse. Although 2–5 years is a typical range for bag life, it may be much shorter or longer, depending on the application and the conditions. Common modes of bag failure include the following:

- Blinding by particles embedded in the fabric
- Blinding by sticky compounds
- Holes from abrasion

- Holes from weakened fibers that suffered chemical attack or high temperature excursions
- Burn holes from hot material
- Broken seams

In some cases, sticky compounds that blind bags are soluble in water or solvent and can be removed by laundering the bags.

23.7.2 Inlet Design

The inlet design configuration can have a direct bearing on bag life because of abrasion from high velocity dust impinging on the bags. It also can affect dust re-entrainment from the hopper. The inlet typically is near, or directed toward, the bottom of the baghouse because bags fill most of the compartment and the high-velocity inlet cannot be directed straight at the bags. A simple baffle plate often is used. However, a baffle plate will not distribute the flow evenly over the horizontal cross section of the baghouse by the time it reaches the bags. Dust entrained in the high-velocity streamlines can abrade the bags, especially at the bottom.

Some vendors offer inlet designs that diffuse the inlet gas flow. The additional expense may warrant the benefits of reduced entrainment and reduced abrasion.

23.7.3 Startup Seasoning

It is common practice for baghouse vendors to specify an initial coating of a large diameter, multidimensional dust to *season* the bags. The intent is to avoid filling the interstices between the fibers with fine dust that blinds the bags at the very beginning of their life and to create an initial dust layer that has good permeability. Larger particles may embed within the fabric also; indeed, a portion of the initial seasoning particulate should stay on the bags for it to be of benefit. Large particles do not blind the fabric completely as small particles are able to do. There is some question as to the necessity and the lasting benefit of pre-seasoning, but that depends largely on the application. It could be very beneficial in applications where very fine particles or fume could embed in the fibers.

23.8 BAGHOUSE DESIGN THEORY

Baghouses are constructed of compartments; each of which contains a certain number of the compartments. The bags are hung in the compartments, usually the same number in each compartment. Initially, the efficiency is low in each bag because dust will pass through the cloth. Owing to impaction, interaction, and diffusion, particles will build up on the fibers and bridge the gap. When a particulate layer has formed, efficiency will increase. Design is based on the pressure drop through the fabric. The pressure drop is defined by Equation 23.4:

$$\Delta P = \Delta P_f + \Delta P_P + \Delta P_S \quad (23.4)$$

where:

ΔP_f is the pressure drop through the fabric
ΔP_P is the pressure drop through the particulate layer
ΔP_S is the pressure drop through the baghouse structure

The pressure drop in each case is defined by Darcy's equation for flow through porous media.

$$\Delta P_f = \frac{D_f \mu V}{60 K_f} \tag{23.5}$$

$$\Delta P_f = \frac{D_P \mu V}{60 K_P} \tag{23.6}$$

Pressure drop through the structure ΔP_S is small and neglected.

In this case, the units are

ΔP_f and ΔP_P are the pressure drop in N/m^2
D_f and D_P are the depth in the direction of the filter and particulate layer, respectively, m
μ is the gas viscosity, kg/m^{-s}
K_f and K_P are the permeability of the filter and particulate layer, m^2
60 is the conversion factor from s/min

As described before, V is the superficial velocity also known as the *air-to-cloth ratio*.

Therefore,

$$V = \frac{Q}{A} \tag{23.7}$$

where:

Q is the air flow rate
A is the cloth area

For example, cfm/ft^2 results in ft/min or [(m^3/min)/m^2] results in m/min.
It is assumed that D_P increases linearly with time for constant V and dust loading.

$$D_P \propto t, \quad \text{thus} \quad D_P = \frac{LV}{\rho_{BL}}$$

Here,

L is the dust loading in kg/m^3
ρ_{BL} is the bulk density of particulate layer in kg/m^3

The filter drag model is defined as

$$\Delta P_t = \Delta P_f + \Delta P_P \tag{23.8}$$

$$\frac{\Delta P_t}{V} = \frac{D_f \mu V}{60 K_f} + \left(\frac{\mu}{60 K_P \rho_{BL}} \right) LVt \tag{23.9}$$

Note that t is time in min.

Define areal dust as W in the unit of kg/m² then

$$W = LVt \tag{23.10}$$

And filter drag S in matching units then

$$S = \frac{\Delta P_t}{V} \tag{23.11}$$

$$S = K_1 + K_2 W \tag{23.12}$$

$$K_1 = \frac{D_f \mu}{60 K_f},\ K_2 = \frac{\mu}{60 K_P \rho_{BL}}$$

It is hard to evaluate K_1 and K_2; therefore, Equation 21.13, an empirical equation, was formed based on Equation 21.12.

$$S = K_e + K_S \cdot W \tag{23.13}$$

Then, K_e and K_S are determined experimentally from the linear portion of a plot of W versus S similar to Figure 23.6. The units of all four terms of Equation 21.13 can be metric or English, but they must be consistent. Note that the units of S and K_e are the same.

For example, in metric units,

> With L in kg/m³, V in m/min, and t in min, then the units of W are kg/m², and the units of K_S are [(N/min)/(kg/m)], then the units of S and K_e are [(N/min)/m³].

And in English units,

> With L in gr/ft³, V in m/min, and t in mins, then the units of W are gr/ft², and the units of K_S are [(lbf-min)/(gr/ft)], then the units of S and K_e are (lbf-min)/ft³.

Note that in this case gr is the unit of grains, a mixed unit that is equivalent of 0.064799 g. K_S has been found to vary with the square root of the superficial velocity V. When pilot plant data were taken at a different V, or when the analysis of a proposed change in flow rate is carried out, the following expression is used:

$$K_{S2} = K_{S1} \left(\frac{V_1}{V_2} \right)^{1/2} \tag{23.14}$$

23.8.1 Design Considerations

To carry out a design of a baghouse, data are required about the characteristic of the fabrics used to make the bags. Table 23.1 contains information about the operating temperatures, chemical resistance, and abrasion resistance of materials used to

TABLE 23.2
Maximum Filtering Velocities for Various Dusts in Shaker or Reverse Air Baghouses

Dusts	Maximum Filtering Velocity, cfm/ft² or ft/min
Aluminum oxide, carbon, fertilizer, graphite, iron ore, lime, paint pigments	2.0
Aluminum, clay, coke, charcoal, cocoa, lead oxide, mica, soap, sugar, talc	2.25
Bauxite, ceramics, chrome ore, feldspar, flour, flint, glass, gypsum, plastics	2.50
Asbestos, limestone, quartz, silica	2.75
Cork, feeds and grain, marble, oyster shell	3.0–3.25
Leather, paper, tobacco, wood	3.50

make the bags. Table 23.2 discusses the maximum filtering velocities for various types of dusts in shaker or reverse air baghouses. Tables relating more to the design variables such as number of compartments needed for a certain cloth area and filtering velocity related to the number of compartments are included and discussed later. One publication that provides this kind of information is that by Danielson.[5]

23.8.2 Number of Compartments

Compartments are costly. More compartments than needed increase cost and maintenance becomes more difficult. Too few compartments mean excessive number of bags per compartment. Once the net cloth area is determined, then Table 23.3 can be used to determine the number of compartments required. When it comes time to clean the bags, one compartment is isolated from the dusty gas flow. In reverse air system, air is blown in the opposite direction to the usual flow. In a shaker baghouse, the bags are shaken to remove the accumulated dust. The number of compartments depends on the total flow to be filtered. Equation 23.15 determines the total filtration time t_f.

$$t_f = N(t_r + t_C) - t_C \tag{23.15}$$

where:

N is the number of compartments
t_r is the run time
t_C is the time to clean one bag

Therefore,

$$t_r = \left(\frac{t_f - t_C}{N}\right) - t_C \tag{23.16}$$

Then, t_f is the elapsed time from the moment one compartment is returned to service until that same compartment is removed for cleaning again after all other

TABLE 23.3
Number of Compartments as a Function of Net Cloth Area

Net Cloth Area, ft²	Number of Compartments
1–4,000	2
4,000–12,000	3
12,000–25,000	4–5
25,000–40,000	6–7
40,000–60,000	8–10
60,000–80,000	11–13
80,000–110,000	14–16
110,000–150,000	17–20
>150,000	>20

compartments have been cleaned in rotation. Figure 23.7 illustrates the pressure drop as a function of time during the cleaning of the bag house.

In this figure, ΔP_M is the maximum pressure drop through the bag house during cleaning. The choice of the best values for the various indicated times is a matter of experience of operation of the cleaning process. Assume the flow rates through all compartments are the same, then the flow through one compartment Q_N is given by Equation 23.17:

$$Q_{N-1} = \frac{Q}{N-1} \tag{23.17}$$

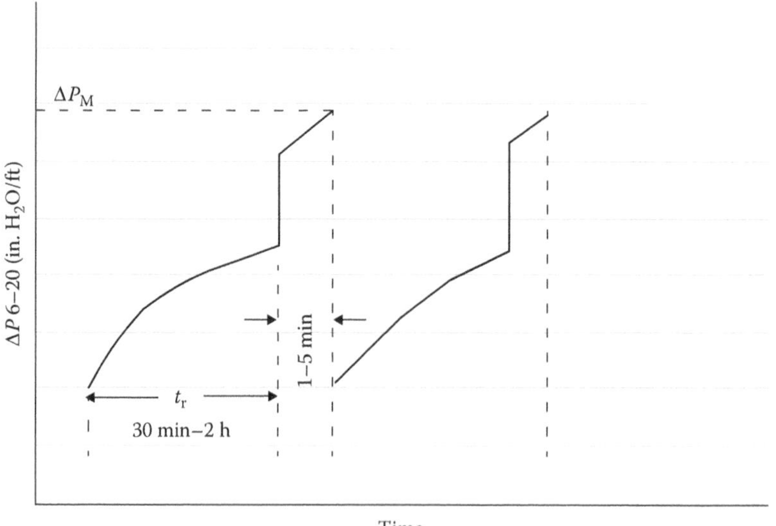

FIGURE 23.7 Pressure drop versus time during cleaning of baghouse compartments.

where:

Q is the total flow rate

With one compartment offline for cleaning, flow rate through each of the other compartments is given by Equation 23.18, and the filtering velocities V_N and V_{N-1} given by Equations 23.18 and 23.19.

$$V_N = \frac{Q_N}{A_C} = \frac{Q}{NA_C} \tag{23.18}$$

$$V_{N-1} = \frac{Q_{N-1}}{A_C} = \frac{Q}{(N-1)A_C} \tag{23.19}$$

where:

A_C is the bag area of one compartment

The maximum pressure drop occurs at the end of cleaning of compartment j just before j–1 is put back in service. At that time, compartment j, next to be cleaned, has been online for time t_j, where

$$t_j = t_f - t_r \tag{23.20}$$

Then from Equation 23.15, substitute for t_f in Equation 23.20:

$$t_f = N(t_r + t_C) - t_C - t_r = (N-1)(t_r + t_C) \tag{23.21}$$

Referring to compartment j,

$$W_j = (N-1)(V_N L\, t_r + V_{N-1}\, L\, t_C) \tag{23.22}$$

where:

L is the particulate loading

W_j is the areal dust density in compartment j

Then, the filter drag in j is given by

$$S_j = K_e + K_S W_j \tag{23.23}$$

In calculating the pressure drop, the actual velocity V_j in compartment j. V_j is lower than the average velocity V_{N-1} because compartment j has the most dust at time t_j. The value of t_j can be determined by the use of Table 23.4.

$$V_j = F_N V_{N-1} \tag{23.24}$$

$$\Delta P_S = \Delta P_M = S_j V_j \tag{23.25}$$

where:

F_N is obtained from Table 23.4

TABLE 23.4
Ratio of Actual Filtering Velocity V_j to Average Filtering Velocity V_{N-1} in a Multicomponent Baghouse

Total Number of Compartments	$F_N = V_j/V_{N-1}$
3	0.87
4	0.80
5	0.76
7	0.71
10	0.67
12	0.65
15	0.64
20	0.62

Then the pressure drop for compartment j is the max allowable.

23.8.3 Example Problem for a Baghouse Design

Dusty air flowing at 40,000 acfm from a flour mill must be treated in a shaker baghouse. The air contains 10 grains of flour/ft^3 of air. The supplier has recommended that a five-compartment baghouse be built. The area of the filter cloth and the number of bags need to be determined to be able to determine the size of the fan needed. The number of compartments should be verified and the maximum pressure drop through the baghouse should be determined. The bags will be 8 ft long and 6 in. in diameter. The pressure drop should be determined for a cleaning time of 4 min and a filtration time of 60 min. A summary of data needed follows:

Flow of dirty air Q = 40,000 acfm from the flour mill
Dust loading L = 10 gr/ft^3
Bags: length = 8 ft and diameter = 6 in. = 0.5 ft
Cleaning time t_c = 4 min
Filtration time t_f = 60 min

In order to begin the design, the following test data have been obtained for a clean fabric:

Time, min	ΔP, in. of H_2O
0	0.400
5	0.550
10	0.700
20	1.000
30	1.300
60	2.200

From Table 23.2, maximum filtering velocity V for flour can be found to be 2.5 ft/min. Then, the values listed below can be calculated from the test data.

Time, min	ΔP, in. of H_2O	$S = \Delta P/V$, in. of H_2O-min/ft	$W = LVt$, gr/ft^2
0	0.400	0.1600	0
5	0.550	0.2200	125
10	0.700	0.2800	250
20	1.000	0.4000	500
30	1.300	0.5200	750
60	2.200	0.8800	1500

These data are plotted as S versus V in the following Figure 23.8.

From Figure 23.8, K_C and K_S can be determined based on Equation 23.13.

$$S = K_C + K_S \cdot W$$

$$K_C = 0.1600 \text{ in. of } H_2O/\text{min/ft} \quad \text{and} \quad K_S = 0.00048 \text{ in. of } H_2O/\text{min/ft/gr/ft}^2$$

Processing flour Table 23.3 requires four or five compartments. The provider suggests we use five compartments, and it is suggested that $V = 2.5$ cfm/min or 2.5 ft/min. The net cloth area required for operation would then be

$$\frac{40{,}000 \text{ acfm}}{2.5 \text{ cfm/ft}^2} = 16{,}000 \text{ ft}^2$$

With one compartment off, we must still be able to operate. Therefore, for each compartment, we must have

$$\frac{16{,}000 \text{ ft}^2}{5-1} = 4000 \text{ ft}^2$$

of bag area needed. Thus, with five compartments, the total amount of fabric needed will be

$$4000 \times 5 = 20{,}000 \text{ ft}^2$$

With all five compartments on line

$$V_N = \frac{40{,}000 \text{ cfm}}{20{,}000 \text{ ft}^2} = 2.0 \text{ ft/min}$$

With only four compartments on line

$$V_{N-1} = \frac{40{,}000 \text{ cfm}}{16{,}000 \text{ ft}^2} = 2.5 \text{ ft/min}$$

Then, we can determine the maximum allowable filter drag from Equation 23.23.

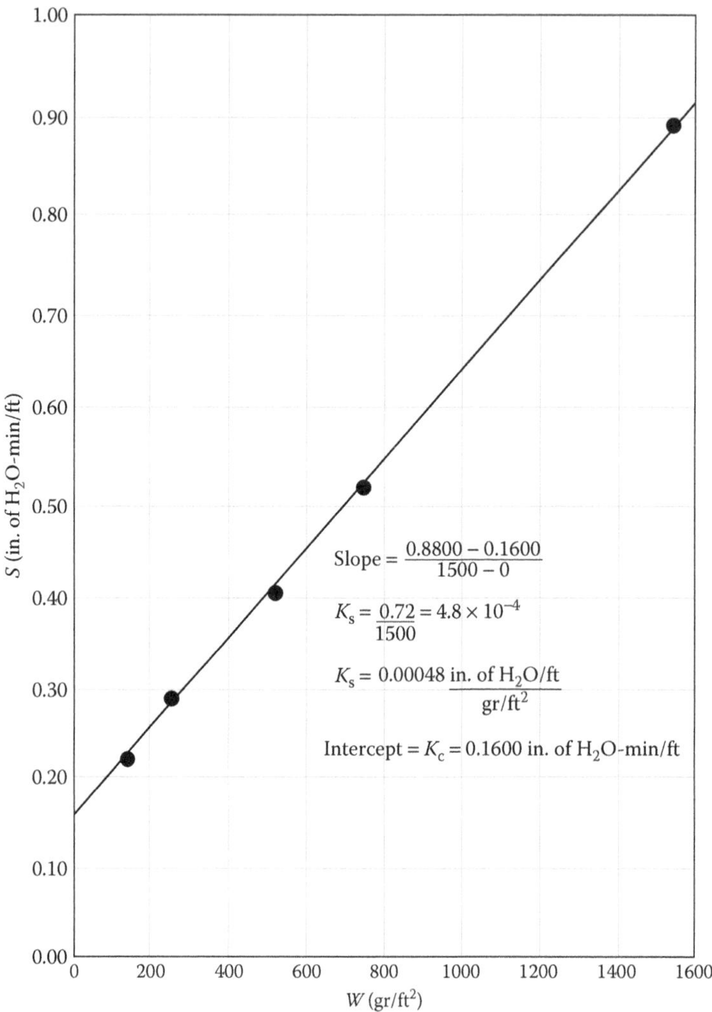

FIGURE 23.8 *S* versus *W* for process data for the baghouse.

$$S_j = K_C + K_S W_j$$

First, we must determine the areal dust density in each compartment W_j.

The total filtration time t_r is found from Equation 23.16.

$$t_r = \left(\frac{t_f - t_C}{N}\right) - t_C$$

$$t_r = \left(\frac{60 + 4}{5}\right) - 4 = 8.8 \text{ min}$$

Then, W_j can be found from Equation 23.22.

$$W_j = (5-1)\left(2.0\text{ ft/min} \times 10\text{ gr/ft}^3 \times 8.8\text{ min} + 2.5\text{ft/min } 10\text{gr/ft}^3 \times 4\text{min}\right) = 1104\text{ gr/ft}^2$$

$$S_j = 0.1600 + 0.00048 \times 1104 = 0.690\text{ in. of }H_2O\text{/min/ft}$$

The compartment velocity V_j can be determined from Equation 23.24, where f_N is found from Table 23.4. For five compartments, f_N is 0.76.

$$V_j = f_N V_{N-1}$$

$$V_j = 0.76 \times 2.5 = 1.90\text{ ft/min}$$

The pressure drop is then determined by Equation 23.25.

$$\Delta P_M = S_j V_j$$

$$\Delta P_M = 0.690\,\frac{\text{in. of }H_2O\text{/min}}{\text{ft}} \times 1.90\text{ ft/min}$$

$$\Delta P_M = 1.31\text{ in. of }H_2O$$

24 Electrostatic Precipitators

24.1 EARLY DEVELOPMENT

The phenomenon of electrostatic attraction amuses children who like to stick balloons to their heads. That opposite charges attract and like charges repel is a basic law of physics. It was noted as early as 600 B.C. that small fibers would be attracted by a piece of amber after it had been rubbed. Modern knowledge of electrostatics was developed throughout the last 400 years, including the work of Benjamin Franklin on the effect of point conductors in drawing electric currents. The first demonstrations of electrostatic precipitation to remove aerosols from a gas were conducted in the early 1800s with fog and tobacco smoke.

The first commercial electrostatic precipitator (ESP) was developed by Sir Oliver Lodge and his colleagues, Walker and Hutchings, for a lead smelter in North Wales in 1885. Unfortunately, this application was unsuccessful because of problems with the high-voltage power supply and the high resistivity of the lead oxide fume. As will be discussed in this chapter, resistivity is an extremely important factor affecting ESP performance. In the United States, Dr. Frederick Cottrell, professor of chemistry at the University of California at Berkeley, and his colleagues developed and improved the technology for industrial application. Cottrell established the nonprofit Cottrell Research Corporation, which supported the experimental studies that formed the fundamental basis of precipitator technology. The technology was applied successfully to control sulfuric acid mist in precious metal recovery kettles. Cottrell installed the next commercial system at a lead smelter. Although the high resistivity of the dust again made it a difficult application, the high-voltage power supply issues were resolved sufficiently well, so that the ESPs could operate at about 80%–90% removal efficiency. Within a few years, ESPs were being installed in Portland cement plants, pulp and paper mills, and blast furnaces. The first installation on a coal-fired boiler was at Detroit Edison Company's Trenton Channel Station in 1924. Eventually, ESPs were specified for most coal-fired boilers until there were more than 1300 installations servicing about 95% of the coal-fired boiler applications.

24.2 BASIC THEORY

An ESP controls particulate emissions by: (1) charging the particles, (2) applying an electric field to move the particles out of the gas stream, and then (3) removing the collected dust. Particles are charged by gas ions that are formed by corona discharge from the electrodes. The ions become attached to the particles, thus providing the charge.

In a typical ESP, vertical wires are used as the negative discharge electrode between vertical, flat, grounded plates. The dirty gas stream passes horizontally between the plates and a dust layer of particulate collects on the plates. The typical spacing between

FIGURE 24.1 Tubular collection electrode. (Courtesy of Geoenergy International, Telford, PA.)

the discharge electrode and the collector plate is 4–6 in. The dust layer is removed from the plates by *rapping*, or in the case of a wet ESP, by washing with water.

An alternative to the plate and wire design is the tube and wire design, in which the discharge electrode wire is fixed in the center of a vertical tubular collection electrode. In this configuration, the gas flow is parallel to the discharge electrode. This configuration, shown in Figure 24.1, is common for wet ESP.

24.2.1 Corona Formation

An electrical potential of about 4000 V/cm is applied between the wires (discharge electrodes) and collecting plates of the ESP. In most cases, the wires are charged at 20–100 kV below ground potential, with 40–50 kV being typical. For cleaning indoor air, the wires can be charged positively to avoid excessive ozone formation. However, the negative corona is more stable than the positive corona, which tends to be sporadic and cause sparkover at lower voltages, so negative corona is used in the large majority of industrial ESP. In the intense electric field near the wire, the gas breaks down electrically, producing a glow discharge or *corona* without sparkover, as depicted in Figure 24.2.

In a negative corona, ionized molecules are formed from the corona glow caused by the high electrical gradient around the discharge wire. The space outside the corona is filled with a dense cloud of negative ions. The dust particles will collide with some of the ions giving them a negative charge. These charged particles will be driven by the electric field toward the plates where they are collected.

24.2.2 Particle Charging

As particles move through the electric field, they acquire an electrostatic charge by two mechanisms, namely bombardment charging and diffusion charging, as illustrated in Figure 24.3. Both types of particle charging act simultaneously, but

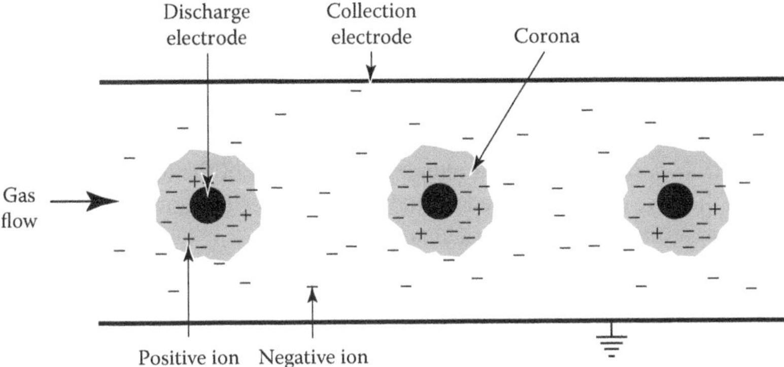

FIGURE 24.2 Corona formation—plan view plate and wire configuration.

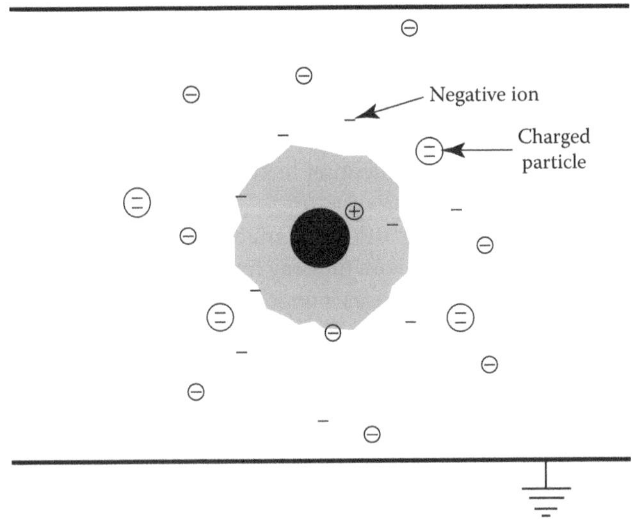

FIGURE 24.3 Particle charging.

bombardment charging is of greater importance for larger particles and diffusion charging is more important for submicron particles. The magnitude of the charging for both mechanisms is lowest for particles in the size range of 0.1–1 μm; therefore, the minimum collection efficiency will occur for this size range. However, a well-designed ESP will be capable of collecting greater than 90% of even these difficult to collect particles.

Bombardment charging is of primary importance for particles greater than 1 μm. Ions and electrons move along the lines of force between the electrodes normal to the direction of flow of particles in the gas stream. Some of the ions and electrons are intercepted by uncharged particles, and the particles become charged. Because the particles are now charged, ions of like charge are now repulsed by the particle, thus reducing the rate of charging. After a time, the charge on the particles will reach a maximum that is proportional to the square of the particle diameter.

Because extremely small particles (<0.1 μm) have an erratic path in the gas stream, due to Brownian motion, they can acquire a significant charge by diffusion charging. Thus, an ESP can be an efficient collection device for submicron particles. However, these particles represent only a small fraction of the mass of dust entering an ESP, so they are often neglected in studies of ESP performance, even though they can be of great importance to particulate emissions.

24.2.3 Particle Migration

Most charged particles migrate under the influence of the electric field toward the plate, although a few particles in the vicinity of corona discharge will migrate toward the wire. The presence of charged particles in the gas space affects the overall electric field. Near the plate, the concentration of charged particles will be high, and interparticle interferences can occur. Finally, particles will collect as a dust layer on the plates, and a portion of their charge may be transferred to the collecting electrode. Ideally, charged particles will migrate to the plate before exiting the ESP, as illustrated in Figure 24.4, and will stick to the dust layer on the collecting electrode until it is cleaned. When the plate is rapped, the dust layer should fall as a sheet into dust collection hoppers without re-entraining into the gas stream.

The velocity at which charged particles migrate toward the plate can be calculated by balancing the electrical forces with the drag force on the particle moving through the flue gas. The electric field produces a force on the charged particle proportional to the magnitude of the field and the charge:

$$F_e = qE \tag{24.1}$$

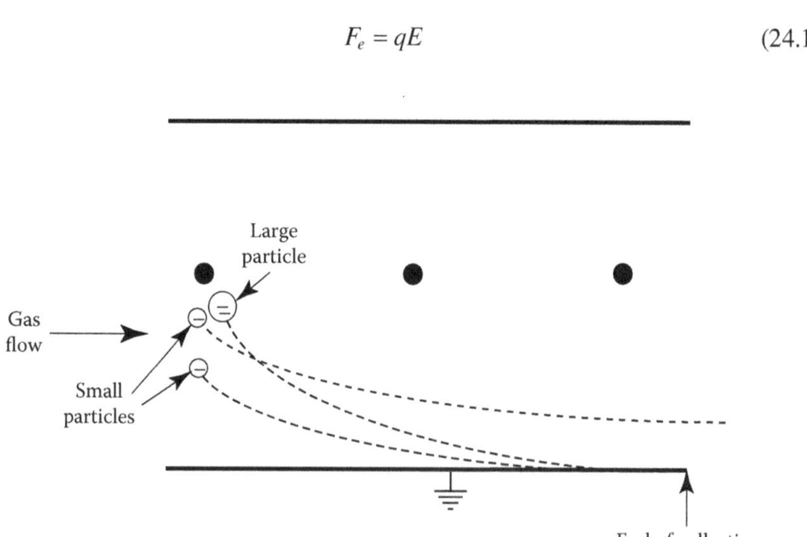

FIGURE 24.4 Migration velocity versus treatment time.

where:

F_e is the force due to electric field
q is the charge on particle
E is the strength of the electric field (V/cm)

However, several simplifying assumptions are needed for calculation of balancing electrical force with drag force:

- Repulsion effects between particles of like charge are neglected.
- The effect of the movement of gas ions (electric wind) is neglected.
- Gas flow within the ESP is turbulent.
- Stokes' law can be applied for drag resistance in the viscous flow regime.
- Particles have been fully charged by bombardment charging.
- There are no hindered settling effects in the concentrated dust near the plate.

After applying these simplifying assumptions, the migration velocity for particles larger than 1 μm charged by bombardment charging is calculated using Equation 24.2:

$$\omega = \frac{(3D/D+2)\varepsilon_o E_c E_p d_p}{3\mu_g} C' \tag{24.2}$$

where:

D is the dielectric constant for the particle
ε_o is the permittivity, 8.854×10^{-12} coulombs/V/m
E_c is the strength of the charging electric field
E_p is the strength of precipitating (collecting) electric field
d_p is the particle diameter
μ_g is the gas viscosity
C' is the Cunningham slip correction factor

Note that the migration velocity is proportional to the square of the electrical field strength, directly proportional to the particle diameter, and inversely proportional to the gas viscosity.

24.2.4 Deutsch Equation

Using the migration velocity to complete a material balance for particles moving toward the ESP plates and particles being carried through the ESP with the gas flow, a common description of particle collection efficiency for monodisperse (same size) particles can be derived:

$$\eta = 1 - \exp\left(-\omega \frac{A}{Q}\right) \tag{24.3}$$

where:

η is the fractional collection efficiency
ω is the migration velocity
A is the plate area
Q is the volumetric gas flow

Any consistent set of units can be used for ω, A, and Q. The expression A/Q is the specific collection area of the ESP, commonly expressed as square feet per thousand actual cubic feet per minute (ft^2/kacfm). When calculating plate area, remember that the surface area of interior plates includes area exposed to gas flow on both sides of the plate, while the two exterior plates are exposed on only one side.

An inherent assumption in the Deutsch equation is that when particles reach the plates, they are permanently removed from the gas stream. This assumption works reasonably well for low efficiency ESPs. However, when the collection efficiency is high (greater than 99%), mechanisms other than balancing migration velocity with treatment time dominate the particle emissions. Sneakage, rapping re-entrainment, scouring re-entrainment, low-resistivity re-entrainment, and poor gas distribution can become controlling nonideal effects that limit collection efficiency.

For very high efficiency ESPs, empirical modifications of the Deutsch equation have been used to fit observed data. These include the Hazen equation:

$$\eta = 1 - \left(1 + \frac{\omega A}{nQ}\right)^{-n} \tag{24.4}$$

where:

n is the empirical constant with typical values of 3–5 to fit most data

and the Matts–Ohnfeldt equation:

$$\eta = 1 - \exp\left(\frac{-\omega A}{Q}\right)^{x} \tag{24.5}$$

where:

x is the empirical constant typically set at 0.5

A more rigorous approach to calculating ESP efficiency uses a computer model that is based on the Deutsch equation, but is applied to individual small band widths of the particle size distribution and accounts for the nonideal effects of sneakage, rapping re-entrainment, nonrapping re-entrainment, space charge, and flow distribution. These factors are accounted for by either experience factors or modeling of fundamental mechanisms. Modeling also can account for changes in electrical conditions as particles are collected.[1]

24.2.4.1 Sneakage

Sneakage occurs when gas bypasses the electric field by sneaking under or over the field in the space between the ends of the plates and the ESP enclosure. The high-voltage wires and grounded plates must be electrically insulated, and some gas flows above the plates by the insulators and some gas flows through the dust collection hoppers beneath the plates. Proper baffling minimizes sneakage.

24.2.4.2 Rapping Re-Entrainment

Another nonideal effect in a dry ESP is rapping re-entrainment. The dust layer of collected particles on the collection plates is knocked loose periodically by *rapping* or knocking the plates, often with a trip hammer. Most of the dust falls as a sheet into collection hoppers, but some particulate is re-entrained into the gas stream. Factors affecting rapping re-entrainment include the aspect ratio of the ESP (length of the ESP divided by plate height), rapping intensity, dust cohesivity, and dust cake thickness (rapping frequency). With a low aspect ratio, dust has further to fall to reach the hopper before it would exit the ESP.

Particles in a cohesive dust cake will tend to stick together as a falling sheet when the plates are rapped. This minimizes re-entrainment. The rapping intensity needs to be strong enough to shear the dust cake from the plate, but not strong enough to produce a cloud. Increasing dust cohesivity with conditioning additives is one of the primary mechanisms for improving fine particle collection.

The frequency of rapping should be adjusted to allow a sufficient dust layer to accumulate so that the layer will fall as a cohesive sheet. Experimental studies with fly ash have shown that a re-entrainment cloud forms when the plate loading is below 0.1 g/cm^2, while the dust layer develops a more cohesive sheet when rapped at a higher loading. However, if the dust layer becomes too thick, it can act as an insulator and cause a potential gradient to build up within the layer. This reduces the electric field strength in the gas space, and could lead to sparking within the dust layer with subsequent re-entrainment.

24.2.4.3 Particulate Resistivity

Once particles reach the dust layer on the collecting electrode, they must stick to the surface until it is cleaned. This is not a problem in a wet ESP because the particle sticks to the wet collection surface until they are washed off by flushing. But in a dry ESP, re-entrainment resulting from dust resistivity that is either too high or too low can reduce the collection efficiency of the ESP. To achieve high collection efficiency when re-entrainment is a factor, the ESP must be oversized to allow particles to be captured again. The forces that hold particles onto the plate include molecular adhesive forces of the London-van der Waals type and electrostatic forces. The optimum resistivity for good removal in a dry ESP is approximately 1×10^9 to 1×10^{10} ohm/cm.

When charged particles arrive at the plate, they are partially discharged. The extent of electrostatic adhesion depends on the rate at which charge leaks away from the particles, which depends on the resistivity of the dust layer. The resistivity of some dusts, including lead smelter fume and coal-fired-boiler fly ash from low-sulfur or alkaline coals, is relatively high. When the resistivity is high, the rate of discharge from the collected particle layer is low. A potential gradient builds up within the layer of collected particles.

Figure 24.5 illustrates the potential gradient as voltage versus distance between the discharge and collection electrodes. Two points on the curve are fixed. The discharge

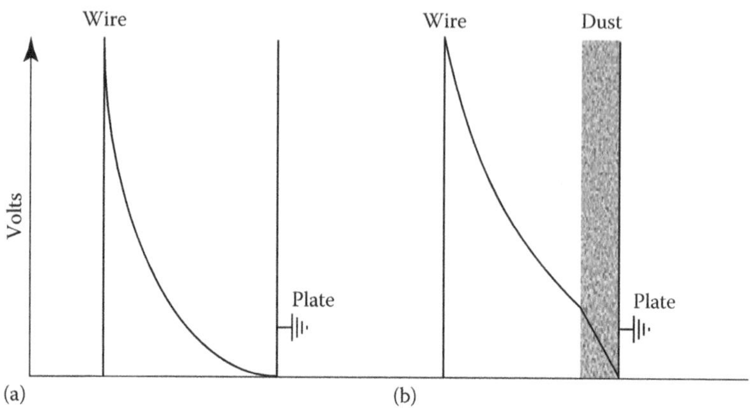

FIGURE 24.5 Voltage gradient in dust layer: (a) clean plate and (b) plate with dust layer.

electrode is charged to the maximum voltage for the limits of the power supply. The collection electrodes are grounded. Without a resistive dust layer, the potential gradient will appear as in Figure 24.5a, with the greatest gradient at the discharge electrode where corona is formed. Figure 24.5b illustrates the effect of a highly resistive dust layer. A substantial portion of the voltage drop occurs across the dust layer, leaving a reduced potential gradient across the gas space. With the lower gradient, the driving force for particle migration is reduced. If the dust resistivity is sufficiently high, the steep potential gradient within the dust layer itself can begin to breakdown of the gases between the dust particles. This is *back corona.* Ions of both charges, including the opposite charge from the discharge electrode, are formed and charge particles. These opposite-charge particles are re-entrained as they migrate back toward the discharge electrodes. Sometimes, the potential gradient within the dust layer can be severe enough to cause a spark within the dust layer, which violently re-entrains some dust and can limit the maximum voltage that can be maintained by the power supply.

Low-resistivity dust also can result in a re-entrainment problem. The particle charge is lost quickly when the dust has low resistivity. A dust layer of uncharged particles is not held against the collecting plate by the potential gradient from the discharge electrode.[2] Carbon dust and moist, low-temperature particles are examples of dusts that have a low resistivity.

Two gas properties that have a significant effect on particle resistivity are temperature and humidity. At high temperatures, above about 400°F, volume conduction of electric charge through the particles tends to control resistivity. Such passage obviously depends upon the temperature and composition of the particles. For most materials, the relationship between resistivity and temperature is given by an Arrhenius-type equation:

$$\rho_e = A \exp\left(\frac{-E}{kT}\right) \tag{24.6}$$

where:

ρ_e is the resistivity
A is the constant
E is the electron activation energy (a negative value)
k is the Boltzmann's constant
T is the absolute temperature

Thus, resistivity decreases as temperature increases.

At lower temperatures, less than 200°F, surface conduction is the predominant mechanism of charge transfer. Electric charges are carried in a surface film adsorbed on the particulate. The presence of moisture increases surface conduction. Humidification of the flue gas upstream of an ESP both decreases temperature and increases moisture content, which reduces particle resistivity.

24.2.4.4 Gas-Flow Distribution

An idealized assumption that is used when applying the Deutsch equation is that the gas flow and the particulate concentration in the gas are distributed uniformly. Customized flow vanes, baffles, and/or perforated-plate gas distributors often are used at the inlet to produce uniform flow. Sometimes, these devices are used at the outlet also. A typical specification for uniform flow distribution requires that 85% of the velocity distribution is within 1.15 times the average velocity, and 99% of the velocity distribution is within 1.40 times the average velocity.[3]

Two approaches are used to ensure uniform velocity distribution: scale-model studies and computational fluid dynamics modeling. computational fluid dynamics modeling is relatively new, but is becoming common as software, computing power, experience, and availability have enabled this tool to be used in a variety of fluid-flow applications.

Although uniform gas distribution is generally accepted as the ideal gas flow distribution, computer modeling and a full-scale demonstration at a coal-fired power station in South Africa show that a skewed distribution reduced particulate emissions by more than 50%.[4,5] In this patented configuration, the inlet flow distribution is skewed with low flow at the top of the precipitator and higher flow at the bottom. At the outlet, the gas flow is skewed with high flow at the top of the precipitator and low flow at the bottom, as shown in Figure 24.6. This distribution utilizes the fact that collected dust exits the precipitator by falling to the bottom. Dust cake dislodged by rapping is less likely to be re-entrained when the distance that it must fall is short. Thus, particles near the bottom are more likely to be removed than particles near the top of the precipitator. At the inlet, a low velocity gives those particles near the top more treatment time for better collection. At the outlet, particles still near the top have not yet worked their way down the precipitator, so are likely to be emitted anyway, so it is better to maximize the collection efficiency of those particles that have been worked toward the bottom and still can be collected.

During operation, flow distribution can be affected by deposits that accumulate on the gas distribution devices. Sometimes, rappers or vibrators are used to remove these deposits.

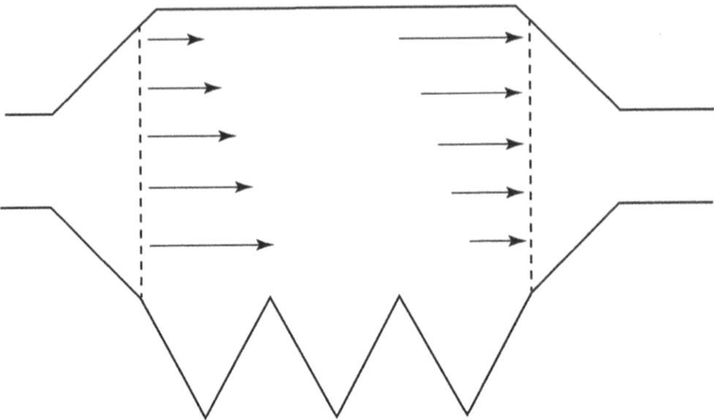

FIGURE 24.6 Skewed gas flow distribution.

24.3 PRACTICAL APPLICATION OF THEORY

24.3.1 Effective Migration Velocity

In most cases, it is far more practical and reliable to determine an *effective migration velocity* from operating experience than it is to calculate the migration velocity from Equation 24.2. Then the effects of many unknown properties, including particle size distribution, and simplifying assumptions are buried in the measured performance. The fractional particulate removal efficiency is determined by measuring the inlet and outlet loading in either a pilot-scale, or better a full-scale, ESP. The effective migration velocity, ω, is calculated after rearranging Equation 24.3:

$$\omega = -\frac{Q}{A}\ln(1-\eta) \tag{24.7}$$

Having the effective migration velocity enables sizing the required collection area for the desired efficiency under similar conditions, bearing in mind the simplifying assumptions and limitations of the Deutsch equation discussed previously.

24.3.2 Automatic Voltage Controller

The potential gradient is the source of corona formation and the driving force for charged particle migration. Equations 24.2 and 24.3 can be used to estimate the effect of increasing voltage for a given application.

The automatic voltage controller increases the secondary (i.e., DC) voltage that is applied to an ESP field until either the spark rate limit is reached or the current limit of the power supply is reached.

A spark is an electrical breakdown of the gas space. Modern electronics are able to detect an incipient spark and immediately reduce the voltage to quench the spark, after which the voltage is slowly increased again.

24.4 FLUE GAS CONDITIONING

Flue gas conditioning has been successfully used to lower fly ash resistivity in cold-side ESP for a number of years and, more recently, has been demonstrated effective in hot-side precipitator applications as well. There are several mechanisms by which conditioning may affect ESP performance, including

- Reducing back corona and increasing potential gradient in the gas space by lowering fly ash resistivity
- Reducing re-entrainment by increasing the cohesive properties of fly ash
- Increasing agglomeration or growth of small particles to form larger particles
- Increasing space charge to produce a more uniform potential gradient in the interelectrode space, thereby allowing operation at higher voltages

In cold-side ESP applications, conditioning agents that have been effective include

- Water vapor
- Sulfuric acid or SO_3
- Ammonia
- Ammonium sulfate
- Triethylamine
- Proprietary additives (e.g., ADA-23)

24.4.1 HUMIDIFICATION

Cooling a hot flue gas stream by evaporating a spray of fine water droplets produces a net decrease in the gas volumetric flow rate, increases the gas density, increases the moisture content of the gas stream, and reduces the particulate resistivity. Increasing both the gas density and the moisture content, increases the spark-over voltage at which the gas breaks down. Therefore, higher field strength can be achieved before spark rate limits further voltage increases.

A key concern with spraying water droplets is that the water droplets are completely evaporated before they impact the interior surfaces in ductwork. If water wets the walls of the ductwork, particulate may stick and accumulate and corrosion may be accelerated rapidly. Generally, at least one second residence time for drying in a straight section of duct is desired for a fine spray with a Sauter mean diameter of 50 μm.[6] Dual-fluid atomizing nozzles are typically used to achieve the fine spray, because high-pressure hydraulic nozzles would require 500–5000 psi with very small orifices that are prone to plugging and wear.

24.4.2 SO_3

Addition of sulfuric acid or SO_3 is a common method of lowering particulate resistivity. Sulfamic acid has been used as a source of SO_3 as it decomposes at flue gas temperatures upstream of the air heater. The conditioning effect is attributed to the formation of sulfuric acid as SO_3 combines with water and condenses on the particulate surfaces, and/or to the direct adsorption of SO_3 on the surface of the particles with the subsequent formation of sulfates, which act as electrolytes and enhance surface conduction.

Because the mechanism of SO_3 conditioning relies on condensation of H_2SO_4, the method is not effective at temperatures above 375°F. Therefore, it is not used for *hot-side* ESPs, which are located upstream of the air heater in a power boiler. Also, neutralization of acid by basic components such as calcium oxide in the fly ash may produce insoluble sulfate salts having a high resistivity. Thus, fly ash alkalinity is another factor in determining the effectiveness of SO_3 conditioning.

24.4.3 Ammonia

Addition of ammonia or ammonia compounds is another method of conditioning that has been used to improve ESP performance, particularly in the petroleum refining industry on fluid catalytic cracking units. Catalyst dust from fluid catalytic cracking units typically has a high resistivity.

Several hypotheses for the mechanism for improved ESP performance by NH_3 conditioning have been proposed. Explanations include neutralization of sulfuric acid in the case of low-resistivity fly ash[7]; reduction of back corona as characteristics of the gas are changed[8]; and higher voltage operation due to an enhanced potential gradient resulting from space charge effects.[9,10] The presence of very fine ammonium sulfate particles produced by the reaction of ammonia and SO_3 increases space charge, which would decrease the electric field strength near the corona discharge and allow a steeper gradient near the collecting plate. Although excessive space charge can lead to corona quenching, moderate space charge produces a higher driving force for particle migration outside of the corona discharge region. The space charge effect can be beneficial when the enhancement in the collecting electric field more than compensates for the decreased corona discharge. Another explanation is that ammonia reacts with sulfuric acid mist in the flue gas, forming ammonium bisulfate, NH_4HSO_4, and ammonium sulfate, $(NH_4)_2SO_4$. Ammonium bisulfate is deliquescent (absorbs water readily, then dissolves in the absorbed water) and has a melting point of 296.4°F.[10] Therefore, it would form a cohesive, sticky material in the ESP.

Recent work with fluid catalytic cracking units indicates that ammonia increases the breakdown strength of the gas space between particles in the dust layer. Sparking within the dust layer occurs when the electric field in the dust layer exceeds the breakdown strength of the gas in the interparticle spaces. Since the dust layer in fluid catalytic cracking unit applications has a high resistivity, the potential gradient within the dust layer is high. It has been demonstrated that ammonia conditions the gas rather than the dust as evidenced by the lack of residual ammonia in collected dust and a rapid response to changes in concentration (Durham, M. D., pers. Comm.).

Unfortunately, ammonia conditioning generates emissions of unreacted ammonia. This disadvantage has become more apparent in recent years with public attention focused by the Toxic Release Inventory reporting requirements and the emphasis to reduce hazardous air pollutants.

24.4.4 SO_3 and Ammonia

The addition of both SO_3 and ammonia would provide ESP performance enhancements by most of the mechanisms discussed above for ammonia alone, but allows control over the amount of each additive and their ratio. The normal flue gas SO_3 concentration produced by coals with the same sulfur content may vary considerably due to basic components in the coal such as calcium oxide, and to trace metal catalysts such as vanadium, which would convert SO_2 to SO_3.

24.4.5 Ammonium Sulfate

The addition of ammonium sulfate rather than ammonia and SO_3 gases separately also can improve ESP performance. In this case, the primary mechanism was attributed to adsorption of SO_3 on the fly ash after ammonium sulfate decomposed to SO_3 and NH_3. Supplemental benefits may have resulted from the coprecipitation of undecomposed ammonium sulfate particles with the fly ash, although there was no indication of agglomeration or particle growth or generation of fine fume to enhance space charge, and rapping emission reductions were attributed to improved collection in the inlet fields rather than to increased cohesion.[11]

24.4.6 Proprietary Additives

A new approach to flue gas conditioning is the use of proprietary phosphaste-based additive solutions that have been developed and commercialized to overcome dust resistivity issues. They have demonstrated dramatic reductions in resistivity of a lignite fly ash over a wide temperature range in a full-scale coal-fired application. Resistivity was reduced by as much as four orders of magnitude at 380°F, and as much as two orders of magnitude at 700°F.[12] Opacity from the stack of a coal-fired boiler with a cold-side (downstream of the air heater) ESP was reduced from 15% to less than 5%.[13] Advantages to this method of conditioning include low capital cost for equipment to inject the additive solution, and that the chemicals are relatively nontoxic compared to traditional conditioning agents SO_3 (which forms H_2SO_4) and ammonia.

24.5 USING V-I CURVES FOR TROUBLESHOOTING

Much troubleshooting of ESP problems can be accomplished by evaluating the electrical characteristics of the ESP. A powerful troubleshooting tool is the V-I curve, which is simply a plot of the secondary current that is produced as the secondary voltage is increased. To generate a V-I curve, the voltage on one bus section is reduced to zero, then manually increased in about 10% increments. Record the secondary voltage and current, and continue stepping up the voltage until 100% capacity is reached. The entire procedure takes only a few minutes, but the section will be operating at less than full power during this time.

The secondary current is plotted on rectangular coordinates as a function of secondary voltage. Example V-I curves are shown in Figure 24.7. Curve 1 is typical

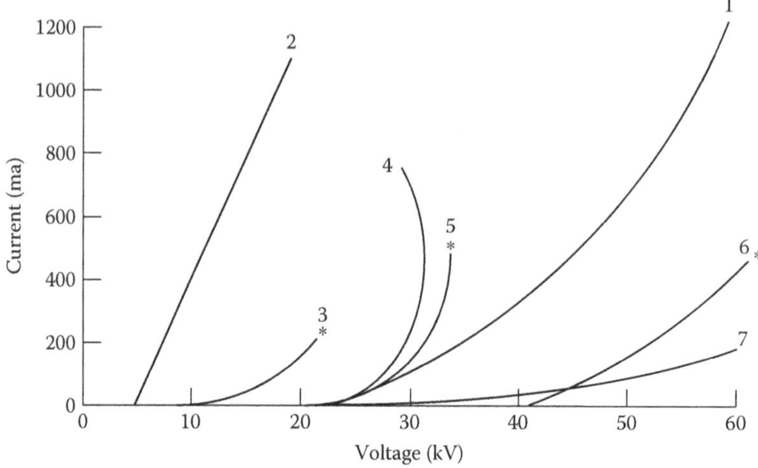

FIGURE 24.7 V-I curves. * indicates spark occurrence.

of normal operation. As the voltage is increased from zero, there is no current. At some voltage, current flow will begin. This occurs at the onset of corona discharge. Current increases with increasing voltage until the current limit of the power supply is reached.

Curves 2 through 7 in Figure 24.7 indicate problems. Curve 2 shows current flow commencing immediately as the voltage is increased, and current increasing with a steep slope until the current limit is reached. This is symptomatic of a direct short to ground, which may be due to a cracked insulator or bridging of dust between the electrodes.

Curve 3 shows corona forming at a low voltage, then voltage and current increasing until sparking occurs at a relatively low voltage. Remember that corona is induced by a steep potential gradient. Early onset of corona is due to an unusually steep gradient, which occurs when the physical spacing between the electrodes is reduced. Also, gas breakdown or sparking occurs when the gap between the electrodes is low. Thus, curve 3 represents a severe misalignment between the emitting electrodes and the collecting plates.

Curve 4 shows normal corona formation, but an unusual current increase that actually can have a backward curve. This unusual shape results from back-corona that forms in the dust layer in addition to the corona formed at the emitting electrodes. Back-corona is formed in high-resistivity dust by the steep potential gradient within the dust layer. Curve 5 is similar to curve 4, but ends with a spark in the dust layer.

Curve 6 is a result of corona not forming until the voltage is very high, then a spark may occur with a small increase in voltage. A likely cause is deposits buildup on the emitting electrodes, which can blunt the sharp edges of the emitting electrode. Since corona forms at the high potential gradient at sharp edges, corona formation is suppressed.

Curve 7 shows normal onset of corona, but the current does not increase with increasing voltage. In this case, severe space charge by charged submicron particles in the gas space flattens the potential gradient sufficiently to suppress additional corona formation.

Air load V-I curves on a new unit can ensure that the unit is installed and aligned correctly. Baseline V-I curves with flue gas at operating temperature will be different than air-load curves. These baseline curves will be helpful in monitoring changes in the curves as the precipitator ages. At higher temperatures, the flue gas is less dense and will break down at lower voltages. However, high flue gas moisture content increases the breakdown voltage of the gas, in addition to affecting the dust resistivity.

References

CHAPTER 1

1. Boubel, R. W., Fox, D. L., Turner, D. B., and Stern, A. C., *Fundamentals of Air Pollution*, 3rd ed., Academic Press, San Diego, CA, 1994.
2. Schrenk, H. H., Heiman, H., Clayton, G. D., Gafafer, W. M., and Wexler, H., *Air Pollution in Donora, PA*, Public Health Bulletin No. 306, Federal Security Agency, Public Health Service, Division of Industrial Hygiene, Washington, DC, 1949.
3. Seinfeld, J. H., *Air Pollution*, McGraw-Hill, New York, 1975.
4. Goldsmith, J. R., Effects of air pollution on human health, in *Air Pollution*, Stern, A. C., Ed., Academic Press, New York, 1977.
5. U.S. Environmental Protection Agency, *Causes of Climate Change*, available at: http://www.epa.gov/climatechange/science/causes.html.
6. Williamson, S., *Fundamentals of Air Pollution*, Addison-Wesley, Reading, MA, 1973.

CHAPTER 2

1. Chambers, A., Fanning the flames, *Power Eng.*, 104(1), 16, 2000.
2. Hazardous Air Pollutants. 42 U.S.C. § 7412 (b)(1)(B), 1988.
3. Boubel, R. W., Fox, D. L., Turner, D. B., and Stern, A. C., *Fundamentals of Air Pollution*, 3rd ed., Academic Press, San Diego, CA, 395, 1994.
4. *General Overview of the CAAA*, www.epa.gov/ttn/oarpg/gen/overview.txt, EPA OAQPS Technology Transfer Network, 1993.
5. Coffin, D. L., Blommer, E. J., Gardner, D. E., and Holzman, R. S., Effects of air pollution on alteration of susceptibility to pulmonary infection, in *Proceedings of the 3rd Annual Conference on Atmospheric Contaminants in Confined Spaces*, Aerospace Medical Research Laboratory, Wright Patterson Air Force Base, Dayton, OH, 71, 1968.
6. *American Trucking Associations, Inc. v. U.S. EPA*, 195 F.3d 4, 1999.
7. *Whitman v. American Trucking Association, Inc.*, 531 U.S. 457, 2001.
8. *Massachusetts v. EPA*, 549 U.S. 497, 2007.
9. *Utility Air Regulatory Group v. EPA*, 134 S.Ct. 2427, 2014.
10. *Final National Emission Standards for Hazardous Air Pollutants*, available at: http://www.epa.gov/airtoxics/mactfnlalph.html.
11. *Sierra Club v. Browner*, D.C. District Court, 93-0124.

CHAPTER 3

1. *New Source Review Workshop Manual*, U.S. Environmental Protection Agency, Draft, October 1990.
2. Leonard, R. L., *Air Quality Permitting*, Lewis Publishers, Boca Raton, FL, 68, 1997.
3. Evans, B. and Weiss, K., Summary of progress on EPA's NSR reform initiative, *EM*, 27, June 2000.
4. Hawkins, B. and Ternes, M. E., *Clean Air Act Handbook*, 3rd ed., American Bar Association, Section of Environment, Energy and Resources, Chapter 6, 2011.

CHAPTER 4

1. Turner, D. B., *A Workbook of Atmospheric Dispersion Estimates*, 2nd ed., Lewis Publishers, Boca Raton, FL, 1994.
2. Schnelle, K. B., Jr. and Dey, P. R., *Atmospheric Dispersion Modeling Compliance Guide*, McGraw-Hill, New York, 1999.

CHAPTER 5

1. *Code of Federal Regulations*, 40 CFR Part 60, Appendix A, available at www.ecfr.gov, current as of February 2, 2015.

CHAPTER 6

1. Lodge, J. P., Jr., Ed., *Methods of Air Sampling and Analysis*, Lewis Publishers, Chelsea, MI, 1989.
2. Glasastone, S., Laidler, K. J., and Eyring, H., *The Theory of Rate Processes*, McGraw-Hill, New York, 1941.

CHAPTER 7

1. Seider, W., Seader, J., Lewin, D., and Widagdo, S., *Product and Process Design Principles: Synthesis, Analysis and Design*, 3rd ed., 544, Wiley, Hoboken, NJ, 2009.
2. Vatavuk, W. M., *Escalation Indexes for Air Pollution Control Costs*, U.S. Environmental Protection Agency, EPA-452/R-95-006, Research Triangle Park, NC, 1995.
3. Humphreys, K. K. and Katell, S., *Basic Cost Engineering*, Marcel Dekker, New York, 108, 1981.
4. Peters, M. S. and Timmerhaus, K. D., *Plant Design and Economics for Chemical Engineers*, 3rd ed., McGraw-Hill, New York, 183, 1980.
5. Perry, R. H., Green, D. W., and Maloney, J. O., *Perry's Chemical Engineers' Handbook*, 7th ed., McGraw-Hill, New York, 9–64, 1997.
6. *OAQPS Control Cost Manual*, 5th ed., U.S. Environmental Protection Agency, EPA 453/B-96-001, Research Triangle Park, NC, 1996.
7. *Process Plant Construction Estimating Standards*, Richardson Engineering Services, San Marcos, CA, 2001.
8. Vatavuk, W. M., *Estimating Costs of Air Pollution Control*, Lewis Publishers, Chelsea, MI, 1990.
9. Keeth, R. J., Miranda, J. E., Reisdorf, J. B., and Scheck, R. W., *Economic Evaluation of FGD Systems*, GS-7193, Vol. 1 and 2, Electric Power Research Institute, Palo Alto, CA, 1991.
10. Keeth, R. J., Ireland, P. A., and Radcliffe, P. T., Economic evaluations of 28 FGD processes, presented at 1991 EPRI/EPA/DOE SO_2 Control Symposium, Washington, DC, December 3–6, 1991.

CHAPTER 8

1. Seider, W. D., Seader, J. D., and Lewin, D. L., *Process Design Principles—Synthesis, Analysis, and Evaluation*, John Wiley & Sons, New York, 1999.
2. Turton, R., Bailie, R. C., Whiting, W. B., and Shaeiwitz, J. A., *Analysis, Synthesis, and Design of Chemical Processes*, Prentice Hall, Englewood Cliffs, NJ, 1998.
3. Phillips, H. W., Select the proper gas cleaning equipment, *Chem. Eng. Prog.*, 96(9), 19, 2000.

4. Nair, C., Letters—Gas cleaning equipment, *Chem. Eng. Prog.*, 97(1), 10, 2001.
5. Smith, J. M., Van Ness, H. C., and Abbott M. M., *Introduction to Chemical Engineering Thermodynamics*, 6th ed., McGraw-Hill, New York, 2001.
6. Felder, M. R. and Rousseau, R. W., *Elementary Principles of Chemical Processes*, 3rd ed., John Wiley & Sons, New York, 2000.
7. El-Halwagi, M. M., *Sustainable Design through Process Integration: Fundamentals and Applications to Industrial Pollution Prevention, Resource Conservation, and Profitability Enhancement*, Elsevier, Oxford, UK, 2012.
8. El-Halwagi, M. M. and Manousiouthakis, V., Synthesis of mass-exchange networks. *AIChE J.*, 35, 1233–1244, 1989.
9. El-Halwagi, M. M., Srinivas, B. K., and Dunn, R. F., Synthesis of heat-induced separation networks. *Chem. Eng. Sci.*, 50, 81–97, 1995.
10. Dunn, R. F., Zhu, M., Srinivas, B. K., and El-Halwagi, M. M., Optimal design of energy-induced separation systems for VOC recovery. *AIChE Symp. Ser.*, 90, 74–85, 1995.

CHAPTER 9

1. Peters, M. S. and Timmerhaus, K. D., *Plant Design and Economics for Chemical Engineers*, 4th ed., McGraw-Hill, New York, 1991.
2. Guthrie, K. M., Capital cost estimating, *Chem. Eng.*, March 24, 76, 114, 1969.
3. Vatavuk, W. M. and Neveril, R. B., Factors for estimating capital and operating costs, *Chem. Eng.*, November 3, 157–162, 1980.
4. Seider, W. D., Seader, J. D., Lewin, D. R., and Widagdo, S., *Product and Process Design Principles: Synthesis, Analysis and Design*, 3rd ed., Wiley, Hoboken, NJ, December 22, 2008.

CHAPTER 10

1. *U.S. EPA Handbook: Control Technologies for Hazardous Air Pollutants*, EPA/625/6-91/014, U.S. EPA, Cincinnati, OH, 1991.
2. El-Halwagi, M. M. and Manousiouthakis, V., Simultaneous synthesis of mass exchange and regeneration networks, *AIChE J.*, 36(8), 1209–1219, 1990.
3. Dunn, R. F. and El-Halwagi, M. M., Process integration technology review: background and applications in the chemical process industry, *J. Chem. Tech. Biotech.*, 78, 1011–1021, 2003.
4. El-Halwagi, M. M., *Sustainable Design through Process Integration: Fundamentals and Applications to Industrial Pollution Prevention, Resource Conservation, and Profitability Enhancement*, Elsevier, Oxford, UK, 2012.

CHAPTER 11

1. Higbie, R., The rate of absorption of pure gas into a still liquid during short periods of exposure. *Trans. AIChE*, L1, 365–389, 1935.
2. Ruddy, E. N. and Carroll, L. A., Select the best VOC control strategy, *Chem. Eng. Prog.*, 89(7), 28, 1993.
3. U.S. Environmental Protection Agency, *Control Technologies for Hazardous Air Pollutants*, EPA/625/6-91/014, U.S. EPA, Washington, DC, 1991.
4. Dunn, R. F. and El-Halwagi, M. M., *Design of Cost-Effective VOC Recovery Systems*, Published by Tennessee Valley Authority Environmental Research Center/EPA Center for Waste Reduction, Muscle Shoals, AL, 82 pp, 1996.

5. Rees, R. L., The removal of oxides of sulfur from flue gases. *J. Inst. Fuel*, 25, 350–357, 1953.
6. Crocker, B. B., Capture of hazardous emissions, in *Proceedings, Control of Specific Toxic Pollutants, Conference*, February 1979, Gainesville, FL, Air Pollution Control Association, Pittsburgh, PA, 414–433, 1979.
7. Crocker, B. B. and Schnelle, Jr., K. B., Air pollution control for stationary sources, in *Encyclopedia of Environmental Analysis and Remediation*, Meyers, R. A., Ed., John Wiley & Sons, New York, 151–213, 1998.
8. Teller, A. J., Control of Gaseous Flouride Emissions, *Chem. Eng. Prog.* 63, 75, 1967.
9. Smith, J. M., van Ness, H. C., and Abbott, M. M., *Introduction to Chemical Engineering Thermodynamics*, 7th ed., McGraw-Hill, New York, 2005.
10. Danckwerts, P. V., Significance of liquid film coefficients in gas absorption, *Ind. Eng. Chem.*, 41, 1460–1467, 1951.
11. Whitman, W. G., The two film theory of gas absorption, *Chem. Metal.*, 2(E), 146–148, 1923.
12. Dunn, R. F. and El-Halwagi, M. M., Optimal recycle/reuse policies for minimizing the wastes of pulp and paper plants, *J. Environ. Sci. & Health*, A28(1), 217–234, 1993.
13. Srinivas, B. K. and El-Halwagi, M. M., Synthesis of reactive mass-exchange networks with general nonlinear equilibrium functions. *AIChE J.*, 40, 1994, 463–472.
14. El-Halwagi, M. M. and Manousiouthakis, V., Synthesis of mass-exchange networks. *AIChE J.*, 35, 1233–1244, 1989.
15. El-Halwagi, M. M., *Sustainable Design through Process Integration: Fundamentals and Applications to Industrial Pollution Prevention, Resource Conservation, and Profitability Enhancement*, Elsevier, Oxford, UK, 2012.
16. El-Halwagi, M. M. and Manousiouthakis, V., Automatic synthesis of mass-exchange networks with single component targets. *Chem. Eng. Sci.*, 45, 2813–2831, 1990.
17. Strigle, R. F., *Packed Tower Design and Applications, Random and Structured Packings*, 2nd ed., Gulf Publishing, Houston, TX, 1994.
18. Kister, H. Z., *Distillation Design*, McGraw-Hill, New York, 1992.
19. Rukovena, F. and Cai, T., Achieve good packed tower efficiency, *Chemical Processing*, Schaumburg, IL, November, 23–31, 2008.
20. Sherwood, T. K., Shipley, G. H., and Holloway, F. A. L., Flooding velocities in packed columns, *Ind. Eng. Chem.*, 30, 765, 1938.
21. Parkinson, G. and Ondrey, G., Packing towers, *Chem. Eng., Newsfront*, 39, December 1999.
22. Kister, H. J., Scherffius, J., Afshar, K., and Abkar, E., CEP website, www.aicheorg/cep July, 2007.
23. Bravo, J. L., Rocha, J. A., and Fair, J. R., *Hydrocarbon Proc.*, 56(3), 45, 1986.
24. Fair, J. R. and Bravo, J. L., *Chem. Eng. Prog.*, 86(1), 19, 1990.
25. Bravo, J. L., Rocha, J. A., and Fair, J. R., *I. Chem. E. Symp. Ser*, No. 128, A439, 1992.
26. Colburn, A. P., Simplified calculation of diffusional processes, *Ind. Eng. Chem.* 33, 459–467, 1961.
27. Perry, R. H. and Green, D. W., *Chemical Engineer's Handbook*, 7th ed., McGraw-Hill, New York, Section 14, 1997.
28. Yaws, C., Yang, H., and Pan X., Henry's Law Constant for 362 organic compounds in water, *Chem. Eng.*, 98, 179–185, November 1991.
29. Carroll, J. J., Use Henry's Law for multicomponent mixtures, *Chem. Eng. Prog.*, 88, 53–58, August 1992.

CHAPTER 12

1. Berg, A. H., Hypersorption design: Modern advancements, *Chem. Eng. Prog.*, 47(11), 585–590, 1951.
2. Ruddy, E. N. and Carroll, L. A., Select the best VOC control strategy, *Chem. Eng. Prog.*, 89(7), 28, 1993.

3. U.S. Environmental Protection Agency, *Control Technologies for Hazardous Air Pollutants*, EPA/625/6-91/014, U.S. EPA, Cincinnati, OH, 1991.
4. Stenzel, M. H., Remove organics by activated carbon adsorption, *Chem. Eng. Prog.*, 89(4), 36–43, 1993.
5. Dunn, R. F. and El-Halwagi, M. M., *Design of Cost-Effective VOC Recovery Systems*, Published by Tennessee Valley Authority Environmental Research Center/EPA Center for Waste Reduction, Muscle Shoals, AL, 82 p., 1996.
6. Humphrey, J. L., Separation processes: Playing a critical role, *Chem. Eng. Prog.*, 91(10), 31, 1995.
7. Perry, R. H. and Green, D. W., *Chemical Engineer's Handbook*, 7th ed., McGraw-Hill, New York, Section 16, 1997.
8. Freundlich, H., M. F., Uber die Adsorption in Losungen, *Z. Phys. Chem.*, 57(A), 385–470, 1906.
9. Langmuir, I., The constitution and fundamental properties of solids and liquids, part 1 solids, *J. Am. Chem. Soc.*, 38, 11, 2221–2295, 1916.
10. Brunauer, S., Emmett, P., and Teller, E., Adsorption of gases in multimolecular layers, *J. Am. Chem. Soc.*, 60, 309, 1938.
11. Goldman, F. and Polyani, M., *Z.* Adsorption von Dämpfen an Kohle und die Wärmeausdehnung der Benetzungsschicht, *Phys. Chem.*, 132, 321, 1928.
12. Grant, R. J. and Manes, M., Correlation of some gas adsorption data extending to low pressures and supercritical temperatures, *Ind. Eng. Chem. Fundam.*, 3(3), 221–224, 1964.
13. Ergun, S., Fluid flow through packed columns, *Chem. Eng. Prog.*, 48, 89–94, 1952.
14. Hougen, O. A. and Marshall, Jr., W. R., Adsorption from a fluid stream flowing through a stationary granular bed, *Chem. Eng. Prog.*, 43(4), 197, 1947.
15. Fair, J. R., Sorption processes or gas separation, *Chem. Eng.*, 90–110, July 14, 1969.
16. Ufrecht, R. H., Sommerfeld, J. T., and Lewis, H. C., Design of adsorption equipment in air pollution control, *J.A.P.C.A.*, 30(42), 1348, 1980.
17. McCleod, H. O. and Campbell, J. M., Mechanisms by which pentane and hexane adsorb on silica gel, *Soc. Petrol. Eng. J.*, 166, June 1966.
18. LeVan M. D. and Schweiger T. A. J., Steam generation of adsorption beds: Theory and Experiments, *Third International Conference on Fundamentals of Adsorption*, Sonthofen, Germany, 487–495, May 5–9, 1989.
19. El-Halwagi, M. M., *Sustainable Design through Process Integration: Fundamentals and Applications to Industrial Pollution Prevention, Resource Conservation, and Profitability Enhancement*, Elsevier, Oxford, UK, 2012.
20. El-Halwagi, M. M. and Manousiouthakis, V., Simultaneous synthesis of mass-exchange and regeneration networks. *AIChE J.*, 36, 1209–1219, 1990.

CHAPTER 13

1. Leite, O. C., Safety, noise, and emissions elements round out flare guidelines, *Oil Gas J.*, 24, 68, 1992.
2. Leite, O. C., Design alternatives, components key to optimum flares, *Oil Gas J.*, 23, 70, 1992.
3. American Petroleum Institute, *Guide for Pressure-Relieving and Depressuring Systems, API Recommended Practice 521*, 4th ed., American Petroleum Institute, Washington, DC, 1997.

CHAPTER 14

1. Smith, J., van Vess, H. C., and Abbott, M. M., *Introduction to Chemical Engineering Thermodynamics*, 6th ed., McGraw-Hill, New York, 2000.

2. Perry, R. H. and Green, D. A., *Perry's Chemical Engineers' Handbook*, 7th ed., McGraw-Hill, New York, 1997.
3. Colburn, A. P. and Hougen, O. A., Design of cooler condensers for mixtures of vapors with noncondensing gases, *I&E Chem.*, 26(11), 1178–1182, 1934.
4. Silver, L., Gas cooling with aqueous condensation, *Trans. Inst. Chem. Eng.*, 25, 30–42, 1947.
5. Bell, K. G. and Ghaly, M. A., An approximate generalized design method for multicomponent partial condensers, *Chem. Eng. Prog. Symp. Ser.*, 131, 72–79, 1973.
6. El-Halwagi, M. M., Srinivas, B. K., and Dunn, R. F., Synthesis of Optimal Heat-Induced Separation Networks. *Chem. Eng. Sci.*, 50(1), 81–97, 1995.
7. Dunn, R. F. and El-Halwagi, M. M., Selection of Optimal VOC-Condensation Systems, *Waste Mgt.*, 14(2), 103–113, 1994a.
8. Dunn, R. F. and El-Halwagi, M. M., Optimal Design of Multi-Component VOC Condensation Systems, *J. Haz. Mtls.*, 38, 187–206, 1994b.
9. Dunn, R. F., Zhu, M., Srinivas, B. K., and El-Halwagi, M., Optimal design of energy-induced separation networks for VOC recovery, *AIChE J. Symp. Ser.*, 90(103), 74–85, 1995.

CHAPTER 15

1. Williams, T. O. and Miller, F. C., Odor control using biofiltration, *BioCycle*, 72–77, October 1992.
2. Leson, G. and Winer, A. M., Biofiltration: An innovative air pollution control technology for VOC emissions, *J. Air Waste Manage. Assoc.*, 41(8), 1045–1054, 1991.
3. Speece, R. E., Biofiltration of gaseous contaminants, State-of-the-art report to Weyerhaeuser Corporation, Federal Way, WA, May 24, 1995.
4. Kampbell, D. H. et al., Removal of volatile aliphatic hydrocarbons in a soil bioreactor, *J. Air Poll. Contr. Assoc.*, 37(10), 1236–1240, 1987.
5. Williams, T. O. and Miller, F. C., Biofilters and facility operations, *BioCycle*, 75–79, November 1992.
6. Bohn, H. L., Soil and compost filters of malodorous gases, *J. Air Poll. Contr. Assoc.*, 25(9), 953–955, 1975.
7. Bohn, H. L. and Bohn, R. K., Soil beds weed out air pollutants, *Chem. Eng.*, 95(6), 73–76, 1988.
8. Adler, S. F., Biofiltration—A primer, *Chem. Eng. Progr.*, 97(4), 33–41, April 2001.

CHAPTER 16

1. Lokhandwala, K. A. and Jacobs, M., New membrane applications in gas processing, *Presented at the 79th Annual Gas Processors Association Convention*, Atlanta, GA, March 2000.
2. Simmons, V. et al., Membrane systems offer a new way to recover volatile organic air pollutants, *Chem. Eng.*, 101(9), 92, September 1994.
3. McInnes, R. G., Explore New Options for Hazardous Air Pollution Control, *Chem. Eng. Progr.*, 91(11), 36–48, 1995.
4. Dunn, R. F. and El-Halwagi, M.M., *Design of Cost-Effective VOC Recovery Systems*, Published by Tennessee Valley Authority Environmental Research Center/EPA Center for Waste Reduction, Muscle Shoals, AL, 82 pp, 1996.
5. Jacobs, M., Gottschlich, D., and Buchner, F., Monomer recovery in polyolefin plants using membranes—An update, *Presented at the 1999 Petrochemical World Review*, Houston, TX, March 23–25, 1999.
6. Baker, R. W. and Jacobs, M., Improve monomer recovery from polyolefin resin degassing, *Hydrocarbon Process*, 75(3), 49, March 1996.

7. Baker, R. W. et al., Recover feedstock and product from reactor vent streams, *Chem. Eng. Progr.*, 99(12), 51, 2000.
8. Crabtree, E. W., Dunn, R. F., and El-Halwagi, M. M., Systematic synthesis of VOC recovery membrane-condensation hybrid systems, *J. Air and Waste Mgt. Assoc.*, 48, 616–626, 1998.

CHAPTER 17

1. Zeldovich, Y. A., *Oxidation of Nitrogen in Combustion*, Academy of Sciences of USSR, Institute of Chemical Physics, Moscow-Leningrad, 1947.
2. Pershing, D. W. and Wendt, J. O. L., Relative contributions of volatile nitrogen and char nitrogen to NO_x emissions from pulverized coal flames, *Ind. Eng. Chem. Proc. Des. Dev.*, 18(60), 60–67, 1979.
3. Gregory, M. G., Cochran, J. R., Fischer, D. M., and Harpenau, M. G., The cost of complying with NO_x emission regulations for existing coal fueled boilers, *Presented at EPRI-DOE-EPA Combined Utility Air Pollutant Control Symposium*, Washington, DC, August 25–29, 1997.
4. Booth, R. C., Kosvic, T. C., and Parikh, N. J., The emissions, operational, and performance issues of neural network control applications for coal-fired electric utility boilers, *Presented at EPRI-DOE-EPA Combined Utility Air Pollutant Control Symposium*, Washington, DC, August 25–29, 1997.
5. Wood, S. C., Select the right NO_x control technology, *Chem. Eng. Progr.*, 90(1), 32, 1994.
6. John Zink Company, Burner design parameters for flue gas NO_x control, Technical Paper 4010B, 1993.
7. Himes, R., Scharnott, M., and Hoyum, R., Fuel system modifications and boiler tuning to achieve early election compliance on a 372-MWe coal-fired tangential boiler, *Presented at EPRI-DOE-EPA Combined Utility Air Pollutant Control Symposium*, Washington, DC, August 25–29, 1997.
8. Wendt, J. O. L. et al., Reduction of sulfur trioxide and nitrogen oxides by secondary fuel injection, in *Proceedings of the 14th International Symposium Combustion*, University Park, PA, 897, 1973.
9. Folsom, B. A. et al., Field experience—Reburn NO_x control, *Presented at EPRI-DOE-EPA Combined Utility Air Pollutant Control Symposium*, Washington, DC, August 25–29, 1997.
10. Garg, A., Specify better low-NO_x burners for furnaces, *Chem. Eng. Progr.*, 90(1), 46, 1994.
11. Sivy, J. L., Sarv, H., and Koslosky, J. V., NO_x subsystem evaluation of B&W's advanced coal-fired low emission boiler system at 100 Million BTU/hr, *Presented at EPRI-DOE-EPA Combined Utility Air Pollutant Control Symposium*, Washington, DC, August 25–29, 1997.
12. Bortz, S. J. et al., Ultra-low NO_x rapid mix burner demonstration at Con Edison's 59th Street Station, *Presented at EPRI-DOE-EPA Combined Utility Air Pollutant Control Symposium*, Washington, DC, August 25–29, 1997.
13. Chen, W. et al., Deviation and application of a global SNCR model in maximizing NO_x reduction, *Presented at EPRI-DOE-EPA Combined Utility Air Pollutant Control Symposium*, Washington, DC, August 25–29, 1997.
14. Ferrell, R., Controlling NO_x emissions: A cooler alternative, *Poll. Eng.*, 4, 50, 2000.
15. Reyes, B. E. and Cutshaw, T. R., $SCONO_x$™ catalytic absorption system, *Western Energ.*, 13, March 1999.
16. Haythornthwaite, S. et al., Stationary source NO_x control using pulse-corona induced plasma, *Presented at EPRI-DOE-EPA Combined Utility Air Pollutant Control Symposium*, Washington, DC, August 25–29, 1997.

CHAPTER 18

1. Brown, C., Pick the best acid-gas emission controls for your plant, *Chem. Eng. Progr.*, 94(10), 63, 1998.
2. Lunt, R. R. and Cunic, J. D., Profiles in Flue Gas Desulfurization, *AIChE*, New York, 2000.
3. Brown, C. A., Blythe, G. M., Humphries, L. R., Robends, R. F., Runzan, R. A., and Rhudy, R. G., Results from the TVA 10-MW spray dryer/ESP evaluation., *Presented at EPA/EPRI First Combined FGD and Dry SO_2 Control Symposium*, St. Louis, MO, October 25–28, 1988.
4. Hedenhag, J. G. and Goss, W. L., Gas suspension absorption: A new approach in FGD systems for boilers and incinerators, *Presented at the 88th Annual Meeting Air and Waste Management Association*, San Antonio, TX, June 18–23, 1995.
5. Lepovitz., L. R., Results of the recent testing of the 10-MW gas suspension absorption flue gas desulfurization technology, *Presented at 88th Annual Meeting of the Air and Waste Management Association*, San Antonio, TX, June 18–23, 1995.
6. Moore, S. R., Toher, J. G., and Sauer, H., Update of the EEC/Lurgi high efficiency, dry circulating fluid bed scrubber experience, *Presented at EPRI Conf. on Improved Technology for Fossil Fueled Power Plants—New and Retrofit Application*, Washington, DC, March 1, 1993.
7. Bland, V. V. and Martin, C. E., *Full-Scale Demonstration of Additives for NO_2 Reduction with Dry Sodium Desulfurization*, Electric Power Research Institute, Palo Alto, CA, EPRI GS-6852, 1990.
8. Baldwin, A. B., Trona use in dry sodium injection for acid gas removal, *Presented at Society of Mining Engineers*, Denver, CO, February 1995.
9. Gmitro, J. I. and Vermeulen, T., Vapor-liquid equilibria for aqueous sulfuric acid, *AIChE*, 10, 740, 1964.
10. Haase, R. and Borgmann, H. W., *Korrosion*, 15, 47, January 1981.
11. Verhoff, F. H. and Banchero, J. T., Predicting dew points of flue gases, *Chem. Eng. Progr.*, 70(8), 71, 1974.
12. Okkes, A. G. and Badger, B. V., Get acid dew point of flue gas, *Hydrocarbon Proc.*, 66(7), 53, 1987.
13. Jones, C. and Ellison, W., SO_3 tinges stack gas from scrubbed coal-fired units, *Power*, 142(4), 73, 1998.
14. *Emergency Planning and Community Right-to-Know Act—Section 313: Guidance for Reporting Sulfuric Acid*, U.S. EPA, Office of Pollution Prevention and Toxics, Washington, DC, EPA-745-R-97-007, 1998.

CHAPTER 19

1. Davies, C. N., Definitive equations for the fluid resistance of spheres, *Proc. Phys. Soc.*, 57, 259, 1945.

CHAPTER 20

1. Dalla Valle, J. M., *Exhaust Hoods*, Industrial Press, New York, 1952.
2. Hemeon, W. E. L., *Plant and Process Ventilation*, Industrial Press, New York, 1954.
3. American Conference of Governmental Industrial Hygienists, Committee on Industrial Ventilation, *Industrial Ventilation, A Manual of Recommended Practice*, 9th ed., ACGIH, Lansing, MI, 1966.
4. American Society of Heating, Refrigerating, and Air Conditioning Engineers, *ASHRAE Handbook—Heating, Ventilating, and Air Conditioning Systems and Application*, Atlanta, GA, 1987.

5. Goodfellow, H. D., Auxiliary equipment for local exhaust ventilation systems, in *Air Pollution Engineering Manual*, Buonicore, A. J. and Davis, W. T., Eds., Air and Waste Management Association, Van Nostrand Reinhold, New York, 1992, chap. 6.
6. King, R. H., Correctly install fans for optimal system performance, *Chem. Eng. Progr.*, 93, 60–69, May, 1997.
7. Urandt, A. D., A summary of design data for exhaust systems, heating and ventilation, 42, 73, May, 1945.
8. McCabe, W. L., Smith, J. C., and Harriott, P., *Unit Operations in Chemical Engineering*, 4th ed., McGraw-Hill, New York, 1985.
9. Crawford, M., *Air Pollution Control Theory*, McGraw-Hill, New York, 1976.
10. Cooper, C. D. and Alley, F. C., *Air Pollution Control*, 2nd ed., Waveland Press, Prospect Heights, IL, 1994.

CHAPTER 21

1. Lapple, C. E., Processes use many collector types, *Chem. Eng.*, 58, 5, May 1951.
2. Leith, D. and Licht, W., The collection efficiency of cyclone type particle collectors—A new theoretical approach, *AIChE Symp. Ser.*, 126(68), 196–206, 1972.
3. Stairmand, C. J., The design and performance of cyclone separators, *Trans. Ind. Chem. Eng.*, 29, 356–573, 1951.
4. Swift, P., Dust control in industry, *Steam Heating Eng.*, 38, 453–456, 1969.
5. Miller and Lissman, Calculation of cyclone pressure drop, *Presented at the Meeting of American Society of Mechanical Engineering*, New York, December 1940.
6. Shepherd, C. B. and Lapple, C. E., Flow pattern and pressure drop in cyclone dust collectors, *Ind. Eng. Chem.*, 32(9), 1246–1248, 1940.
7. Kalen, B. and Zenz, F., Theoretical empirical approach to saltation velocity in cyclone design, *AIChE Symp. Ser.*, 70(137), 388–396, 1974.
8. Koch, W. H. and Licht, W., New design approach boosts cyclone efficiency, *Chem. Eng.*, 84(24), 80–88, 1977.

CHAPTER 22

1. Crocker, B. B. and Schnelle, Jr., K. B., Air pollution control for stationary sources, in *Environmental Analysis and Remediation*, Meyers, R. A., Ed., John Wiley & Sons, New York, 151–213, 1998.
2. Hesketh, H. E., Atomization and cloud behavior in wet scrubbers, *US-USSR Symposium Control of Fine Particulate Emissions* (1), 15, 1974.
3. Calvert, S., Goldshmidt, F., Leith, D., and Mehta, D., *Scrubber Handbook*, NTIS Publication PB-213016 and PP-213017, U.S. EPA, Washington, DC, July–August 1972.
4. Porter, H. F., Schurr, G. A., Wells, D. F., and Semrau, K. T., *Chemical Engineers' Handbook*, 6th ed., McGraw-Hill, New York, 20–77 to 20–110, 1984.
5. Sargent, G. D., Dust collection equipment, *Chem. Eng.*, 76, 130–150, 1969.
6. Celenza, G. F., Designing air pollution control systems, *Chem. Eng. Prog.*, 66(11), 31, 1970.
7. Semrau, K. T., Correlation of dust scrubber efficiency, *J. Air Poll. Contr. Technol. Assoc.*, 10(3), 200, 1960.
8. Semrau, K. T., Dust scrubber design—A critique on the state of the art, *J. Air Poll. Contr. Technol. Assoc.*, 13(12), 587, 1963.
9. Calvert, S., Engineering design of fine particle scrubbers, *J. Air Poll. Contr. Assoc.*, 24(10), 929, 1974.
10. Calvert, S., How to choose a particulate scrubber, *Chem. Eng.*, 39, 54–68, August 29, 1977.

Additional References

Licht, W., *Air Pollution Control Engineering, Basic Calculations for Particulate Collection*, 2nd ed., Marcel Dekker, New York, 1988.

Loftus, P. J., Stickler, D. B., and Diehl, R. C., A confined vortex scrubber for fine particulate removal from flue gases, *Environ. Prog.*, 11, 27–32, 1992.

Blackwood, T. R., An evaluation of fine particle wet scrubbers, *Environ. Prog.*, 7, 71–75, 1988.

CHAPTER 23

1. Schimmoller, B. K., Tuning in to acoustic cleaning, *Power Eng.*, 103(107), 19, 1999.
2. Plinke, M. et al., Catalytic filtration—Dioxin destruction in a filter bag, in *Recent Developments in Air Pollution Control, Topical Conference Proceedings AIChE Spring National Mtg.*, Atlanta, GA, 167, 2000.
3. Wagner, R. A., Ceramic composite hot-gas filter development, *Presented at Symposium on High-Temperature Particulate Cleanup*, Birmingham, AL, April 20–23, 1998.
4. Bustard, C. J. et al., Demonstration of novel additives for improved fabric filter performance, *Presented at EPRI/DOE International Conference on Managing Hazardous and Particulate Air Pollutants*, Toronto, Canada, August 15–18, 1995.
5. Danielson, J., *Air Pollution Engineering Manual*, 2nd ed., AP40, US EPA, Research Triangle Park, NC, 1973.

CHAPTER 24

1. McDonald, J. R., Computer simulation of the electrostatic precipitation process, in *Proceedings of the International Conference on Electrostatic Precipitation*, Air Pollution Control Association, Pittsburgh, PA, October 1981.
2. Katz, J., Factors affecting resistivity in electrostatic precipitation, *J. Air Poll. Contr. Assoc.*, 30(1), 165, 1980.
3. *Gas Flow Model Studies*, Industrial Gas Cleaning Institute, Pub. No. EP-7, Rev I, Washington, DC, 1969.
4. Hein, A. G., Dust reentrainment, gas distribution and electrostatic precipitator performance, *J. Air Poll. Contr. Assoc.*, 39(5), 766, 1989.
5. The McIlvaine Company, Skewed gas flow technology demonstrated in South Africa, *Precip. Newsl.*, 241, 1996.
6. Butz, J. R., Baldrey, K. E., and Sam, D. O., Computer modeling of ESP performance improvement due to spray cooling, *Presented at EPRI/DOE International Conference on Managing Hazardous and Particulate Air Pollutants*, Toronto, Canada, August 15–18, 1995.
7. Reese, J. T. and Greco, J., Electrostatic precipitation, *Mech. Eng.*, 90(10), 13–7, 1968.
8. Watson, K. S. and Blecher, K. J., Investigation of electrostatic precipitators for large P.F. fired boilers, *Presented at Clean Air Conference*, Sydney, Australia, 1965.
9. Dismukes, E. B., *Conditioning of Fly Ash with Sulfur Trioxide and Ammonia*, Environmental Protection Agency, Washington, DC, EPA-600 / 2-75 - 015, TVA-F75 PRS-5, 1975.
10. Krigmont, H. V. and Coe, Jr., E. L., Experience with dual flue gas conditioning of electrostatic precipitators, *Presented at Eighth Symposium on Transfer and Utilization of Particulate Control Technology*, San Diego, CA, March 20–23, 1990.

11. Gooch, J. P., Bickelhaupt, R. E., and Dismukes, E. B., *Investigation of Ammonium Sulfate Conditioning for Cold-Side Electrostatic Precipitators—Volume 1: Field and Laboratory Studies*, Electric Power Research Institute, Washington, DC, EPRI CS-3354, 1, 1984.
12. Durham, M. D., Bustard, C. J., and Martin, C. E., New ESP additive controls particulates, *Power Eng.*, 101(06), 44, 1997.
13. Durham, M. D. et al., Full-scale experience with ADA-34 2nd generation flue gas conditioning for hot-side and cold-side ESPs, in *EPRI-DOE-EPA Combined Utility Air Pollution Control Symposium. The MEGA Symposium*, EPRI, Washington, DC, 1999.

Index

Note: Locator followed by '*f*' and '*t*' denotes figure and table in the text

A

Abrasive dusts, handling, 315
Absorption process
  and adsorption, 121–125
    fluid mechanics terminology, 123–124
    removal of HAP and VOC, 124–125
  advantages, 128
  aqueous systems, 128–129
  catalytic, 282–283
  description, 127–128
  disadvantages, 128
  equipment types and arrangements, 129–132
    baffle tray tower, 129, 134*f*
    cyclonic spray chamber, 129, 132*f*
    packed column, 129, 130*f*
    plate column, 129, 133*f*
    spray chamber, 129, 131, 131*f*
  nonaqueous systems, 129
  sample design calculation, 174–184
    dumped packing, 174–179
    flooding, 179–180
    structured packing, 180–184
Acceleration, 329
Acclimation process, 263
Acid deposition control, 34–35, 35*f*
Activated adsorption, 192
Adsorbents
  with activated carbon, 192
  nature of, 190–192
  pore structure, 192
  properties of, 190, 190*t*–191*t*
  type of, 185, 186*t*
  water-saturated, 189
Adsorber
  effectiveness, 204–208
    hot air/gas regeneration, 205, 207–208
    reactivation, 208
    steam regeneration, 205, 206*f*, 218–219
  fixed-bed, 200–203
    arrangement of, 188, 189*f*
    canister, 201
    carbon adsorption system, 207, 207*f*
    design considerations, 201–203
    modular, 202
    regenerative, 202
    safety measures, 202–203
    types of, 201–202
    unsteady-state, 188–189, 200–201
  moving-bed, 188
  pressure drop through, 203–204
  rotary wheel, 189
Adsorption isobar, 194
Adsorption isotherm, 194
Adsorption process, 186*t*, 187–190
  absorption and, 121–125
    fluid mechanics terminology, 123–124
    removal of HAP and VOC, 124–125
  activated, 192
  carbon, *see* Carbon adsorption
  chemical, 192
  chromatographic, 189
  data of, 194–195
  equilibrium, 194, 195*f*
  fixed-bed, 188–189
  of hexane
    from air, 217*f*
    from natural gas, 213, 214*f*
  isotherms, 195–198
    BET, 196–198
    Freundlich's equation, 195
    Langmuir's equation, 196
  newer technologies, 189–190
    chromatographic, 189
    pressure swing, 190
    rotary wheel adsorber, 189
  operations, 185–187
    advantages, 186
    description, 185–186
    disadvantages, 186
  phenomenon, 187
  physical, 187, 192–193
  pressure swing, 190
  stagewise, 188
  temperature-swing, 186
  theories of, 192–194
  with/without capillary condensation, 196–198
Advection inversion, 52
Aerodynamic diameter, 310, 343, 357
Air
  chemical composition of, 5–6, 6*t*
  as noncondensable, 251–252
  polluted atmosphere *vs.* pure, 6, 7*t*
  quality analysis, 48–50
    full, 50
    preliminary, 49–50, 49*t*
Air load V-I curves, 405

Air permit application, elements, 39–44
  applicability, 40–41
  emissions netting, 43–44
  modification, 42–43
  significant emission rates, 41–42
Air pollution
  effects of, 10–12, 12*f*
  incidents, 2, 3*t*
  parameters affecting transport of pollutants, 8–10
  particle size distribution, 305–309, 306*f*
    data, 305, 306*t*
    on linear graph paper, 308, 308*f*
    on linear-probability plot, 305, 307*f*
    on log-probability plot, 305, 307*f*
    number *vs.* mass fraction, 305, 306*t*
    on semi-logarithmic graph paper, 308, 308*f*
  problem
    federal involvement in control of, 3–4
    history, 1–3, 6, 7*t*
    recipe for, 6–12
  process, 7, 8*f*
  sources of, 8, 55–57, 56*t*
  in the United States, 1–2
Air Pollution Control Act Amendments of 1960, 3
Air Pollution Control Act of 1955, 3, 13
Air pollution control systems
  cost estimates
    annualized capital cost, 93
    detail and accuracy, 95
    escalation factors, 93
    resources, 95–96
    time value of money, 91–93, 92*f*
    types of, 94–95
  equipment cost, 95–96
    factors affect, 117
    OAQPS control cost manual, 95
  process design
    decisions, 98–99
    flowsheets, 99
    mass and energy balances, 99–104, 100*f*, 102*f*, 103*f*
    overview, 97
    process of, 97, 98*f*
    strategy of, 97–99
    systems-based approaches, 104–107
  separation processes, 119
Air Quality Act of 1967, 4
Air quality control regions (AQCRs), 4
Air Quality Modeling Group (AQMG), 59
Air sampling train, 78–80
Air-to-cloth ratio, 376–377, 381
Amagat's law, 136
Ambient air-quality sampling program, 73
American Meteorological Society/EPA Regulatory Model (AERMOD), 61
American Trucking Associations, Inc. v. U.S. EPA, 20
Aqueous systems, absorption, 128–129
Area model for mixture condensing
  algorithm for, 257–258
  derivation of, 256–257
Arrhenius-type equation, 398
Atmosphere, characterizing, 4–6, 5*f*
Atmospheric-diffusion models, 57–58
Attainment areas, 16
Automatic voltage controller, 400

B

Back corona, 398, 404
Baffle tower design, 131, 134*f*
Baghouse design theory, 380–389
  compartments, 383–386, 384*t*
  considerations, 382–383
  example, 386–389
Baghouse, fabric filter, 367
  bag life
    failure modes, 379–380
    inlet design, 380
    startup seasoning, 380
  cleaning mechanisms, 368–372
    pulse jet (high/low pressure), 370–372, 372*f*
    reverse air, 369–370, 370*f*, 371*f*, 383*t*
    shake/deflate, 368–369
    sonic horns, 372
  design, 367–368
  design theory, 380–389
    compartments, 383–386, 384*t*
    considerations, 382–383
    example, 386–389
  pressure drop, 377–379, 378*f*, 384*f*
  properties, 373–376, 373*t*
    catalytic membranes, 375
    ceramic candles, 376
    felted fabric, 374
    membrane fabrics, 375
    pleated cartridges, 375–376
    surface treatment, 374
    weight, 374–375
    woven bags, 374
  shaker, 368, 369*f*, 383, 383*t*
  size, 376–377
    air-to-cloth ratio, 376–377
    can velocity, 377
  S vs. W for process data, 387, 388*f*
Base case design, 110
Baseline V-I curves, 405
Batch operation, 123
Berl saddles random packing, 154, 155*f*
Best Available Control Technology (BACT), 44–48
  control technologies

evaluate, 47–48
identify, 45–46
cost-effectiveness, 45
rank remaining options, 46–47
select, 48
technically feasible, 46
BET theory, 194
Binary mixture process, 247, 248*f*
condensation of, 248–251
Biofilter technology, 259
case studies, 266
components of, 261
considerations, 266
design parameters and conditions, 261–265
biofilter bed, 262, 262*f*
inorganic nutrient supply, 263
loading and removal rates, 264–265
microorganisms, 262–263
moisture content, 263–264
oxygen supply, 263
pH, 264
pressure drop, 265
pretreatment of gas streams, 265
temperature, 264
mixture of, 260
open, 261, 261*f*
overloading, 260
start-up of, 263
theory of, 260
in the United States and Europe, 259
vs. other control technology, 265
Biotransformations, 260
Bombardment charging, 392–393
Breakthrough curve, 200
Breakthrough model, 208–217
curve example, 211–217
mass transfer, 209–211, 211*f*
Brunauer, Emmett, Teller (BET) isotherm, 196–198
Bubble cap tower, 130
Budget cost estimates, 94
Buoyancy effect in plumes, 62

C

CALPUFF model, 61
Calvert cut diameter design technique, 357–362
Canister adsorber, 201
Canopy hoods, 318
Can velocity, 377
Capital investment and product cost, 115–117
Carbon adsorption
beds, 186
fixed-bed activated, 207, 207*f*
technology, 265
Catalytic absorption, 282–283
Catalytic incineration, 235–237, 236*t*
Centrifugal force mechanism, 313, 329, 331
Chemical adsorption, 192
Chemical composition of air, 5–6, 6*t*
Chemiluminescence process, 87
Chemisorption, 187, 192
vs. physical adsorption, 193
Chromatographic adsorption, 189
Circulating lime reactor process, 296–298, 297*f*
Citizen suits, 15, 37
Claus process, 287
Clausuis-Clapeyron equation, 193
Clean Air Act (CAA), 3–4
GHG, 21
history, 13–17
in the United States, 13
Clean Air Act Amendments (CAAA)
1970, 13–14
citizen suits, 15, 37
HAP, 14–15
NAAQS, 14
NSPS, 14
1977, 15–17
nonattainment areas, 16–17
PSD, 16
1990, 17–37, 17*t*
acid deposition control, 34–35, 35*f*
attainment and maintenance of NAAQS, 18–21
disadvantaged business, 37
employment transition assistance, 37
enforcement, 36–37
HAP program, 22–34
miscellaneous provisions, 37
mobile sources, 21
operating permits, 36
research, 37
stratospheric ozone and global climate protection, 36
The Clean Air Act of 1963, 4
The Clean Air Amendments Act of 1970, 4
Cleaning mechanisms, baghouse, 368–372
pulse jet (high/low pressure), 370–372, 372*f*
reverse air, 369–370, 370*f*, 371*f*, 383*t*
shake/deflate, 368–369
sonic horns, 372
Code of Federal Regulations (CFR) test methods, 65, 66*t*–68*t*, 76
Collection mechanisms, 311–314
diffusion, 313
impaction, 312–313
interception, 313
other, 313–314
centrifugal force, 313
diffusiophoresis, 314
electrostatic attraction, 313
gravity, 313
thermophoresis, 314

Collection mechanisms (*Continued*)
  particle, 312, 312*f*
  wet scrubbers
    and efficiency, 342
    and particle size, 342–343
Column design, absorption, 132–150
  dilute solution case, 146–147
  equilibrium relationships, 135–136
  ideal solutions, 136–138
  mass separating agent, 147–150
  mass-transfer coefficients, 140–141, 143–144
  transfer unit method, 145–146
  Whitman two-film theory, 142–143, 142*f*
Combustion control techniques, 274–279
  flue gas recirculation, 275, 276*f*
  low excess air firing, 275
  NO_x burners
    low, 277–278, 277*f*, 278*f*
    out of service, 276
    ultra-low, 278–279
  overfire air, 275
  reburn, 276
  reduce air preheat, 275
  reduce firing rate, 275
  water/steam injection, 276
Combustion processes, 221
  flame temperature, 223
  flares, 223–229
    burner tip, 225, 225*f*
    elevated, open, 224–225, 224*f*
    ground, 227–228
    height, 226–227
    hydrocarbon liquids, 229
    refinery, 226
    safety features, 228–229
    smokeless, 225–226, 226*f*
  heat of, 221, 222*t*–223*t*
  incineration, 229–237
    catalytic, 235–237, 236*t*
    classifications, 229
    direct flame, 230–231
    energy recuperation in, 237, 237*f*
    thermal, 231–235, 231*f*, 232*t*
Concentration–time effect, 11
Concurrent/co-current flow, 121, 122*f*
Condensation mechanism, 345
  based heat-induced network, 252, 252*f*
  of binary mixture, 248–251
  dropwise, 239
  energy-induced separating network, 254, 254*f*, 255*f*
  film-type, 239
  pinch diagram for VOC, 253, 253*f*
  surface, 193
  systems-based approach, 252–255
Condensed sulfuric acid, 302
Condensers, 239
  contact, 241
  HAP, 239
  in refrigerated systems, 241
  surface, 240–243, 241*f*
  vapor pressure curves, 242, 242*f*
  vertical upflow total, 249*f*
  VOC, 239–243
Condensing film, heat transfer, 244, 256*f*
Coning plume, 54
Contact condensers, 241
Contacting power, scrubber, 345
Contemporaneous time period, 43
Continuous air-quality monitors, 81–90, 82*f*
  chemilumenescence for ozone/nitrogen oxides, 86–88
  device, 81
  hydrocarbons by flame ionization, 86
  nondispersive infrared method for CO, 84–85
  SO_2
    coulometric analyzer, 83–84, 83*f*
    electroconductivity analyzer, 82–83
    flame photometric detection of sulfur and, 85–86
    fluorescent monitor, 86
  specifications, 88
  steady-state calibrations, 88–90
Continuous emissions monitoring (CEM), 81
Continuous monitoring method, 75–76, 76*t*
Continuous *vs.* integrated sampling, 74
Controlling film concept, 169
Control technologies
  evaluate, 47–48
    economic impacts and cost-effectiveness, 48
    energy impacts, 47
    environmental impacts, 47
  identify, 45–46
Conventional scrubbers, 341
Corona formation, 392, 393*f*
Corona-induced plasma, 283
Cost estimates
  approaches for, 117
  resources, 95–96
  types of, 94–95
Coulometric analyzer for SO_2, 83–84, 83*f*
Countercurrent flow, 121, 121*f*, 122*f*
  absorption column design, 132–150
    dilute solution case, 146–147
    equilibrium relationships, 135–136
    ideal solutions, 136–138
    mass separating agent, 147–150
    mass-transfer coefficients, 140–141, 143–144
    transfer unit method, 145–146
    Whitman two-film theory, 142–143, 142*f*
  absorption tower design, 151–174
    baffle, 129, 134*f*
    controlling film concept, 169

diameter, 165–168, 166*f*
equations, 138–140, 139*f*
graphical absorption, 165, 165*f*
Henry's law, 171–174, 172*t*
L/G ratio on packing height, 169–171, 171*f*
liquid–gas flow ratio, 164–165
mass-transfer information, 171–173
operations of, 151–153, 151*t*
overview, 151
packed tower internals, 162–164
structured packing, 168–169
tower packings, 153–161
Countercurrent spray chambers, 347–356
Criteria pollutant, 14, 271
Cross-flow absorber, 122–123, 122*f*
Cunningham correction factor, 343
Cunningham slip correction, 310–311, 311*f*
Cut–power relationship, scrubbers, 362, 362*f*
Cyclone device, 329
collection efficiency, 329–337
comparison, 336–337
factors affecting, 331–333
Lapple's, 334–335, 334*f*
Leith and Licht efficiency model, 335–336
theoretical, 333
configuration parameter, 336, 336*t*
curves, 336, 337*f*
dimensions, 331, 332*f*, 333*t*
efficiency relationships, 331, 332*f*
inlet piping configuration, 330*f*
pressure drop, 337–338
saltation, 338–339
slope parameter, 335, 335*f*
standard, 330*f*
Cyclonic spray towers, 341, 347*f*

D

Decarbonation process, 298
Declining balance method, 114
Definitive cost estimates, 95
Degrees of severity for nonattainment, 18, 18*t*
Density/molecular seal, 228, 229*f*
Depreciation, effect of, 113–115
Desulfurization process, 128
Deutsch equation, 395–399
gas-flow distribution, 399
inherent assumption in, 396
particulate resistivity, 397–399, 398*f*
rapping re-entrainment, 397
sneakage, 396–397
Diffusion mechanism, 313
Diffusiophoresis mechanism, 314
Direct flame incineration, 230–231
Dispersion of pollutants, 63
Diurnal cycle, 52–53
Donora, Pennsylvania, 10
Dropwise condensation, 239
Dry adiabatic lapse rate, 9
Dry sodium-based systems, 290
Dual alkali, 294
Dual-fluid atomizing nozzles, 401
Duct design, 318–325
circular, 322, 322*f*
fan power, 325–326
mechanical energy balance, 320–325
friction head, 321–325
velocity head, 321
minimum velocity, 319, 319*t*
rectangular and circular, 323, 323*f*
Dust cake removal systems, 368
Dynamic gas seal, 228, 229*f*
Dynamic response testing, 88

E

Effective migration velocity, 400
Effective source height, 62
Electroconductivity analyzer for SO_2, 82–83
Electrostatic attraction mechanism, 313
Electrostatic precipitator (ESP), 391
basic theory, 391–399
corona formation, 392, 393*f*
Deutsch equation, 395–399
particle charging, 392–394, 393*f*
particle migration, 394–395, 394*f*
controls particulate emissions, 391
flue gas conditioning, 401–403
ammonia, 402
ammonium sulfate, 403
humidification, 401
proprietary additives, 403
SO_3, 401–402
SO_3 and ammonia, 403
hot-side, 402
practical application, 400
vertical wires, 391
V-I curves for troubleshooting, 403–405, 404*f*
Elevated, open flare system, 224–225, 224*f*
Emission rate, significant, 41–42, 42*t*
Enclosing hoods, 317
Energy-induced separation network, 106, 254
Energy-separating agent (ESA), 105, 120
Environmental Protection Agency (EPA), 4, 13
Air Trends website, 8
computer programs for regulation of industry, 59–61
complex model, 60
new models, 61
screening models, 60–61
GHG emissions, 21
PSD, 16
SCRAM programs, 59
source categories, 22

Equilibrium adsorption, 194, 195*f*
Escalation factors, 93
Evaporation inversion, 52

F

Fabric filter baghouse, 367
  bag life, 379–380
    failure modes, 379–380
    inlet design, 380
    startup seasoning, 380
  cleaning mechanisms, 368–372
    pulse jet (high/low pressure), 370–372, 372*f*
    reverse air, 369–370, 370*f*, 371*f*, 383*t*
    shake/deflate, 368–369
    sonic horns, 372
  design, 367–368
  design theory, 380–389
    compartments, 383–386, 384*t*
    considerations, 382–383
    example, 386–389
  pressure drop, 377–379, 378*f*, 384*f*
  properties, 373–376, 373*t*
    catalytic membranes, 375
    ceramic candles, 376
    felted fabric, 374
    membrane fabrics, 375
    pleated cartridges, 375–376
    surface treatment, 374
    weight, 374–375
    woven bags, 374
  shaker, 368, 369*f*, 383, 383*t*
  size, 376–377
    air-to-cloth ratio, 376–377
    can velocity, 377
  S vs. W for process data, 387, 388*f*
Fabric filtration, 367
Fanning fashion, 53
Fan power, ductwork, 325–326
Federal reference methods, 75–76, 75*t*
  for NSR, 39, 41, 42*t*, 44
  regulated air pollutants
    continuous monitor methods, 75, 76*t*
    determination of, 76–78
Film-type condensation, 239
Filter drag model, 378, 381, 385
Fire triangle, 221, 222*f*
Fixed-bed adsorber, 200–203
  arrangement of, 188, 189*f*
  carbon adsorption system, 207, 207*f*
  design considerations, 201–203
  safety measures, 202–203
  types of, 201–202
  unsteady-state, 188–189, 200–201
Fixed *vs.* mobile sampling, 74
Flame photometric detection, 85–86
Flares, combustion process, 223–229
  burner tip, 225, 225*f*
  elevated, open, 224–225
  ground, 227–228
  height, 226–227
  hydrocarbon liquids, 229
  refinery, 226
  safety features, 228–229
  smokeless, 225–226
FLEXIPAC® HC™ structured packing, 160, 160*f*
FLEXIPAC® structured packing, 160, 160*f*
FLEXIRING random packing, 154, 156*f*
Flow parameter, 182
Flue gas conditioning, 401–403
  ammonia, 402
  ammonium sulfate, 403
  humidification, 401
  proprietary additives, 403
  SO_3, 401–402
  SO_3 and ammonia, 403
Flue gas recirculation, 275, 276*f*
Flue gas treatment techniques, 279–283
  absorption
    catalytic, 282–283
    low temperature oxidation, 281–282, 282*f*
  corona-induced plasma, 283
  SCR, 280–281, 280*f*
  SNCR, 279, 280*f*
Fluid mechanics terminology, 123–124
Fluorescence process, 87
Fluorescent SO_2 monitor, 86
Forced oxidation process, 293
Freundlich's equation, 195
Friction pressure/head, 321
Fuel NO_x mechanism, 274
Fugitive emissions, 41
Full impact analysis, 50

G

Gaseous pollutants, 65, 68
  condensers for removal of, 120
Gas-flow distribution, 399
Gas seals, 228, 229*f*
Gas streams
  humidification, 264
  pretreatment, 265
Gas-to-cloth ratio, 377
Gaussian concentration model, 63
Gaussian plume model, 58, 60
Gibbs–Duhem equation, 135
Gore-Tex®, 375
Gravity mechanism, 313
Greasko Limited plant, 97
Greenhouse gases (GHGs), 21
Ground flare system, 227–228

H

H_2S control, 285–287, 286*f*
Hazardous air pollutants (HAP), 14–15, 15*t*
  condensers, 239
  MACT standards, 24, 33–34
  penalties, 36–37
  program, 22–34
  release of, 36
  removal of VOC and, 124–125
  risk management plans, 34
  source categories, 22, 23*t*–24*t*
Hazen equation, 396
Heat exchangers
  coolant and, 244–247
  plate- and frame-type, 241
  process, 241
    gas–liquid temperature difference, 250, 251*f*
    temperature profiles, 250, 250*f*
  shell and tube
    advantages/disadvantages, 244, 246*t*
    surface condenser, 240, 241*f*, 243*f*
    vertical, 244, 245*f*
  vertical, 249*f*, 250
Heat-induced network system, 252, 252*f*
Heat-induced separation networks (HISENs), 105–107, 106*f*
Heat-induced separator, 105, 106*f*, 252
Heel, 204
Henry's law, 136–138, 137*f*
  constants, 171–174, 172*f*
  for multicomponent solutions, 174
Heuristic-based design approach, 126
High-efficiency/throughput cyclones, 331
Hood design, 316–318
  control velocity, 316, 316*t*
  effect of entrance into, 325, 325*t*
  flow relationship
    canopy, 318
    enclosing, 317
    rectangular or round, 317
    slot, 317–318
Hood–duct system, 326–328, 326*f*
Hot-gas regenerative system, 205, 206*f*
Hot-side ESP, 402
Humidifier process, 264
Hydraulic radius concept, 322–323
Hypersorption process, 185

I

Impaction mechanism, 312–313
IMTP™ random packing, 154, 157*f*
Incineration, combustion process, 229–237
  catalytic, 235–237, 236*t*
  classifications, 229
  direct flame, 230–231
  energy recuperation in, 237, 237*f*
  thermal, 231–235, 231*f*, 232*t*
Incremental cost-effective approach, 48
Industrial source complex model, 60
Inertial impaction, 342, 350
Inlet design configuration, 380
Intalox™ high capacity structured packing, 158, 159*f*
Intalox™ SNOWFLAKE™ random packing, 154, 157*f*
Integrated sampling, 74
  devices, 80–81
Interception mechanism, 313
Inversion, 9
  advection, 52
  evaporation, 52
  subsidence, 52
  surface/radiation, 51–52
Ionosphere, 5
ISC-PRIME model, 61
ISCST3 model, 61
Isokinetic sampling, 70–72, 72*f*
Isopropyl alcohol (IPA)—water system, 248, 249*f*
Isotherms, adsorption, 195–198
  BET, 196–198
  Freundlich's equation, 195
  Langmuir's equation, 196

J

Job Training Partnership Act, 37

L

Langmuir's equation, 196
Lapple's efficiency correlation, 334–335, 334*f*
Lean streams, 149
Leith and Licht efficiency model, 335–336
Lewis and Randall fugacity rule, 136, 137*f*
Lime spray drying process, 294–296, 295*f*
Liquid-seal drums, 228
Log-mean temperature difference (LMTD) method, 247–248
London smog *vs.* Los Angeles smog, 10–12, 11*t*
Looping plumes, 54
Los Angeles, California, 9–10
Lower explosive limit (LEL), 221, 230
Low-resistivity dust, 398
Low temperature oxidation process, 281–282, 282*f*
Luminescence process, 87
Lung-damaging dust, 305

M

MACCOOKER huge meat smoker, 232–235, 233*f*
MACT floor, 24, 33

Mass exchange network (MEN), 104, 105*f*, 219
Mass separating agent (MSA), 104–105, 119, 219
Mass transfer, 209–211, 211*f*
  process, 119
Mass-transfer coefficients
  data, 172, 173*t*
  origin of volume-based, 140–141
  overall, 143–144, 143*f*
  volume-based, 144
Matts–Ohnfeldt equation, 396
Maximum achievable control technology (MACT) standards, 33–34
  final, 24, 25*t*–33*t*
  floor, 24, 33
  technology-based, 33
Membrane separation technology, 267
  advantages, 268
  applications, 269–270
  description, 267
  disadvantages, 268
  performance, 268–269
  polymeric membranes, 268
  systems design, 270
  for VOC gaseous emissions, 267, 268*f*, 269*t*
Micrometeorological processes, 5
Microorganisms, biofilter, 262–263
Minimum composition driving force, 148
Mobile sampling, 74
Modular adsorbers, 202
Moisture content, biofilter, 263–264
Molecular diffusion, steady-state, 140–141
Monitoring systems, 73–75
  continuous *vs.* integrated sampling, 74
  fixed *vs.* mobile sampling, 74
  instrumentation and methods, 75
  selection of, 74
Montreal Protocol, 36
Motor Vehicle Air Pollution Control Act of 1965, 4, 13
Moving-bed adsorbers, 188

N

National Ambient Air Quality Standards (NAAQS), 4, 14
  attainment and maintenance, 18–19
  revisions, 19–21, 19*t*
National Emission Standards for Hazardous Air Pollutants (NESHAP), 14, 22
New Source Performance Standards (NSPS), 14
New source review (NSR), 39
  reform, 50
Nonaqueous systems, absorption, 129
Nonattainment areas
  degrees of severity, 18, 18*t*
  offsets in, 16–17
Nondispersive infrared method for CO, 84–85
Nonthreshold pollutant, 20, 20*f*
NorPro®Proware™Ceramic saddles, 154, 156*f*
NO_x (nitrogen oxides)
  burners
    low, 277–278, 277*f*, 278*f*
    out of service, 276
    ultra-low, 278–279
  from combustion
    fuel, 274
    prompt, 274
    thermal, 271–274
  control techniques, 274–283
    combustion, 274–279
    flue gas treatment, 279–283
  emission
    sources, 271, 272*f*
    in the United States, 271, 272*f*
  equilibrium concentrations/constants, 273, 273*t*

O

OAQPS control cost manual, 95
Office of Air Quality Planning and Standards (OAQPS), 95
Opacity, 76
Open biofilter configuration, 261, 261*f*
Operating line, 140, 143
Order of magnitude cost estimate, 94
Organic vapors, mixtures of, 247–251
Ozone nonattainment, 18, 18*t*

P

Packed absorption tower design, 151–174
  baffle, 131, 134*f*
  controlling film concept, 169
  diameter, 165–168
    random dumped packing, 165–167, 166*f*
    structured packing, 168–169, 168*f*
  equations, 138–140, 139*f*
  graphical absorption, 165, 165*f*
  Henry's law, 171–174, 172*t*
  L/*G* ratio on packing height, 169–171, 171*f*
  liquid–gas flow ratio, 164–165
  mass-transfer information, 171–173
  operations of, 151–153, 151*t*
  overview, 151
  packed tower internals, 162–164
    bed limiter, 163–164, 164*f*
    liquid distributors, 162, 163*f*
    liquid redistributors, 163, 163*f*
    packing support plate, 162, 162*f*
  structured packing, 168–169
  tower packings, 153–161
    grid-type, 161, 161*f*
    random or dumped, 153–155, 155*f*, 156*f*
    structured, 156–161

Packing factor, 165
for Intalox structured packing, 169, 169*t*
for random dumped packings, 167, 167*t*
Pall ring random packing, 154, 156*f*
Particle charging, 392–394, 393*f*
Particle migration, 394–395, 394*f*
Particulate resistivity, 397–399
Particulate traverses, 68–70, 70*f*, 71*f*
Permeation tube, 89
Perry's Chemical Engineers' Handbook (book), 95–96
pH of biofilter, 264
Phosphorescence process, 87
Physical adsorption, 187, 192
vs. chemisorption, 193
Pinch point, 253
Pitot tube for velocity measurement, 68–69, 69*f*
Plumes, smoke, 52–53
assumptions, 60
buoyancy effect in, 62
coning, 54
looping, 54
types, 53
Polanyi potential theory, 198–200, 199*f*
Pollutants
annual production of, 8, 9*f*
bulk transport of, 63
concentrations, 74
criteria, 14, 271
destruction efficiency of VOC, 223, 224*t*
dispersion of, 63
gaseous, 65, 68, 120
H_2S, 285–287, 286*f*
hazardous air, *see* Hazardous air pollutants (HAP)
nonthreshold, 20, 20*f*
organic, 120
parameters affecting transport of, 8–10
threshold, 20, 20*f*
Polymeric membranes, 268
Pore structure, adsorbent, 192
Potential to Emit (PTE), 40–41
Preliminary dispersion modeling analysis, 49–50
Pressure swing adsorption, 190
Prevention of Significant Deterioration (PSD), 16
source categories, 40, 40*t*
Primary capture mechanism, 350, 353*f*
Principal smoke-plume models, 53–54
Process design
decisions, 98–99
flowsheets, 99
mass and energy balances, 99–104, 100*f*, 102*f*, 103*f*
overview, 97
process of, 97, 98*f*
strategy of, 97–99
systems-based approaches, 104–107
Process Plant Construction Estimating Standards (book), 96
Process stacks, 57
Process synthesis technology, 125–126
Profitability analysis
mathematical methods, 109–110, 110*f*
measure of, 110–113
process, 116
profit goal, 109
Prompt NO_x mechanism, 274
PSD trigger, 41
Pulse jet baghouse, 370–372, 372*f*

R

Rapping, ESP, 392
dust cake dislodged by, 399
re-entrainment, 397
Raschig ring random packing, 154, 155*f*
Reasonably available control technology (RACT), 42, 45
Rectangular/round hoods, 317
Regeneration modeling, 119, 218–219
Regenerative adsorbers, 202
Relative humidity, effect of, 192
Return on investment (ROI)
example, 111–113
incremental rate of, 110–113
Reverse air baghouse, 369–370, 370*f*, 371*f*, 383*t*
Rich streams, 149
Risk management plans, HAP, 34
Rotary wheel adsorber, 189
Rules of thumb, 57

S

Saltation concept, 338–339
Sample design calculation, absorption, 174–184
dumped packing, 174–179
flooding, 179–180
structured packing, 180–184
Sampling program
ambient air-quality, 73
objectives of, 73
Sampling techniques, representative, 65, 68–72
gaseous pollutants, 65, 68
isokinetic sampling, 70–72, 72*f*
velocity and particulate traverses, 68–70, 69*f*, 70*f*, 71*f*
$SCONO_x^{TM}$ technology, 282
SCREEN model, 49, 60–61
Scrubbers, 151
Secondary emissions, 41
Selective catalytic reduction (SCR), 280–281, 280*f*
Selective noncatalytic reduction (SNCR), 279, 280*f*
Self-induced spray tower, 344, 348*f*
Semrau principle, 344–347, 352*t*

Separation processes, 119
Shake/deflate baghouse, 368–369
Shaker baghouse, 368, 369*f*, 383, 383*t*
Shell and tube heat exchangers
  advantages/disadvantages, 244, 246*t*
  surface condenser, 240, 241*f*, 243*f*
  vertical, 244, 245*f*
Single mass exchanger, 104, 104*f*
Six-tenths factor rule, 116
Skewed gas flow distribution, 399, 400*f*
Slot hoods, 317–318
Smokeless flare system, 225–226, 226*f*
SO_2 (sulfur dioxide)
  coulometric analyzer, 83–84, 83*f*
  electroconductivity analyzer, 82–83
  flame photometric detection of sulfur and, 85–86
  fluorescent monitor, 86
SO_3 mechanism, 402
Soda ash wet scrubber, 293–294, 294*f*
Sodium bicarbonate process, 298–299, 299*f*
Solvent-recovery process, 205
Sonic horns, 372
Source testing, 65
Source-transport-receptor problem, 61–63
  receptor, 63
  source, 61–62
  transport, 62–63
    bulk transport of pollutants, 63
    dispersion of pollutants, 63
    effective emission height, 62
Space charge effect, 402
Spray towers, 344, 346*f*
  application to, 351–356
  cross-current flow, 361, 364*f*
  cyclonic, 341, 347*f*
  mass balance, 353, 354*f*
  parameters on penetration, 355, 356*f*
  self-induced, 344, 348*f*
  vertical countercurrent flow, 361, 363*f*
Stability array (STAR), 60
Staged air low NO_x burner, 277, 277*f*
Staged fuel low NO_x burner, 277, 278*f*
Stairmand cyclone design, 336–338
Startup seasoning, baghouse, 380
Steady-state calibrations, 88–90
Steam regeneration, 205, 206*f*, 218
Stratosphere, 5
Subsidence inversion, 9, 52
Sulfur dioxide (SO_2) emissions
  capital *vs.* operating costs, 290–291
  example evaluation, 300, 301*t*
  and HCL removal, 287–300
  reagents, 287–290
    calcium-based reactions, 287–289
    chemicals, 287, 288*t*
    sodium-based reactions, 289–290
  removal processes, 291–300, 292*t*
    circulating lime reactor, 296–298, 297*f*
    lime spray drying, 294–296, 295*f*
    other, 300
    sodium bicarbonate/sodium sesquicarbonate injection, 298–299
    wet limestone, 291–293, 293*f*
    wet soda ash/caustic soda, 293–294, 294*f*
  sources, 285, 286*f*
  trend, 285, 286*f*
Sulfur trioxide (SO_3) emissions
  and H_2SO_4 formation, 300–304, 302*f*
  toxic release inventory, 304
Superficial velocity, 376
Superfund Amendments and Reauthorization Act (SARA), 304
Super Intalox™ Saddles® random packing, 155, 158*f*
Support Center for Regulatory Air Models (SCRAM), 59
Surface condensation, 193
Surface condensers, 240–243, 241*f*
Surface/radiation inversions, 51–52
Surface reaction, 193
Surveillance and control system, 78, 79*f*
Suspended particulate matter, 80–81, 80*f*
Swift cyclone design, 336–338
Systems-based approach, 252–255

T

Tail gas, 287
Tall stack, 54
  definition of, 56–57
  size of tall things, 54, 54*t*
Temperature inversion, 9
Temperature-swing adsorption, 186
Tennessee Valley Authority (TVA), 54, 55*f*
Theoretical efficiency, cyclone, 333
Theories of adsorption, 192–194
Thermal incineration process, 231–235, 231*f*, 232*t*
Thermal NO_x mechanism, 271–274
Thermal reactivation, 208
Thermoluminescence process, 87
Thermophoresis mechanism, 314
Threshold pollutant, 20, 20*f*
Top down methodology, 45
Tower design, packed absorption, 151–174
  baffle, 131, 134*f*
  controlling film concept, 169
  diameter, 165–168
    random dumped packing, 165–167, 166*f*
    structured packing, 168–169, 168*f*
  equations, 138–140, 139*f*
  graphical absorption, 165, 165*f*
  Henry's law, 171–174, 172*t*
  L/*G* ratio on packing height, 169–171, 171*f*

liquid–gas flow ratio, 164–165
mass-transfer information, 171–173
operations of, 151–153, 151*t*
overview, 151
packed tower internals, 162–164
bed limiter, 163–164, 164*f*
liquid distributors, 162, 163*f*
liquid redistributors, 163, 163*f*
packing support plate, 162, 162*f*
structured packing, 168–169
tower packings, 153–161
grid-type, 161, 161*f*
random or dumped, 153–155, 155*f*, 156*f*
structured, 156–161
Toxic Release Inventory, 304, 402
Troposphere, 5
Troubleshooting ESP, 403–405, 404*f*
Tubular collection electrode, 392, 392*f*

U

Uniform gas distribution, 399
Upper explosive limit (UEL), 230
Upper loading point, 152

V

Valve tray, 131
van der Waals adsorption, 192
Vatavuk Air Pollution Control Cost Index, 93, 94*t*
Velocity traverses, 68–70, 71*f*
Venturi scrubbers, 342, 350*f*, 357
hydrophobic particles, 361, 366*f*
jet, 344, 351*f*
model for, 357
performance curve, 346, 352*f*
Vertical heat exchanger, 249*f*, 250
Volatile organic compounds (VOCs), 14, 105–107, 221
biofilters, 259
condensers, 239–243
control technologies, 119, 120*t*
coolant and heat exchanger type, 244–247, 244*t*, 245*f*
destruction efficiency of, 223, 224*t*
gaseous emission streams, 239–240, 252–255
membrane separation, 267, 268*f*, 269*t*
organic vapors mixtures, 247–251
pinch diagram for condensation, 253, 253*f*
process synthesis technology for, 125–126
recovery process, 267
removal of HAP and, 124–125
Vortex finder tube, 329

W

Wet limestone process, 291–293, 293*f*
Wet limestone scrubbing, 287
Wet scrubbers
characteristics of, 344, 345*t*
collection mechanisms
and efficiency, 342, 345
and particle size, 342–343, 344*t*
Semrau principle, 344–347, 352*t*
contacting power, 345
conventional, 341
countercurrent spray chambers, 347–356
cut diameter method of, 357–362
cut–power relationship, 362, 362*f*
devices for, 344
disadvantages of, 342
overview, 341–342
plate, 344, 349*f*, 361, 365*f*
selection and design of, 344
soda ash, 293–294, 294*f*
spray tower, 344, 346*f*
cyclonic, 344, 347*f*
mass balance, 353, 354*f*
parameters on penetration, 355, 356*f*
self-induced, 344, 348*f*
technology, 265
venturi, 342, 350*f*, 357
hydrophobic particles, 361, 366*f*
jet, 344, 351*f*
model for, 357
performance curve, 346, 352*f*
Wet sodium-based scrubbers, 289–290
Whitman two-film theory, 142–143, 142*f*
Wind-profile law, 60
Wire gauze structured packing, 158, 159*f*